PERSPECTIVES ON TWENTIETH-CENTURY PHARMACEUTICALS

PERSPECTIVES ON TWENTIETH-CENTURY PHARMACEUTICALS

Edited by Viviane Quirke
and Judy Slinn

PETER LANG
Oxford · Bern · Berlin · Bruxelles · Frankfurt am Main · New York · Wien

Bibliographic information published by Die Deutsche Nationalbibliothek.
Die Deutsche Nationalbibliothek lists this publication in the Deutsche Nationalbibliografie; detailed bibliographic data is available on the Internet at http://dnb.d-nb.de.

A catalogue record for this book is available from the British Library.

Library of Congress Cataloguing-in-Publication Data:

Perspectives on twentieth-century pharmaceuticals / [edited by] Viviane Quirke and Judy Slinn.
p. ; cm.
Includes bibliographical references and index.
ISBN 978-3-03910-920-3 (alk. paper)
1. Pharmaceutical industry--History--20th century. I. Quirke, Viviane. II. Slinn, Judy.
[DNLM: 1. Drug Industry--history. 2. Drug and Narcotic Control--history. 3. History, 20th Century. 4. Technology, Pharmaceutical--history. QV 711.1 P466 2009]
HD9665.5.P47 2009
338.4'761510904--dc22
2009049320

ISBN 978-3-03910-920-3

Hochfeldstrasse 32, CH-3012 Bern, Switzerland
info@peterlang.com, www.peterlang.com, www.peterlang.net

Printed in Germany

Foreword

This volume of papers represents our choice from those delivered at an international conference we organized in Oxford on 14–16 July 2005. Its purpose was to bring together some of the significant strands in the current studies of pharmaceuticals and their development in the twentieth century. The diversity of the approaches as well as the vibrancy of the subject are, we believe, reflected in this collection. It includes papers by some who were then new entrants in the field and whose careers have moved on while this volume was in preparation, as well as established scholars from different countries. We hope that its cross-national and multidisciplinary perspectives will further and enrich our understanding of an industry, of an object, and of a process at the heart of highly medicalized contemporary societies. Our thanks to all the contributors for their hard work and patience during the time it has taken to put this volume together. We are also grateful to the Wellcome Trust and the Economic History Society for their financial support of the conference and of this book. We thank the School of Arts and Humanities (particularly Carol Beadle) and the Business School at Oxford Brookes University for their help and support.

VIVIANE QUIRKE and JUDY SLINN, Oxford, September 2009

Contents

List of Tables and Figures

Tables

Figures

VIVIANE QUIRKE AND JUDY SLINN

Perspectives on Twentieth-Century Pharmaceuticals: an introduction

One of the most striking features of the twentieth century has been the rapid growth of the pharmaceutical industry and the large increases in the use and consumption of its products, particularly in North America, Europe and Japan. By the end of the century, worldwide sales by drug companies were valued at approximately $350bn, and were expected to rise to a figure of $500bn within five years.[1] This trend began in the first half of the century, but accelerated most sharply after the Second World War, when the creation of national systems of healthcare created mass markets for drugs. The industry then assumed a major economic, social and political significance, and became one of the most highly regulated sectors of the economy.

These changes attracted the attention of industry analysts and academics, and have been reflected in the research and writing of the last few decades. As well as an expanding literature on the subjects of regulation and industrial policy,[2] there have been studies of the structure

1 IMS health 2000.

2 L. Hancher, *Regulating for Competition: government, law and the pharmaceutical industry in the United Kingdom and France* (Oxford: Clarendon Press, 1990); L.G. Thomas, 'Implicit industrial policy: the triumph of Britain and the failure of France in global pharmaceuticals', *Industrial and Corporate Change*, 3 (1994): 451–489; J. Abraham and H. Lawton Smith (eds), *Regulation of the Pharmaceutical Industry* (Basingstoke: Palgrave Macmillan, 2003); A.A. Daemmrich, *Pharmacopolitics: drug regulation in the United States and Germany* (Chapel Hill/London: the University of North Carolina Press, 2004); M.N.G. Duke, *The Law and Ethics of the Pharmaceutical Industry* (Amsterdam and Oxford: Elsevier, 2005); J. Slinn, 'Price Controls or Control through Prices? Regulating the cost and consumption of prescription pharmaceuticals in the UK 1948–1967', *Business History*, 47 (2005): 352–366.

of the industry and its importance to the economy;[3] histories of individual companies and of national drug industries;[4] accounts of drug dis-

3 C.J. Thomas, 'The pharmaceutical industry', in D. Burn (ed.), *The Structure of British Industry* (Cambridge: Cambridge University Press, 1958) vol. 2, pp. 331–375; K. Blunden, *Etude sur l'évolution de la concentration dans l'industrie pharmaceutique en France* (Luxembourg: Office des Publications Officielles des Communautés Européennes, 1975); W.D. Reekie, *The Economics of the Pharmaceutical Industry* (London: Macmillan, 1975); idem, 'Pharmaceuticals', in P.S. Johnson (ed.), *The Structure of British Industry* (London: Granada, 1980), pp. 106–130; G. Owen, *From Empire to Europe* (London: HarperCollins, 1999), chapter 13, pp. 360–387.

4 For example: H.G. Lazell, *From Pills to Penicillin: the Beechams story* (London: Heinemann, 1975); J. Slinn, *May & Baker, 1834–1984* (Cambridge: Hobsons Ltd, 1984); G. Tweedale, *At the Sign of the Plough: 275 years of Allen & Hanburys' and the British pharmaceutical industry, 1715–1990* (London: Murray, 1990); A. Blondeau, *Histoire des laboratoires pharmaceutiques en France: et de leurs médicaments* (Paris: Le Cherche Midi, 1992) vols. 1 and 2; R.P.T. Davenport-Hines and J. Slinn, *Glaxo: a history to 1962* (Cambridge: Cambridge University Press, 1992); E. Jones, *The Business of Medicine* (London, Profile Books: 2001) R.P. Amdam and K. Sogner, *Wealth of Contrasts: Nyegaard & Co. – a Norwegian pharmaceutical company, 1874–1985* (Oslo: Ad Notam Gyldendal, 1994); L. Galambos with J. Eliot Sewell, *Networks of Innovations: vaccine development at Merck, Sharpe & Dohme, and Mulford, 1895–1995* (Cambridge, Mass.: Cambridge University Press, 1995); M. Ruffat, *175 Ans d'industrie pharmaceutique française: histoire de Synhélabo* (Paris: La Découverte, 1996); C. Kobrak, *National Cultures and International Competition: the experience of Schering AG, 1851–1950* (Cambridge, Cambridge University Press, 2002); W. Abelshauser, W. von Hippel, J. Allen and R.G. Stokes, *German Industry and Global Enterprise BASF: the history of a company* (Cambridge: Cambridge University Press, 2004); R. Church and E.M. Tansey, *Knowledge, Trust, Profit: a history of Burroughs Wellcome & Co. and the transformation of the British Pharmaceutical Industry* (Lancaster: Crucible Publishing, 2007). There has also been a growing number of histories of national pharmaceutical industries: P. Starr, *The Social Transformation of American Medicine: the rise of a sovereign profession and the making of a vast industry* (New York: Basic Books, 1982); J. Liebenau, *Medical Science and Medical Industry: the formation of the American pharmaceutical industry* (Baltimore: Johns Hopkins University Press, 1987); idem, 'The twentieth-century British pharmaceutical industry in international context', in J. Liebenau, G.J. Higby and E.C. Stroud (eds), *Pill Peddlers: essays on the history of the pharmaceutical industry* (Madison, Wis.: American Institute of the History of Pharmacy, 1990), pp. 123–133; M. Robson, 'The French pharmaceutical industry, 1919–39', in ibid., pp. 107–122; C. Davis, *The Pharmaceutical Industry and the Market in the USSR and its Successor States: from reform to fragmentation to transition* (Richmond: PjP Books, 1993); D.A. Rajimwale, *The Indian Pharmaceutical Industry* (New Delhi: the People's Pub. House, 1996); S. Chauveau, *L'Invention pharmaceutique: la pharmacie entre l'Etat et la société* (Paris: Institut d'Edition Sanofi-Synthélabo, 1999); T.A.B. Corley, 'The British pharmaceutical industry since 1851', in L. Richmond, J. Stevenson and A. Turton (eds), *The*

covery,[5] academic-industrial relations,[6] and pharmaceutical R&D and innovation.[7] Recent work has also explored the growth of the biotechnology

---

Pharmaceutical Industry: a guide to historical records (Aldershot: Ashgate, 2003), pp. 14–32; A.D. Chandler, *Shaping the Industrial Century: the remarkable story of the evolution of the modern chemical and pharmaceutical industries* (Cambridge, MA: MIT Press, 2005).

5 M. Weatherall, *In Search of a Cure: a history of pharmaceutical discovery* (Oxford: Oxford University Press, 1990); R. Vos, *Drugs Looking for Diseases: innovative drug research and the development of the beta-blockers and the calcium antagonists* (Dordrecht: Kluwer Academic, 1991); W. Sneader, *Drug Discovery: the invention of modern medicine* (Chichester: John Wiley & Sons, 1996); V. Quirke, 'Making British cortisone: Glaxo and the development of corticosteroid drugs in Britain in the 1950s and 1960s', *Studies in History and Philosophy of Biology and Biomedical Sciences*, 36 (2005): 645–674; idem, 'From alkaloids to gene therapy: a brief history of drug discovery in the twentieth century', in S. Anderson (ed.), *A Brief History of Pharmacy* (London: the Pharmaceutical Society 2005), pp. 177–201; idem, 'Putting theory into practice: James Black, receptor theory, and the development of the beta-blockers at ICI, 1958–1978', *Medical History*, 50 (2006): 69–92; idem, 'The material culture of British pharmaceutical laboratories in the Golden Age of Drug Discovery, ca. 1935–1975', *International Journal for the History of Engineering and Technology*, 79 (2009): 280–299.

6 J.P. Swann, *Academic Scientists and the Pharmaceutical Industry: cooperative research in twentieth-century America* (Baltimore: Johns Hopkins University Press, 1988); J. Liebenau, 'The MRC and the pharmaceutical industry: the model of insulin', in J. Austoker and L. Bryder (eds), *Historical Perspectives on the Role of the MRC* (Oxford: Oxford University Press, 1989), pp. 163–180; J. Liebenau and M. Robson, 'L'Institut Pasteur et l'industrie pharmaceutique', in M. Morange (ed.), *L'Institut Pasteur: contributions à son histoire* (Paris: Presses Universitaires de France, 1991), pp. 52–61; T. Jones, 'The value of academia/industry links in R&D', in S.R. Walker (ed.), *Creating the Right Environment for Drug Discovery* (Lancaster: Quay Publishing, 1991), pp. 77–84; N. Oudshorn, 'United we stand: the pharmaceutical industry, laboratory and clinic in the development of sex hormones into scientific drugs, 1920–1940', *Science, Technology and Human Values*, 18 (1993): 5–24; J. Goodman, 'Can it ever be pure science? Pharmaceuticals, the pharmaceutical industry and biomedical research in the twentieth-century', in J.-P. Gaudillière and I. Löwy (eds), *The Invisible Industrialist: manufactures and the production of scientific knowledge* (Basingstoke: Macmillan, 1998), pp. 143–165, and other contributions in this volume; N. Rasmussen, 'The moral economy of the drug company-medical scientist collaboration in interwar America', *Social Studies of Science*, 34 (2004): 161–185; V. Quirke, *Collaboration in the Pharmaceutical Industry: changing relationships in Britain and France* (Abingdon/New York: Routledge, 2008); D. Tobbell, 'Allied against reform: pharmaceutical industry-academic physician relations in the United States, 1945–1970', *Bulletin of the History of Medicine*, 82 (2008): 878–912.

7 D. Schwartzman, *Innovation in the Pharmaceutical Industry* (Baltimore: Johns Hopkins University Press, 1976); J. Howells and I. Neary, *Intervention and Technological Innovation*

industry,[8] the globalization of the economy, and the challenges these have posed to the traditional pharmaceutical sector.[9]

At the same time, the growing dependence of the medical profession and the public on pharmaceutical products, and the resulting profits made by drug companies, which have been linked to the increasing role of marketing in firms' activities, have led to mounting criticism and controversies.[10] Many of these have been concerned with the advertising, prescription and

(Basingstoke: Macmillan, 1988); A. Gambardella, *Science and Innovation: the US Pharmaceutical industry during the 1980s* (Cambridge: Cambridge University Press, 1995); J. Slinn, 'Research and development in the UK pharmaceutical industry from the nineteenth century to the 1960s', in R. Porter and M. Teich (eds), *Drugs and Narcotics in History* (Cambridge: Cambridge University Press, 1996), pp. 168–186, and other chapters in this volume; J. Slinn, 'Innovation at Glaxo and May & Baker, 1945–165', *History and Technology*, 13 (1996): 133–147; R. Landau, B. Achilladelis and A. Scriabine (eds), *Pharmaceutical Innovation* (Philadelphia: Chemical Heritage Press: 1999); V. Quirke, 'Standardizing pharmaceutical R&D in the second half of the twentieth century: ICI's Nolvadex Development Programme in historical and comparative perspective', in C. Bonah et al. (eds), *Harmonizing Drugs: standards in 20th Century Pharmaceutical History*, (Paris: Glyphe, 2009), pp. 123–50.

8 M. McKelvey, *Evolutionary Innovations: the business of biotechnology* (Oxford: Oxford University Press, 1996); R. Henderson, L. Orsenigi and G. Pisano, 'The pharmaceutical industry and the revolution in molecular biology: interactions among scientific, institutional and organizational change', in D.C. Mowery and R.R. Nelson (eds), *Sources of Industrial Leadership* (Cambridge: Cambridge University Press, 1999). See also various contributions to M. Mazzucato and G. Dosi (eds), *Knowledge Accumulation and Industry Evolution: the case of pharma biotech* (Cambridge: Cambridge University Press, 2006).

9 R. Ballance, J. Pogany and H. Forstner, *The World's Pharmaceutical Industries* (Aldershot: Edward Elgar, 1992); J.M. Taggart, *The World Pharmaceutical Industry* (London: Routledge, 1993); P. Ramirez, 'The globalisation of research in the pharmaceutical industry: a case of uneven development', *Technology Analysis and Strategic Development*, 18 (2006): 143–167.

10 For example: J.-P. Dupuy and S. Karsenty, *L'Invasion Pharmaceutique* (Paris: Le Seuil, 1974); H. Redwood, *The Price of Health: the link between research in the pharmaceutical industry and health care systems in the developed world* (London: Adam Smith Institute, 1989); J. Abraham, *Science, Politics, and the Pharmaceutical Industry: controversy and bias in drug regulation* (New York: St Martin's Press, 1995); P. Pignarre, *Le Grand secret de l'industrie pharmaceutique* (Paris: La Découverte, 2003); M. Urfalino, *Le Grand méchant loup pharmaceutique: angoisse ou vigilance?* (Paris; Textuel, 2004).

uses of psychiatric drugs.[11] Given the difficulties in accessing corporate archives and information about the industry and its activities, impartial judgement of these issues has become problematic, leading to both factual and fictional publications directed against 'Big Pharma'. This has been particularly so in the USA, the only significant market to permit direct-to-consumer advertising, where public interest and concern has helped to make popular books on the subject into bestsellers.[12]

However, in parallel, considerable scholarly research has been carried out on the development of drugs as medicines in national and international markets, on their regulation in different contexts and at different times, on their role in medical practice, as well as on their representation and use in society more widely.[13] Unsurprisingly for an industry that discovers, develops, manufactures and sells products that have come to occupy such an important place in human and animal health and well-being, the

11 A. Ehrenberg, *La Fatigue d'être Soi: dépression et société* (Paris: Odile Jacob, 1998); D. Healy, *Let them Eat Prozac: the unhealthy relationship between the pharmaceutical industry and depression* (New York and London: New York University Press, 2004); A. Tone, *The Age of Anxiety: a history of America's turbulent affair with tranquillizers* (New York: Basic Books, 2008).

12 For an early example see T. Mahy, *The Merchants of Life: an account of the American pharmaceutical industry* (New York: Harper, 1959). For more recent examples: J. Crawford, *Kill or Cure? The role of the pharmaceutical industry in society* (London: Arc Print, 1988); L. Marsa, *Prescription for Profit: how the pharmaceutical industry bankrolled the unlikely marriage between science and business* (New York: Scribner, 1997); J. Law, *Big Pharma: exposing the healthcare agenda* (Robinson Publishing, 2006); idem, *Big Pharma: how the world's biggest drug companies control illness* (London: Constable, 2006); J. Moran and C. Guerra, *Pill Pushers: a Big Pharma battle for market share* (Booksurge plc, 2007). See also the recent spate of novels portraying the dark world of the corporate drug industry, including E. Jacobs, *The Pawn of Pharma* (Outskirts Press, 2005); J. Prieve, *Big Pharma: a novel* (PublishAmerica, 2006).

13 J. Goodman and V. Walsh, *The Story of Taxol: nature and politics in the pursuit of an anti-cancer drug* (Cambridge/New York: Cambridge University Press, 2001); T. Pieters, *Interferon: the science and selling of a miracle drug* (London: Routledge, 2005); R. Bud, *Penicillin: triumph and tragedy* (Oxford/New York: Oxford University Press, 2007); J.A. Greene, *Prescribing by Numbers: drugs and the definition of disease* (Baltimore, Johns Hopkins University Press, 2007); N. Rasmussen, *On Speed: the many lives of amphetamine* (New York/London: New York University Press, 2008).

drugs themselves have therefore attracted researchers from many different disciplines – from health economics, to medical anthropology, and social studies of science, as well as history.[14] All of this has helped to provide a rich and somewhat better rounded picture.

This volume brings together a collection of papers exploring and reflecting upon some of the significant strands in the current studies of pharmaceuticals in the twentieth century. It is organized in five parts, each containing three chapters by three different contributors, structured thematically and thereby representing the variety and complexity of perspectives on the pharmaceutical industry and its products. To introduce this collection, we begin by giving a brief chronological overview of the development of twentieth century pharmaceuticals, placing each chapter in its historical context. We then set out the themes of each of the five parts of the book, with short summaries of each chapter.

Historical Overview of Twentieth-Century Pharmaceuticals and the Pharmaceutical Industry

In the twentieth century, drugs came to occupy a central place in medical practice, especially in wealthier countries, where they helped to transform the health and life expectancies of individuals. Most of these drugs were developed in corporate laboratories, using scientific knowledge and technical know-how from a number of disciplines, with chemistry playing

14 For example, M. Gilswijt-Hofstra, G.M. Van Heteren and E.M. Tansey (eds), *Biographies of Remedies: drugs, medicines and contraceptives in Dutch and Anglo-American healing cultures* (Amsterdam/New York: Rodopi, 2002); C. Bonah and A. Rasmussen (eds), *Histoire et médicament aux 19e et 20e siècles* (Paris: Editions Glyphe, 2005); J. Collin, M. Otero and L. Monnais (eds), *Le Médicament au coeur de la socialité contemporaine: regards croisés sur un objet multiple* (Sainte-Foy, Quebec: Presses Universitaires du Québec, 2006); A. Tone and E. Siegel Watkins (eds), *Medicating Modern America: prescription drugs in history* (New York: New York University Press, 2007).

a major role for much of the century. The costs of research and development were high, increasing exponentially over time, and the pharmaceutical industry, which came to be dominated by large corporations, sought economies of scale and international markets for its products. Towards the end of the century a number of new trends were emerging, which shaped the markets for medicines and challenged the industry, and are discussed in several papers in this selection. They include: the growth of biotechnology; the changing nature and extent of regulatory systems; the respective roles of research and marketing in drug development; the validity of clinical trials and their reporting; the appearance of patient-activist movements, and the part played by doctor–patient relationships in drug consumption.

Public and private institutions, and the scientific traditions and corporate cultures associated with them, underpin the evolution of drugs and the pharmaceutical industry in the twentieth century, providing it with significant sources of continuity. However, geopolitical events, in particular the two World Wars;[15] regulatory changes, especially in the USA; and last but not least key developments and discoveries, such as the discovery of diphtheria antiserum, antibiotics and, more recently, the development of biotechnology, have created major discontinuities. These help to structure our chronological overview, which is therefore in four parts: 1880s-World War One; the inter-war years; post-World War Two; since the 1970s.

The end of the nineteenth century until World War One

Medicinal plants have always played an important part in the treatment of disease. However, it was only at the beginning of the nineteenth century that the first therapeutically active principle was extracted from a plant. This was the narcotic principle of opium, which was later named 'morphine'.

15 See J. Goodman, 'Pharmaceutical industry', in R. Cooter and J.V. Pickstone (eds), *Medicine in the Twentieth Century* (Amsterdam: Harwood Academic, 2000), pp. 141–154; more particularly on the impact of war on the pharmaceutical industry see V. Quirke, 'War and change in the pharmaceutical industry: a comparative study of Britain and France in the twentieth century', *Enterprises et Histoire*, 36 (2004), 64–83.

The isolation and identification of other plant alkaloids, many of them from tropical areas of the expanding European empires, soon followed. These included emetine from ipecacuanha root for dysentery, and quinine from cinchona bark for malaria. The pharmaceutical businesses that manufactured these drugs often also developed the chemical expertise required for selecting and extracting suitable material. Demand for quinine was especially high, but supplies difficult to obtain. Attempts were therefore made to synthesize it from material that was more readily available, such as coal tar. It was one such attempt, by a young British chemist named William Perkin, which led to the first artificial dye, mauveine, and hence to the synthetic dyestuffs industry.[16] Because of favourable economic and political conditions, the industry grew most rapidly in Germany,[17] and it was in the laboratories of the German chemical companies that many of the first synthetic drugs, including aspirin, were developed.[18]

Towards the end of the nineteenth century, the use of machinery for the manufacture of tablets became common in the industry, enabling the mass-production and facilitating the consumption of drugs such as aspirin. At around the same time, a new approach to the treatment and prevention of disease appeared. Also requiring an 'industrial' style of production, it was based on the use of vaccines and sera, and the first major application of this approach was in the mid-1890s, against diphtheria. In Chapter 1, Simon and Hüntelmann contrast the development of diphtheria antitoxin in France and Germany, highlighting the different organization of research and legal contexts for the production and sale of such medicines in the two countries. Also in the same period, the medical marketplace was re-structured, leading to a sharp distinction being made between 'ethical'

16 A. Travis, *The Rainbow Makers: the origins of the synthetic dyestuffs industry in Western Europe* (Bethlehem: Lehigh University Press, 1993).

17 J.J. Beer, 'The emergence of the German Dye Industry', *Illinois Studies in the Social Sciences*, 44 (1959): Ch. 7; G. Meyer-Thurow, 'The industrialization of invention: a case study from the German chemical industry', *ISIS*, 73 (1982): 363–381; E. Homburg, 'The emergence of research laboratories in the dyestuffs industry', *British Journal for the History of Science*, 25 (1992): 91–111.

18 For aspirin see D. Jeffreys, *Aspirin: the remarkable story of a wonder drug* (London:Bloomsbury, 2004).

and 'proprietary' medicines and their manufacturers.[19] This had important implications for the long-term evolution of the drug industry, helping to establish the connection between business and science that has lasted to this day. This is discussed by Huisman in his study of Dutch pharmaceutical firms in Chapter 2, and by Sismondo in his analysis of the relationship between marketing and research in Chapter 8.

Following the discovery of diphtheria antitoxin many drug companies, in France, Germany, and elsewhere, became involved in the manufacture of biological remedies. However, chemotherapy, rather than biotherapy, became the dominant approach to therapy in the twentieth century, and was used to target a wide range of illnesses, from infectious diseases, to chronic disorders, and cancer. The inventor of chemotherapy – the use of chemical substances for the prevention and treatment of disease – was the German bacteriologist and immunologist Paul Ehrlich, working at the Institute for Infectious Diseases in Berlin. As well as elaborating a theory of drug action (receptor theory, described by Prüll in Chapter 5), in 1910 Ehrlich developed the first chemotherapeutic remedy, Salvarsan, a 'magic bullet' targeting the micro-organisms responsible for syphilis without harming their human hosts.[20] However, in the period that followed the discovery of Salvarsan, other than some successes against tropical diseases caused by protozoa, chemotherapy appeared to have failed its early promise. It was not until the discovery of the sulphonamides in the 1930s that chemotherapy can be said to have come into its own.

19 See J. Liebenau, 'Ethical business: the formation of the pharmaceutical industry in Britain, Germany and the US before 1914', in R.P.T. Davenport-Hines and G. Jones (eds), *The End of Insularity: essays in comparative business history* (London: Cass, 1988), pp. 117–129.

20 J. Parascandola 'The Theoretical Basis of Paul Ehrlich's Chemotherapy', *Journal of the History of Medicine and Allied Sciences*, 36 (1981): 19–43; J. Liebenau, 'Paul Ehrlich as commercial scientist and research administrator', *Medical History*, 34 (1990): 65–78.

The inter-war period

Soon after the discovery of Salvarsan, the First World War interrupted European and American supplies of German imported drugs, leading to the abrogation of German patent rights, and stimulating the production of synthetic drugs in countries at war with Germany. Nevertheless, the innovatory advantage in drug discovery remained principally in Germany, where the close relationship that existed between academia and industry was an important contributory factor. This is well illustrated by the life and career of the malariologist Werner Schulemann, which spanned both the academic and industrial worlds, as well as the time periods before and after the Second World War, and is explored by Hulverscheidt in Chapter 4. Schulemann was involved in the development of the first synthetic anti-malarial, Plasmochin (also known by its generic name, pamaquin), at the Bayer laboratories (IG Farben) in Wuppertal-Elberfeld, before moving in 1938 to the Institute for Pharmacology at Bonn University, where he continued to attract industrial funding for his research. It was in these same Bayer laboratories that, in the mid-1930s, Gerhard Domagk discovered the antibacterial properties of the red dye Prontosil Rubrum, the first broad-spectrum antibacterial sulphonamide drug, and a milestone in the history of drug discovery and the pharmaceutical industry.[21]

Between the wars there were also important achievements in replacement therapy, against deficiency diseases caused by a lack of vitamins or hormones. In this area, many significant developments occurred in Britain and North America, where a physiological approach to drug development provided an alternative to the chemical approach favoured by most German laboratories.[22] Insulin was discovered by researchers at the University of Toronto, in Canada, and knowledge about the hormone and know-how

21 See D. Bovet, *Une Chimie qui guérit: histoire de la découverte des sulfamides* (Paris: Payot, 1988); J.E. Lesch, 'Chemistry and biomedicine in an industrial setting: the invention of the sulfa-drugs', in S. Mauskopf (ed.), *Chemical Sciences in the Modern World* (Philadelphia: University of Philadelphia Press, 1993), pp. 158–215; idem, *The First Miracle Drugs* (Oxford: Oxford University Press, 2007).

22 See Weatherall, *In Search of a Cure*.

concerning the production method were transferred rapidly across the world through the firms chosen to be licensees.[23] As well as the sex and other hormones, in this period several vitamins were identified,[24] and various processes for their manufacture were devised and patented, thereby creating a precedent for the patenting of life-science innovations in the second half of the twentieth century.[25] Thus, in Chapter 13, Bächi recounts how, in an attempt to synthesize vitamin C, the Swiss drug company Hoffman-la Roche (now Roche), working in collaboration with the chemist Tadeus Reichstein, developed a new process that included a biotechnological step. A forerunner of the hybrid processes that became the hallmark of the biotechnological era, it gave much higher yields than either the extraction of the natural vitamin or a purely chemical synthesis, but was controversial at the time.[26]

These early successes and achievements of the biomedical sciences and the pharmaceutical industry led to increasing demand for drugs to treat an ever-growing variety of complaints. Despite the economic recession, in the 1930s there was a rising tide of concern about body weight and image, in particular in the USA. This coincided with the emergence – in an as yet largely unregulated market – of an effective but dangerous medicine, dinitrophenol. As Swann demonstrates in Chapter 10, dinitrophenol enabled weight loss, but its serious side effects created a potentially major

23 M. Bliss, *The Discovery of Insulin* (London: Faber and Faber, 1988). See also C. Sinding, 'Making the unit of insulin: standards, clinical work, and industry, 1920–1925', *Bulletin of the History of Medicine*, 76 (2002): 231–270.

24 See various contributions in S. de Chadarevian and H. Kamminga (eds), *Molecularizing Biology and Medicine: new practices and alliances, 1910s-1970s* (Amsterdam: Harwood Academic, 1998).

25 This would lead to debate and controversy in the post-war period, especially in connection with the patenting of penicillin. See R. Bud, 'Upheaval in the moral economy of science? Patenting, teamwork, and the World War II experience of penicillin', *History and Technology*, 24 (2008): 173–190, and other articles in this special issue. See also G. Dutfield, *Intellectual Property Rights and the Life Science Industries: past, present and future* (World Scientific Publishing, 2nd Edition, 2009).

26 A. Kornberg, 'The Two Cultures: chemistry and biology', *Biochemistry*, 26 (1987): 6888–6891; idem, *The Golden Helix: inside the biotech ventures* (Sausalito, California: University Science Books, 1995).

public health disaster. Although use of the drug persists to this day, the combined efforts of the American Medical Association, the Food and Drug Administration (FDA), and the journalists who publicized its dangers, contributed to a 'tectonic shift' in American drug safety legislation. This occurred in 1938, bringing the distribution of dinitrophenol under control of the FDA, and conferring upon the agency considerable influence. Because of the growing size and importance of the American market for medicines after the Second World War, this influence would be exerted not only in, but also outside the USA.[27]

Post-World War Two

The Second World War resulted in major discontinuities for science and industry, while at the same time providing a significant stimulus for growth during, as well as immediately after the hostilities, especially in the burgeoning field of antibiotics, and in the USA. The development of penicillin during the war is widely recognized as a watershed.[28] Celebrated as the first veritable cure for infectious diseases, demand for the drug was high in the post-war years, as Santesmases shows in the case of Spain, discussed in Chapter 3. Although penicillin itself could not be patented, the deep fermentation process developed in the USA during the war was protected by patents.[29] Therefore European companies wishing to use it – among them Spanish firms – had to pay royalties in return for the technical know-how and a licence to manufacture the antibiotic.

Hard on the heels of penicillin came streptomycin, giving new hope to tuberculosis sufferers, and that was followed by more new antibiotics developed and launched in the 1950s. The war and penicillin had brought

27 For more on the FDA see D. Carpenter, *Reputation and Power: organizational image and pharmaceutical regulation at the FDA* (Princeton: Princeton University Press, forthcoming in 2010).

28 See J. Le Fanu, *The Rise and Fall of Modern Medicine* (London: Abacus, 2000).

29 For patents and their uses see various papers in J-P Gaudillière (ed.), *History and Technology*, Special Issue 24.2. (June 2008).

new companies into the industry, perhaps most notably the American firm Pfizer, which built on its contribution to the deep fermentation process for penicillin and developed the novel antibiotic tetracycline. Like other American companies in the 1950s, Pfizer established manufacturing facilities abroad, at first mainly for the purpose of making and selling antibiotics, and this expansion helped to fuel the growth of the American pharmaceutical industry in the post-war period. Later, in order to capitalize on high-quality education systems, particularly in Europe, American firms also established R&D facilities abroad, for instance Pfizer's research centre at Sandwich in Britain.[30]

War created a special need for anaesthetics, antibiotics, anti-malarials, and other medicinal products. In Britain, and elsewhere, as well as mobilizing traditional drug companies for the war effort, it had helped to establish new firms in the pharmaceutical sector, such as the then largest British chemical group, Imperial Chemical Industries (ICI).[31] In the 1950s, ICI built a new R&D facility at Alderley Edge, South of Manchester, where in the 1960s and 1970s drugs such as tamoxifen and the beta-blockers were developed, the former for the treatment of breast cancer, the latter for the treatment of cardiovascular diseases. Underpinning the discovery of the beta-blockers and the other receptor antagonists or stimulants that followed them, was the concept of receptors. However, for the concept to become well established, particularly among the somewhat sceptical community of pharmacologists, it needed first to be successfully applied. In Chapter 5 Prüll examines the uneasy diffusion of Raymond Ahlquist's theory of alpha- and beta-receptors. Although it would play a key role in the development of remedies such as the anti-ulcer drugs Tagamet and

30 On the role of Anglo-American relations in the post-war growth of the UK pharmaceutical industry, see for example V. Quirke, 'Anglo-American relations and the co-production of American "hegemony" in pharmaceuticals', in H. Bonin and F. de Goey (eds), *American Firms in Europe* (Geneva: Droz, 2009), pp. 363–384. See also Tony Corley's contribution to this volume in Chapter 7. On the globalisation of the pharmaceutical industry more generally, see Ramirez, 'The globalisation of research in the pharmaceutical industry'.

31 W.J. Reader, *Imperial Chemical Industries: a history* (London: Oxford University Press, 1975), vol II; C. Kennedy, *ICI: the company that changed our lives* (London: Hutchinson, 1986).

Zantac, it was accepted mainly after it had been used to develop the first beta-blockers, which were found to be effective in the treatment of hypertension and other cardiovascular diseases.[32]

There were many other new drugs developed and launched in the 1950s and 1960s. These included not only antibiotics and receptor antagonists or stimulants, but also anaesthetics, psychiatric drugs, cortisone and other corticosteroid hormones, non-steroidal anti-inflammatory drugs (NSAIDs), and the contraceptive pill.[33] All contributed to meeting and reinforcing the expectations of the new national health services', of medicine's, and of industry's abilities to discover cures for dreaded diseases, as well as treat minor complaints and effect lifestyle changes. The latter is evidenced in Chapter 9, in Niquette and Buxton's discussion of popular pharmaceuticals, such as antacids, laxatives, and later the 'famous blue pill' (Viagra), as well as in Chapter 12 in Prescott's account of the morning-after pill and other contraceptive measures.

To what extent these expectations were lowered in the early 1960s in the wake of the thalidomide disaster, which contributed to public anxieties about the end of the 'Age of Optimism' and to undermining faith in science and medicine, has not yet been fully explored.[34] But thalidomide did usher in a period of tightening drug safety regulation across the world.[35] At the

32 For more on this, see Quirke, 'Putting theory into practice'.

33 J. Swazey, *Chlorpromazine in Psychiatry: a study of therapeutic innovation* (Cambridge, Mass: MIT Press, 1974); C. Djerassi, 'Steroid research at Syntex: "The pill" and cortisone', *Steroids*, 57 (1992): 631–641; E.M. Tansey, '"They used to call it psychiatry": aspects of the development and impact of psychopharmacology', in M. Gijswijt-Hofstra and R. Porter (eds), *Cultures of Psychiatry and Mental Health Care in Postwar Britain and the Netherland* (Amsterdam: Rodopi, 1998), pp. 79–101; L. Marks, *Sexual Chemistry: a history of the contraceptive pill* (New Haven/London: Yale University Press, 2001); D. Healy, *The Creation of Psychopharmacology* (Cambridge, Mass.: Harvard University Press, 2004).

34 Le Fanu, *The Rise of Modern Medicine*, Part 2; Bud, *Penicillin.*

35 On the impact of thalidomide on British drug safety legislation, see for example E.M. Tansey and L.A. Reynolds (eds), 'The Committee on Safety of Drugs', *Wellcome Witnesses to Twentieth Century Medicine* (London: Wellcome Trust, 2007), vol. 1, pp. 103–132. More specifically on ICI, see J. Abraham and C. Davis, 'Testing times: the emergence of the practolol disaster and its challenge to British drug regulation in the modern period',

same time, growing concerns about the costs, prices and profits made by the industry were voiced, especially in the USA and in the UK. This led to the Kefauver Committee hearings in the former, and the investigations of the Sainsbury Committee in the latter.[36]

Since the 1970s

In the last quarter of the century, it was becoming clear that the number of new drugs launched in the market was decreasing, the rate of innovation was beginning to slow down, and the period of the so-called 'Therapeutic Revolution' (roughly from the 1930s to the 1970s) was drawing to a close. Stricter regulatory procedures and more extensive clinical trials meant that it took much longer for new medicines to be approved, thereby eating into the patent period and reducing profitability. For some firms, increased financial risks, combined with greater scepticism, and sometimes outright opposition from the medical profession and the public towards the products' intended purpose or use, had the effect of discouraging innovative R&D. Prescott argues in Chapter 12 that this was the case with contraceptive drugs and devices in the USA, while Tobbell shows in Chapter 11 that, combined with other factors such as continuing disparities in regulatory practices and the uneven development of disease-based organizations, this could result in very different national therapeutic regimes, as illustrated by the treatment of thalassemia patients in Britain and the USA.

Partly because of increased social, political and economic pressures, in the 1980s and 1990s the industry underwent significant consolidation and restructuring, and despite persisting national differences, much of it was transnational in character. Thus, in Chapter 7, Corley examines the

---

Social History of Medicine, 19 (2006): 127–147; also V. Quirke, 'The impact of thalidomide on the British pharmaceutical industry: the case of Imperial Chemical Industries', in J.-P. Gaudillière and V. Hess (eds) 'Ways of Regulating: therapeutic agents between plants, shops and consulting rooms', (Max Planck Institut für Wissenschaftgeschichte, Preprint 363, Berlin, 2009), pp. 125–141.

36 For more on the latter see Slinn, 'Price controls or control through prices?'.

effects of one of the larger consolidations: the merger of the British-based Beecham Group with the American firm SmithKline, Beckman, which in 1985 produced SmithKline Beecham (SB). By seeking assistance from across the Atlantic in order to overcome deficiencies at home, the merger helped to transform the British company's previously inward-looking ethos into one which made effective use of innovation and marketing to overtake its rivals. This raised Beecham to somewhere near the top of the world's pharmaceutical league, and prepared it for its next big merger with the British industry leader, Glaxo Wellcome, to form GlaxoSmithKline (GSK), in 2000.

Increasingly, large corporations such as SB, and later GSK, were urging their R&D on to search for 'blockbuster' drugs, i.e. those which would find large markets across the world. In the 1970s and 1980s, they included anti-ulcer drugs, such as Glaxo's Zantac, and in the 1990s, Pfizer's Viagra. Viagra was one of the first of what are now called 'lifestyle' drugs, and as such has attracted considerable interest. However, as Niquette and Buxton argue in Chapter 9, the success of Viagra cannot simply be attributed to the industry's promotional campaigns. Indeed, they show that the popularity of drugs like *Viagra* is rooted in the everyday relationships people have with medication, as well as with others through medication. These have shaped the social representation of pharmaceuticals at the same time as reflecting the relational requirements of life in modern, post-industrial societies.

As to prescription medicines, they have reached the consumer through a series of transactions that has become increasingly complex, partly because of the requirement for informed consent that was introduced in many countries after the Second World War.[37] In Chapter 6 Richard and Lussier focus on one specific transaction, that between doctors and patients. The exchange that takes place during medical consultations remains the primary way of informing the patient, to enable him to give his informed consent and take his medication correctly. Knowing that patient participation is associated with better health outcomes, Richard and Lussier examine the

37 On the history of informed consent, see P.J. Weindling, 'The origins of informed consent: the International Commission for the Investigation of Medical War Crimes and the Nuremberg Code', *Bulletin of the History of Medicine*, 75 (2001): 37–71; idem, *Nazi Medicine and the Nuremberg Trials: from medical war crimes to informed consent* (Basingstoke: Palgrave-Macmillan, 2004).

content, attitudes and emotions in audio-recorded primary care encounters, and devise a model to help not only improve the quality of doctor–patient exchanges, but also increase patient participation.

The complexity of transactions through which prescription medicines reach the consumer has, since the beginning of the twentieth century, required a considerable marketing effort directed at the prescribers. However, it would seem that it is only in recent decades that the marketing function in pharmaceutical corporations has become more directly involved in creating a market and, potentially, in playing a decision-making role in R&D. This trend is dissected by Sismondo in Chapter 8. Drawing on books by marketers and cases that have recently come to light, he argues that, in the rational world centred on health which the industry and its various customers have helped to create, there is no intrinsic divide between research and marketing.

It is unclear as yet how far the growth of the marketing function in traditional pharmaceutical firms has been related to the shift that has been taking place in innovation from Big Pharma to biotechnology start-ups. Indeed, since the 1990s, many of the most innovative drugs have not come from the large R&D laboratories of the pharmaceutical industry, but from the much smaller biotech companies. These developed in clusters, first in the USA, then mainly in Europe, and most recently in some countries of the Far East. In Chapter 15, Saives, Mehran, Desmarteau and Garnier discuss the Quebec biotechnology cluster, the largest in Canada, where it has benefited from a number of positive factors, such as geographical proximity to the American market, government policies stimulating innovation in the biopharmaceutical sector, and the presence of Montreal as a world-class research centre in life and health sciences, and analyse the evolution of 100 biotech firms located within it.

What is now regarded as first-generation biotechnology, i.e. brewing and baking, as well as the cross-breeding of plants and animals, has a long history.[38] The second generation evolved in the first half of the twentieth

38 For histories of biotechnology, see R. Bud, 'Biotechnology in the Twentieth Century', *Social Studies of Science*, 21 (1991): 415–457; idem, *The Uses of Life: a history of biotechnology* (Cambridge: Cambridge University Press, 1993); Kornberg, *The Golden Helix*.

century with the development of biological products such as vaccines, vitamins, hormones, and antibiotics. This is studied by Bächi in Chapter 13 through the example of the Swiss company, Hoffman-la Roche, which in the 1930s succeeded in adding a biotechnological step to enhance its yields of synthetic vitamin C. The third generation of biotechnology has developed since the 1970s, based on two fundamental discoveries from the new discipline of molecular biology: Cohen and Boyer's discovery of recombinant DNA in 1973, and the development of hybridoma and monoclonal antibodies by Milstein and Kohler in 1975.[39] However, while the traditional pharmaceutical sector watched these developments with interest, only a few firms entered the biotechnology business themselves.[40] Perhaps in part because of its early experience with vitamin C, one of the first of these firms was Hoffman-la Roche, whose establishment of a molecular biology laboratory in Nutley, New Jersey, contributed to its transition from a chemistry-based company to a life-science enterprise, and is examined by Bürgi and Strasser in Chapter 14.

Many of the developments discussed in this brief chronological overview occurred in the West, which for much of the twentieth century remained the hub of drug discovery and the heart of the world's pharmaceutical sector. At the beginning of the twenty-first century, inevitably the direction the industry and drug development will take is unclear. The

39 L.E. Kay, *The Molecular Vision of Life: Caltech, the Rockefeller Foundation, and the rise of the new biology* (New York: Oxford University Press, 1993); P.G. Abir-Am, 'The Molecular Transformation of Twentieth-Century Biology', in J. Krige and D. Pestre (eds), *Science in the Twentieth Century* (Amsterdam: Harwood, 1997), pp. 495–524; R. Bud, 'Molecular biology and the long-term history of biotechnology', in Arnold Thackray (ed.), *Private Science: biotechnology and the rise of the molecular sciences* (Philadelphia: University of Pennsylvania Press, 1998), pp. 3–19; M. Morange, *A History of Molecular Biology* (Cambridge, MA: Harvard University Press, 1998); S. de Chadarevian and Bruno Strasser (eds), 'Molecular Biology in Postwar Europe', Special issue of *Studies in the History and Philosophy of Biological and Biomedical Sciences*, 33C (2002); P.G. Abir-Am, 'Molecular Biology and its Recent Historiography: a transnational quest for the "Big Picture"', *History of Science*, 44 (2006): 95–118.

40 See L. Galambos and J. Sturchio, 'Pharmaceutical firms and the transition to biotechnology: a study in strategic innovation', *Business History Review*, 72 (1998): 250–278.

effects of the creation and dissemination of new scientific knowledge, as well as the expanding generics industry and rising affluence in the world's two fastest growing economies, India and China, have yet to be seen. Nevertheless, that biotechnology has transformed therapeutic innovation at the end of the twentieth century cannot be doubted. It has encouraged Big Pharma into strategic alliances with start-ups and academia, leading to the growth of what is known as the 'Bioscience industry'. Although the extent to which it has fulfilled its early promise in the field of human healthcare has been questioned, there have been some significant achievements; these include human insulin, human growth hormone, and recombinant human interferon. And at the beginning of the twenty-first century, bioscience continues to promise us new and better therapies, such as therapeutic cloning and gene therapy, and medicines targeting the individual patient rather than the disease.

Themes and Structure in Twentieth-Century Pharmaceuticals

Part 1: Different countries, times and perspectives

The launch of new drugs has varied from one country to another, as also has the use and mode of delivery of a drug. National differences of this type, identifiable in the late nineteenth and early twentieth century, have been persistent in ways not always anticipated, and indeed continue today. In Part 1 these continuities and variations are explored in several European countries over different time periods. In Chapter 1, Simon and Hüntelmann contrast the French and German experience of diphtheria serum. In the late nineteenth century, serum therapy was developed as a new method for the treatment and prevention of disease. The first successful application of this principle was against diphtheria, starting in the mid 1890s. By the beginning of the twentieth century, the serum was being mass-produced around Europe and in the USA. This chapter explores the different models for serum production in France and in Germany, models that were related not only to the contrasting institutional organization of research and development, but also to very different legal contexts for the production and sale of such medicines in the two countries.

In France, the research into potential treatments (principally vaccination) for diphtheria was conducted at the newly founded Pasteur Institute by Roux and Yersin, who in 1888 famously isolated the toxin produced by the diphtheria bacteria. In Berlin, Behring pursued his own line of research that resulted in the discovery that the serum of the blood from animals immunized against diphtheria could confer immunity on ones that had never been exposed to the disease. This discovery formed the basis for the development of a treatment for humans using the serum of blood extracted from animals inoculated with the diphtheria toxin. In France, production was for the most part controlled by the Pasteur Institute, which, with the help of generous contributions from the French public, set up its own dedicated stable/factory for the production of the diphtheria 'antitoxin'. In Germany, the responsibility for mass production was assumed by private pharmaceutical laboratories such as Hoechst, which worked in partnership with government-employed research scientists. The chapter explores the effects that these different configurations of public/private collaboration had on the forms of production, distribution and consumption of the diphtheria antitoxin. The production of the serum therapy was also the subject of different legislation in the two countries. In Germany, the state required that all sera be checked by a national laboratory before they could be sold. Thus the government founded the Institut für Serumprüfung und Serumforschung, which in 1899 became the Königlichen Institut für experimentelle Therapie, under the direction of the illustrious Paul Ehrlich, and which was responsible for providing official approval of the medicine. In France, on the other hand, the issue of quality control was left to the producers of the therapies although they had to apply for approval from a special commission. The chapter concludes with a consideration of the effect that these different legal situations had on the relationship between the manufacturers of these therapies, the government, and the public at the beginning of the twentieth century.

Chapter 2 takes us to the Netherlands, where Huisman explores the development and transformation of the national pharmaceutical industry in the first half of the twentieth century. It is often assumed that the pharmaceutical industry 'took off' only after the Second World War. While the industry certainly enjoyed enormous growth in that period (due to the introduction of new medicines), its expansion was facilitated by structures that had come into existence much earlier, between 1880 and 1940. By using the perspective of the medical market, Huisman argues, it is possible to do justice to the many dimensions of this early transformation of pharmaceutical production. Thus, the market perspective may enrich the historiography on pharmaceuticals, because it does not restrict itself to only either 'scientific' or 'successful' companies. It could even be argued that a similar one-dimensional view on the healthcare system (focusing on theory or success) has impoverished the historiography, because it excludes discussions about the dynamics of its evolution.

Traditionally, pharmacists had been the most important professional producers of medicines. In the course of the nineteenth century, the pharmaceutical industry developed, representing an enormous threat to them. The market was flooded by standardized, cheap, industrial products that were produced on a large scale, meeting a growing demand. In its advertising campaigns the new industry addressed itself directly to the public, disregarding the profession. Pharmacists felt threatened in their economic position as well as in their professional self-esteem. United with physicians and lawyers in the Association against Quackery they tried to do battle with the new industry by drawing a sharp line between ethical and unethical medicines and producers, 'scientificity' being the criterion of demarcation. 'Scientific' stood for ethical, humanitarian, and public; its opposite for unethical, commercial and secret. The construction of this distinction, the chapter suggests, was an important part of the professional strategy of pharmacists and – later on – of the pharmaceutical industry in an attempt to gain legitimacy and obtain a greater share of the market.

Initially the industry defended itself by calling upon the liberal principle of the freedom of trade. Gradually, however, some pharmaceutical entrepreneurs realized that a 'scientific' profile need not stand in the way of commercial success. Furthermore, it could facilitate the cooperation of physicians and pharmacists. As it turned out, pharmacists were more than willing to associate themselves with the 'ethical' part of the industry, appreciating its complementary – and growing – pharmacological expertise. New alliances were shaped, and a new pharmacological landscape developed. However, the success of 'ethical' pharmaceutical companies did not mean that their 'unethical' counterparts disappeared. It would seem that the success of a remedy and its producer is less dependent on its pharmacodynamic 'effects' or its legal status than on its cultural image. Whereas some consumers were attracted to 'neat' products with appealing brand names and fancy packaging, others preferred 'ethical', scientific remedies. While sections of the new industry were transformed from threat into allies, acquiring an unchallenged position, quackery and self-medication never vanished. The added value of the medical market perspective, Huisman concludes, is that it does not selectively focus on medical progress or growing professionalization. Using the market model makes it possible to show how some groups decide to join forces, and exclude others. In the process, the traditional image of a linear development of medicine is replaced by a new, multidimensional, vision of the structuring of the health market.

In Chapter 3 we move to a different time period – the 1950s – and to a different country, Spain, where Santesmases discusses the arrival, the manufacture, and the impact of antibiotics. The discovery of penicillin enjoyed world-wide publicity which led, initially, to demand far outstripping the availability of the drug in Spain, as elsewhere. The first antibiotics made in Spain did not reach the Spanish market until 1955, although Spanish physicians had started to report on the uses of penicillin and streptomycin, and academic researchers had received information about the drug from the Professor of Biology at the University of Madrid, Florencio Bustinza, well before that. The drugs themselves were available before that date, but mainly through the black market, as only small amounts of imported drugs were released by the Spanish government. Strong demand already existed, and fictional accounts (famous novels on the post-war period) placed these very limited supplies – or, when available, the huge prices – of early antibiotics in Spain after the Second World War at the core of public imagination and culture. It was in this context that Franco's government launched a public competition to approve two factories for penicillin production in

Spain. The manufacture of antibiotics in Spain was dependent on foreign patents, and agreements were signed between Spanish pharmaceutical firms and American firms to give them access to American technical processes and expertise. This created a strong dependency on foreign innovations concerning the manufacture of penicillin and on any further innovation concerning production of other antibiotics developed outside Spain.

Thus, the production of antibiotics in Spain lagged behind not only the demand for them generated by physicians and scientists, but also their production abroad. While Fleming and Waksman were becoming public heroes, the production of antibiotics in Spain began and provided the firms involved with long-lasting success in the pharmaceutical field.

Part 2: Different actors: scientists, doctors and patients

The processes by which drugs are discovered, brought to the market, prescribed for patients and eventually consumed by them, involve many actors. The three chapters in this section focus on the scientists working on the research that leads (sometimes) to drug development and, at the other end of the complex series of transactions involved, on doctor–patient relationships.

In Chapter 4, Hulverscheidt explores the life and work of the German malariologist, Werner Schulemann (1888–1975), and offers a perspective on how pharmaceutical research was funded in Germany between 1920 and 1970. Schulemann was, by any standard, an extremely successful researcher over a long period of time. He worked at the Bayer Laboratories (IG Farben) in Wuppertal-Elberfeld until 1936, became a Professor and the Director of the institute for pharmacology at Bonn university in 1938, and was involved in the development of the first synthetic anti-malarial, Plasmochin. For his research he received the Mary-Kingsley-Medal from the Liverpool School of Tropical Medicine. He had a great talent for obtaining money for his work; being able to attract funding from the pharmaceutical industry, as well as from the University, from the state and from the DFG (a foundation which funded scientific research). His applications show how, at different times, different rationales were needed. He was not involved in anti-malarials in the 1950s, since due to the effectiveness of DDT at that time there was little demand for such drugs. But in the early 1960s, when the first cases of resistance against DDT were documented, Schulemann returned to his earlier work using his old recipes to develop anti-malarials. He was then more than 70 years old, and was still being funded by the DFG. The example of Schulemann shows how a scientist can spend his entire lifetime on a single line of research. Funding depends not only on the quality of ideas and outcomes, but also on the contacts and abilities of the individual researcher. The chapter discusses whether, and in what ways, funding regulations influence the direction and progress of research.

In Chapter 5, Prüll presents a very different perspective on the development of influential scientific ideas, more particularly, the receptor concept. In 1948, Raymond P. Ahlquist (1914–1983), then head of the Pharmacological Department of the Medical College of Georgia/Augusta, published a theory on drug binding which has had a decisive influence on the development of scientific pharmacology and drug discovery in the second half of the century. 'Receptors' are proteins on the surface of the cell, which enable the latter to attach to foreign or bodily substances. Ahlquist differentiated between two receptors of the adrenergic system, which is responsible for heart rate and blood pressure. With this research he paved the way for pharmacological research on an ever-growing number of receptors, and for the development of drugs to treat an increasing variety of diseases, from heart, lung, to gastric disorders and cancer. However, the contribution of Ahlquist's theory remained unacknowledged until the 1960s, and he waited in vain for the award of the Nobel Prize.

Based on the thesis that scientific knowledge is socially constructed, Prüll's chapter analyses the fate of Ahlquist's concept. Using printed and unprinted sources as well as interviews with Ahlquist's former friends and colleagues in Augusta, Prüll delivers the first detailed examination of the subject. Several factors caused problems for Ahlquist: misunderstandings about his theory, the ambiguities of the theory itself, Ahlquist's personality and his own attitude towards his 'dual adrenoceptor concept'. Prüll concentrates on difficulties related to the orientation of Ahlquist's research work within pharmacology, and to the fact that Eli Lilly and Co. sponsored much basic research on the adrenergic system in the 1940s and 1950s. Ahlquist and other research groups depended on that sponsorship and were part of a competitive network that tried to unravel the fundamental mechanisms of the adrenergic system.

Although Ahlquist had the practical application of his concept in mind, initially the concept remained in the realm of theoretical pharmacological research. Prüll argues that Ahlquist's research environment was largely responsible for the late recognition of his work. Because of the narrow scope of his research, a vision of how to apply his approach to medicine only began with the work of Sir James Black on the beta blockers in 1957.

With Chapter 6, the focus shifts to examine the roles of very different actors in the transaction outlined above. Richard and Lussier put a microscopic lens on the nature and intensity of discussions between patients and physicians on the subject of medications, and present us with their results. The purpose of their study is to examine the content, attitudes and emotions in audio-recorded primary care encounters. They measure the extent of the dialogue which takes place between patient and physician, and relate it to the content in terms of whether the encounter concerns new prescriptions, repeat prescriptions, or ongoing medication. Their analysis, set out in this chapter, identifies three clusters of themes from their observations. This enables them to suggest a model to be used in further explorations of the patient-physician relationship, in order to improve communication between doctor and patient and in the hope of improving health outcomes for the patient.

Part 3: Developing, selling and representing drugs

As the industry grew rapidly in the second half of the twentieth century, the ways in which drugs were marketed and sold were transformed. In Chapter 7, Corley explores the changes as they were reflected in one corporation, the Beecham Group, in the last two decades of the century. By 1985 the British-owned Beecham Group, lately evolved from a pills and proprietaries manufacturer into the innovator of semi-synthetic penicillins, had clearly lost its sense of direction. It lacked an overall corporate strategy, so that its pharmaceutical and consumer product sub-groups scarcely spoke to each other, and it had diversified into such non-core products as cosmetics and DIY requisites. It had few contacts with the institutions – which held 80 per cent of its shares – or the City analysts. In principle it avoided using management consultants. Instead, it relied on a strong 'Beecham' ethos, nourished by the pre-1968 innovation- and marketing-driven Leslie Lazell. Following some dismal half-year results, the Beecham chairman was sacked in a boardroom coup.

The new chairman, an American, deliberately and relentlessly set out to eradicate that inward-looking ethos. He replaced most of the incumbent directors with executives brought in from more progressive companies, such as Cadburys and BOC, to head the key finance, corporate communications and personnel functions and to act as enforcers of his new strategy, based on Harvard Business School principles. He hired a firm of American consultants, which recommended a 'merger of equals' with a company of similar size, to raise Beecham from 23rd in the world pharmaceutical league to somewhere near the top.

The American corporation SmithKline Beckman was chosen as partner on account of its advanced R&D facilities in Philadelphia and its well-trained sales force. After the merger in 1989, the new SmithKline Beecham (SB) pursued a thorough programme of integrating every level of management across the Atlantic, advised by McKinsey & Co. as consultants. Within a few years SB markedly improved its performance. It licensed some of its drugs and information to other companies, enhanced its corporate image and shifted emphasis from mere curative medicines to the broader health care concept.

SB was soon faced with external pressures such as an accelerating merger movement throughout the industry, the introduction of managed health care and stricter drug regulations, but also the onset of the biotechnology revolution, manifested in the very costly gene sequencing process. In 1998 it therefore sought to merge, again on an equal footing, with the British industry leader, Glaxo Wellcome, a bid which succeeded in 2000. As its head office was in London but the operational headquarters were in Philadelphia, it had become truly outward-looking.

The pharmaceutical industry's marketing function has become in recent years a focus for research and debate. This is reflected in Chapter 8, in Sismondo's examination of the links between R&D and marketing. He starts with the view of the management guru, Peter Drucker, that the overarching goal of a business is 'to create a satisfied customer'. From this formula, it not difficult to see that both the customer and their satisfaction need to be created. In any case, marketing is the prime force behind the creation at least of customers, and perhaps also their satisfaction.

Yet pharmaceutical companies have to portray themselves as research and development organizations involved only by necessity in marketing and sales. Most of their various customers – for prescription drugs these are traditionally physicians, though increasingly drug companies pay attention also to patients, potential patients, pharmacists, health maintenance organizations, hospitals, government agencies, nursing homes, and clinics – want their drugs to be part of a rational world centred on health. Any visible aspect of drug research, development, or promotion that is not part of a logic of health is immediately suspect. Thus ghost-writing of research is hidden, recognition of the pleasure drugs can produce is carefully managed, and marketing often takes an educational form.

Sismondo goes on to describe in very general terms the integration of clinical research and marketing, drawing on books by marketers and recent cases that have come to the public eye. The tools that have been used to accomplish this integration over the past half-century are various, but they all stem from a realization that in a rational world centred on health there need be no intrinsic divide between research and marketing. Most obviously, marketing drugs to physicians, who are professionals acting within their own spheres, depends crucially on research. Physicians respond, and need to see themselves as responding, to facts, figures, and studies. The well-chosen images and vehicles for marketing campaigns must be subordinated to research. Yet at the same time research is a means of increasing sales.

Pharmaceutical companies are, of course, among the most successful of businesses. They have become so by resolving, or at least appearing to resolve, the conflicts between the logic of business and the logic of health. Big Pharma and its representatives easily and often argue that their actions are innocuous: just as research and marketing (and education) are necessarily connected, if the drugs are seen as efficacious then improved sales means improved patient outcomes. Thus, the chapter concludes, we need to look carefully both at conflicts and their management to understand why we should care about pharmaceutical marketing.

Aspects of marketing are also explored in Chapter 9, in which Niquette and Buxton discuss the role that the development of advertising has played in the growth of the pharmaceutical industry in the twentieth century. They begin by noting that advertising itself has been greatly influenced by the promotion of patent medicines, body-care products and pharmaceutical drugs. Drawing on the example of medications that have become part of global popular culture over the last century – such as pain killers, cough syrups, anti-itching ointments, analgesic balms – they offer us a theoretical model through which the representations of pharmaceuticals can be studied in relation to social practices. They begin by describing the modern characteristics that make pharmaceuticals an object radically different from others; secondly, they show how these characteristics have contributed to the transformation of everyday life; and thirdly, how the changes in the role that pharmaceuticals play in society are reflected in the social representations that circulate in the public sphere.

It is commonly thought that the day-to-day uses of pharmaceuticals are closely related to the discourses that are promulgated by the drug industry and disseminated by the medical establishment. However, few studies have addressed the fact that social practices surrounding the use of medications are the product not only of professional and promotional discourses, but also of the very process by which popular images of pharmaceuticals are constituted. In other words, the social representations of pharmaceuticals are rooted in the everyday relationship people have with medication, and with others *through* medication. For instance, the popularity of the Sildenafil Citrate (Viagra), as compared to other pharmaceuticals used for the treatment of symptoms associated with the andropause, cannot simply be attributed to the success of promotional campaigns. The representations used to promote the famous blue pill are inspired by the various ways in which this medication was already involved in society, beyond the quest for more satisfying intercourse. These representations include themes such as social status in a highly competitive environment, male timidity in the doctor's surgery, the desire to experience a constant good mood, the sense of obligation to make one's partner happy, and the ideals of autonomy and spontaneity in sex. All these themes are typical of the relational requirements of modern life in post-industrial societies. Therefore, studying the changes in the relational themes through which the uses of medication have been depicted from the beginning of the twentieth century up to the present allows us to understand the process by which pharmaceuticals, beyond their therapeutic functions, have become reflexive tools by which we relate to one another.

Part 4: Drug regulation and its limits in the USA

Over the course of the twentieth century, the pharmaceutical industry became one of the most highly regulated industries, a contentious issue in Europe and North America especially. The three chapters in this part of the book each illustrate certain aspects of regulation. In Chapter 10 Swann examines the case of the weight-reducing drug, dinitrophenol, in the USA in the 1930s. At that time, he argues, the conjunction of a rising tide of concern for body image and weight consciousness with the emergence of an effective but hazardous medicine to lose weight laid the foundation of a potentially major public health disaster. The public was being told increasingly in advertising, motion pictures, and other venues of the desirability of a lithe, trim, and athletic figure in both women and men. It was in this social context that dinitrophenol, a toxic component used in the munitions industry of the First World War, came to be understood by pharmacologists as capable of raising the metabolic rate to such an extent as to readily dissipate body fat. Its narrow margin of safety, however, prompted many to advise against its indiscriminate use by the public. Two leading researchers who studied dinitrophenol argued that it should be restricted only to those professionals who could monitor a patient's basal metabolic rate. But at this time in the USA there was no mechanism to compel such a limit on distribution of a medicine of this kind. Consequently, dozens of dinitrophenol-containing products were launched on the market for self-medication, often without the active ingredient even being labelled on the package. The American Medical Association and journalists tried to publicize the dangers associated with the use dinitrophenol, which could result in serious injuries such as cataracts, as well as fatalities. The Food and Drug Administration (FDA) joined in this effort, using creative means to alert the public. But there was little else the FDA could do under the then current food and drug law. A tectonic shift in the drug law in 1938 had a swift impact on the distribution of dinitrophenol in the USA, although use of the drug has persisted, even up to the present.

In Chapter 11, Tobbell explores the concept of nationally distinctive pharmaceutical 'cultures', created in part by different national systems of regulation. Arthur Daemmrich has argued that different systems of drug regulation, clinical trials, and post-marketing surveillance of pharmaceutical drugs developed in Germany and the USA because of differences in the 'therapeutic cultures' of the two countries; that is, the relationships among the state, the pharmaceutical industry, the medical profession, and disease-based organizations.[41] Tobbell describes the development of iron chelation therapy for the treatment of thalassemia in Britain and the USA in the second half of the twentieth century as a case study for examining the importance of 'therapeutic cultures' in pharmaceutical development. While British physicians readily incorporated iron chelators such as Desferal (marketed by Ciba-Geigy, now Novartis) into medical practice in the 1960s, physicians in the USA were more reluctant to do so. Despite the publication of several reports from Britain in the early 1970s demonstrating the long-term efficacy of Desferal, American physicians continued to question the clinical value of iron chelation therapy, and Desferal remained on the margins of American therapeutic practice. The reasons for this difference in medical practice include, in particular, the different ways in which thalassemia patients and their families, physicians and researchers, pharmaceutical companies, and the state influenced the development of iron chelation therapy in Britain and the USA in the second half of the twentieth century.

41 Daemmrich, *Pharmacopolitics*.

National regulatory differences in the approach to contraception are the subject of Chapter 12, focussing on the USA. In the late 1980s, Prescott writes, officials from various American and international family planning organizations reported an alarming trend: due to an increasingly hostile legal and regulatory environment precipitated by lawsuits against the contraceptive pill, the Dalkon shield, and other potentially harmful reproductive technologies, women in the USA had fewer birth control options than they did in the previous decade. Indeed, family planners in the 1980s observed that women in Third World countries, which continued to use birth control methods that had been banned or were no longer in use in the USA, actually had more options than did their American counterparts. Drawing on records from the U.S. Food and Drug Administration, Planned Parenthood Federation of America, and the National Institutes of Health, as well as interviews with contraceptive researchers, this chapter explores how the political, legal, and regulatory environment of the 1970s and 1980s hindered contraceptive research and development in the USA. It also shows how recent work by non-government organizations has made new reproductive technologies available by going beyond the venture capital model used by most pharmaceutical companies in the USA.

Part 5: From Pharma to Biotech

The development of the biotechnology industry in the last three decades of the twentieth century has had a significant impact on the pharmaceutical industry and its products. Often seen as a 'new' industry in the last three decades of the century, evidence of an earlier biotechnological process is brought to the fore by Bächi in Chapter 13. In 1933, in a laboratory at the Swiss Federal Institute of Technology (Zurich), the synthesis of l-ascorbic acid (vitamin C) was achieved by Tadeus Reichstein and his collaborators. When the patents concerning the Reichstein procedure, owned by the small Swiss foodstuff company Haco, were offered to the Swiss pharmaceutical company F. Hoffmann-la Roche, Basle, in May 1933, Roche was not really interested in them. That is, not until Reichstein had improved the synthesis by introducing a *biotechnological* step which allowed the use of glucose as a basic raw material. By using bacteria to transform sorbitol into sorbose (an idea already floated in the nineteenth century), he had found that the synthesis of vitamin C brought higher yields compared with the extraction of natural vitamin C. Therefore, Roche bought the patent, a decision that led to Roche's first (traditional) biotechnological production step.

However, there remained much corporate and academic research to be done in order to translate Reichstein's laboratory synthesis into industrial manufacture. Above all, the use of bacteria caused unforeseen troubles, and a number of resistances had to be overcome. First of all, Roche's industrial chemists, who simply had no experience of handling bacteria and would have preferred a purely chemical synthesis, disapproved of the bacteriological technique. Introducing biotechnology into the chemical company therefore required intense cooperation between academia and industry. Moreover, the scaling up of this microbiological step caused new kinds of problems different from the ones experienced in a scientific laboratory. Last but not least, during the economic crisis of the 1930s, Roche's general director hesitated in making such an investment into new, expensive equipment indispensable for this biotechnological step.

In the process of introducing biotechnology into the pharmaceutical industry, corporate traditions, path-dependency, and academic-industry relations have played an important role, but so have different social and national contexts. Thus, while the fear of bacteria helped to sell synthetic vitamin C to consumers, the same fear hampered the diffusion of the Reichstein procedure. In Nazi Germany, I.G. Farben favoured the purely chemical Helferich synthesis for the vitamin, refusing to have bacteria present in their production plants, because, as they told Roche's management, they feared 'poisoning and degeneration'.

In Chapter 14 Bürgi and Strasser examine Roche's approach to drug development over two critical decades, the 1960s and 1970s, more particularly its move into research in molecular biology. Early in 1967, Sidney Udenfriend and Herbert Weissbach, both collaborators of the National Institutes of Health, and John J. Burns, director of research at Hoffmann-la Roche in Nutley (New Jersey, USA), discussed the establishment of a research institute guided by academic scientists and financed entirely by the company. Two months later, the company's top management in Basel approved the project and soon the construction of the Roche Institute of Molecular Biology (RIMB) in Nutley began. In 1972, the new building opened its doors to 128 scientists.

The establishment of the institute meant two major changes to the research policy of Hoffman-la Roche: first, the company embarked on its own 'academic' line of research; second, the emphasis on biological research challenged the position of organic chemistry as the main supplier of scientific knowledge inside the company. The decision to build a corporate research institute devoted to fundamental biological research was reached within a very short time. However, the conditions that made it possible for the company to change its research policy so quickly had been developing since the late 1950s. An analysis of the minutes of the Roche Research Management Group (RRMG), founded in 1956, and the reports composed for its annual meetings, make it possible to identify five determining factors: business success, new conceptions of research management, changes in regard to the legal environment, the growing influence of the company's American subsidiary, and difficulties in recruiting qualified staff.

Soon after it had been founded, the RIMB was recognized as an academic research institution. Its members were integrated into university networks and benefited from the exchange of ideas and materials. In 1970, the RRMG for the first time discussed the opportunities offered by 'genetic bioengineering'. In 1977, together with the industrial research department in Nutley and with Genentech, the RIMB embarked upon developing recombinant human interferon, which has been sold by Hoffmann-la Roche since 1986. In the 1990s, the company closed the RIMB and contracted a large number of start-up companies and university institutions. The RIMB was, in essence, just a phase within Hoffman-la Roche, but what it did was to contribute to the company's transition from a chemistry-based enterprise to a bioscience company.

Across the world the biotechnology industry is generally to be found in clusters, and Canada is no exception to this rule. In Chapter 15, Saives, Mehran, Desmarteau and Garnier present and discuss the results of their exploratory study of the Quebec biotechnology cluster, the largest in Canada. As the literature on innovation management has shown, with the advent of a new scientific paradigm, new players within the biopharmaceutical industry have emerged, i.e. firms dedicated to biotechnology, often regarded as essential partners to traditional pharmaceutical companies. These enterprises focus on the management of innovation with the aim of entering a growth cycle based on R&D projects, on the choice of intellectual property to be protected and traded, and with the aim of managing financial options. The purpose of the study is to obtain a deeper understanding of the technological and organizational development cycle of these firms. It relies on a series of data extracted from semi-structured interviews taken from over 110 biotech firms located within the bio-industrial cluster of Quebec. Close to 30 qualitative variables describing the stages of classical evolution within growing firms are examined using a multi-factorial analysis. This exploratory field study leads to the observation of a number of discrepancies between the organization of knowledge creation and the type of financial governance within biotechnology firms. It concludes that three modes of development are present: that is pre-entrepreneurial, entrepreneurial and managerial, and that the passage from one to the other is marked, in the first instance, by a teleological gap, and in the second by a creativity gap.

Concluding remarks

The papers in this collection illustrate the wealth and variety of perspectives on pharmaceuticals and their development over the course of the twentieth century. They touch upon many of the issues that are matters of concern and debate today, including the pharmaceutical industry, biotechnology, innovation, academic-industrial relations, the interaction between doctors and patients in the Age of Information, as well as the drugs themselves and their uses and representation in society. We hope that this cross-national and multidisciplinary approach will stimulate further debate on the subject.

PART I

Different Countries, Times and Perspectives

JONATHAN SIMON AND AXEL C. HÜNTELMANN

1 Two Models for Production and Regulation: the diphtheria serum in Germany and France[1]

The development of serum therapy for diphtheria, first in Germany and subsequently in other European countries, represents a major innovation in modern medicine. While this treatment was introduced at the end of the nineteenth century, it nevertheless profoundly marked twentieth century European medicine and public health in several respects. In particular, this therapy of biological origin – produced by horses – attracted intense legislative attention, reflecting a perception that a failure to oversee its production and distribution would entail significant public health risks.[2]

In this paper, we look in more detail at the responses to the particular problems posed by the introduction of serum therapy for diphtheria in two neighbouring European countries, France and Germany. By examining the regimes of production and legislation that governed the manufacture and distribution of serum in these two countries in the closing years of the nineteenth century, we elucidate the similarities and differences between them. Then, we examine these comparative elements in order to identify what are only superficial divergences and what are fundamental differences between

1 This article was written as part of the project 'The Industrialisation of experimental knowledge' financed by the German Research Foundation – DFG HE 2220/4–1 and 2. We would like to thank Christoph Gradmann, Anne I. Hardy and Volker Hess for their invaluable contributions to this work.

2 For other studies that treat the history of serotherapy in France, see J. Liebenau and M. Robson, 'L'Institut Pasteur et l'industrie pharmaceutique', in M. Morange (ed.), *L'Institut Pasteur: contributions à son histoire* (Paris: La Découverte, 1991), pp. 52–61, and P. Weindling, 'Emile Roux et la diphtérie', in ibid., pp. 137–143. For a detailed account of this history in Germany, see C. Throm, *Das Diphtherieserum. Ein neues Therapieprinzip, seine Entwicklung und Markteinführung* (Stuttgart: WVG, 1995).

the management of the serum in the two countries. This work allows us to contribute to ongoing historical discussions concerning national traditions or national styles in the domain of public health and the pharmaceutical industry at the dawn of the twentieth century.[3]

The Origins of Serum Therapy for Diphtheria

Despite repeated claims for priority from other French and German researchers, the dawn of serum therapy was ushered in by a paper published in 1890 by Emil Behring (1854–1917) and Shibasaburo Kitasato (1852–1931) in the *Deutsche Medizinische Wochenschrift*.[4] Thus, these two researchers at the Royal Prussian Institute for Infectious Diseases in Berlin announced how they had successfully treated animals infected with a lethal disease (diphtheria and tetanus respectively) using blood serum extracted from individuals of the same species that had been rendered immune to the disease in question.[5] Already in this publication, Behring and Kitasato predicted the use of this technique in the treatment of human disease,

3 For a challenging contribution, as well as a review of classic positions, see P. Baldwin, *Contagion and the State in Europe, 1830–1930* (Cambridge: Cambridge University Press, 1999). The issue of national styles of microbiology is raised by Andrew Mendelsohn in his analysis of the Pastorian and Kochian schools: J.A. Mendelsohn, 'Cultures of Bacteriology: formation and transformation of a science in France and Germany, 1870–1914' (PhD thesis, Princeton University, 1996).

4 The French claim to priority was championed by Héricourt, who maintained that the idea originated in Richet's research from the 1880s. See J. Héricourt, *La Sérothérapie. Historique, état actuel, bibliographie* (Paris: Rueff, 1899). In Germany, Behring was challenged by the Japanese researcher Masanori Ogata and Rudolf Emmerich from Munich for priority in discovering the principle of serotherapy. For more on these priority claims, see D.S. Linton, *Emil von Behring: infectious disease, immunology, serum therapy* (Philadelphia: American Philosophical Society, 2005), pp. 92–98.

5 E. von Behring and S. Kitasato. 'Ueber das Zustandekommen der Diphtherie-Immunität und der Tetanus-Immunität bei Thieren', *Deutsche Medicinische Wochenschrift*, 16 (1890): 1113–1114.

but much work lay ahead in finding a way of producing effective serum in large enough quantities for first experimental, and later commercial use. Despite the beginning of experimental therapy as early as 1892, the success of this new treatment came to the attention of a broad public only at the end of 1894, following the highly publicized intervention by Émile Roux (1853–1933) at the Eighth International Hygiene Congress held in Budapest in September of that year.[6] Soon afterwards, the Pasteur Institute in France joined a number of German companies in producing the serum, which quickly became a standard treatment for diphtheria.

Production in Germany

During the period immediately following the discovery of the therapy, 1894–1895, several serum manufacturers started production in Germany. In Hamburg, for example, two pharmacists set up the company Ruete & Enoch and started to produce serum, as did the well-established pharmaceutical and chemical company, E. Merck in Darmstadt. In the same region, the Farbwerke Hoechst, formerly known as Meister, Lucius and Brüning, produced the serum, along with Schering in Berlin, although only these two companies had invested in the experimental development of the serum. Starting in 1892, Hoechst signed a contract with Emil Behring at the Berlin Institute for Infectious Diseases to provide financial support for his studies in this area. Around 1893, Schering started supporting Hans Aronson (1865–1919) at the Berlin veterinary school in a similar fashion. In March 1894, Schering published an advertisement announcing that they were able to produce and deliver a diphtheria serum.[7] Despite this

6 Roux's intervention was published in full in *La Semaine Médicale* (8 September, 1894): 405–410.

7 See the flyer 'Diphtherie-Antitoxin-Lösung-Schering' in the archive of Emil von Behring, University of Marburg; and in the Federal Archive Berlin (Bundesarchiv – hereafter BA Berlin), R 86/1646.

publicity, however, they were unable to supply either a very high quality serum or the quantities needed.[8] Hoechst, Schering's main competitor, continued their research and, in August 1894, were the first company to make diphtheria serum widely available through the pharmacies. As predicted, the diphtheria serum was a resounding economic success. Following the successful completion of clinical trials in the winter of 1893/1894 and the following summer, Hoechst decided to build a new production plant dedicated to diphtheria serum, with the new buildings being officially inaugurated in November 1894. Hoechst's new serum production plant set the standard, moving its capacity well beyond Schering in terms of both quantity and quality.

The Serum Production Process

The process of manufacturing serum starts with the production of the diphtheria toxin. For this, pure cultures of the diphtheria bacillus (from infected children) are sown on an appropriate culture medium, and, after several days of breeding, are killed using a disinfectant.[9] This was a difficult process, as the amount of toxin produced depended on the strain of bacteria used and the preparation of the culture medium, among other factors, and the details of toxin production differed from company to company. The quality of the toxin was, however, of vital importance as it had to be strong enough to stimulate the immunization process but without causing excessive inflammation at the injection site.

In 1894, the immunization process at Hoechst lasted between four and five months, but by 1900, this time had been reduced to only four weeks.

8 The serum was tested by Paul Ehrlich who found that it was not as effective as claimed, see BA Berlin, R 86/1646.

9 For a detailed description, see the outline for the founding of a state-run control station in the archive of the Paul Ehrlich Institute in Langen (hereafter APEI), Dept. Va, No. 1, Vol. 1. A general account is given in Throm, *Diphtherieserum*, p. 82.

During this period, the horses were inoculated between eight and eleven times at regular intervals using increasing doses of toxin. Test-bleedings indicated when the antitoxin content of the serum was at its maximum level, at which time the horse could be bled regularly. The quantity of blood drawn varied from 4 litres at Hoechst in 1894 to as much as 12 litres a few years later. After bleeding, the blood was left to stand so that the serum separated out. At Hoechst extra steps of filtration and centrifugation ensured the purity of the product. After the quality of the serum had been checked, it was placed in a phial at an appropriate dose (5 or 10 millilitres), labelled with the immunization power and date of preparation, ready for distribution.[10]

Around 1900, Hoechst came to dominate the field, supplying around three quarters of the serum in Germany. Hoechst's product also had the highest value in terms of immunization units and, in contrast to Schering and most other producers, there were never any complaints from the control station about Hoechst's serum.[11] Finally, it should be noted that Hoechst – again in contrast to Schering – enjoyed excellent relations with the Ministry for Cultural Affairs, Robert Koch (1843–1910) and his Institute for Infectious Diseases, with a history of providing the Institute with dye products used for staining. Later on, Schering often complained that only Hoechst was included in the ministerial proceedings.[12] It is interesting to note that in September 1894 the *Berliner Lokal-Anzeiger* tried to mount a public subscription for the benefit of the poor on the model

10 Throm, *Diphtherieserum*, pp. 99–113; A. Eiermann, 'Die Einrichtung zur Darstellung des Diphtherie-Heilserums in den Höchster Farbwerken', *Münchener Medicinische Wochenschrift*, 41 (1894): 1038–1040.

11 The amounts of serum produced and its value, as well as the complaints about it are in the annual reports of the control station in the APEI, Dept. IV, No. 1, Vol. 2.

12 For example, see Schering to the Prussian Ministry for Cultural Affairs, 15.10.1894 and 26.8.1904, Secret Central Archives (Geheimes Staatsarchiv Preußischer Kulturbesitz – hereafter GStA PK), HA 1, Rep. 76 VIII B, No. 3747 and No. 3754. On Schering, see C. Kobrak, *National Cultures and International Competition: the experience of Schering AG, 1851–1950* (Cambridge: Cambridge University Press, 2002).

of the *Figaro* (described below), but they never succeeded in raising very much money.[13]

Production in France

Although the archives at the Pasteur Institute are not complete enough to be certain, it seems likely that French serum research and production consistently lagged behind that of its German neighbours and probably owed much to what the pastorians could read in German publications. This is not surprising, given that the pioneers in the research leading to a stable production process for the serum were Behring, Wernicke (1859–1928), and Aronson in Berlin. Nevertheless, a French group of researchers following the German developments undertook its own experimental program starting in 1892.[14] While Émile Roux was the senior member of the group conducting this research, it is likely that the most active in this area were Edmond Nocard (1850–1903), a professor at the National veterinary school in Alfort and disciple of Pasteur, and Louis Martin (1864–1946) a young doctor who had recently replaced Alexandre Yersin as Roux's assistant. Thus, a series of experiments on large animals conducted at Alfort led to a French method for producing toxin that differed only in detail from the German one described above. In the end, the French favoured the use of an iodine solution to reduce the potency of the toxin for its initial administration to horses, and never adopted the sophisticated methods for evaluating the

13 For more on the financial aspects, see A.C. Hüntelmann, 'Two Cultures of Regulation? The development, production, and state control of diphtheria serum at the end of the nineteenth century in France and Germany', *Hygiea Internationalis*, 6 (Autumn 2007): 99–120.

14 E. Roux and L. Martin, 'Contribution à l'Étude de la Diphtérie (Serum-Thérapie)', *Annales de l'Institut Pasteur*, 8 (1894) : 609–661; E. Roux, 'Sur le sérum antitoxique. Communication faite au Congrès de Budapest', *Annales de l'Institut Pasteur*, 8 (1894): 722–727.

quality of the serum developed by Paul Ehrlich and discussed below. Roux started testing the Pasteur Institute's serum on children in the diphtheria service of the nearby *Hôpital des Enfants Malades* on 1 February 1894, a clinical experience that would provide the results he presented in Budapest. Nevertheless, there was a sudden shift in the scale of demand and the perception of the serum starting in September 1894, a shift for which the Pasteur Institute was radically under-prepared. Roux's presentation at the Congress of Hygiene in Budapest – a congress that Behring did not attend – made the front page of the *Figaro* on 6 September, and in the following weeks two things happened that would determine the form serum production took in France. First, there was suddenly a huge demand that the Pasteur Institute was unable to satisfy, and second, the *Figaro* started a subscription that would finance large-scale serum production in France.[15] The press coverage following the Budapest meeting also helped to found the national myth that the serum was a French discovery, an idea that Roux dismissed in his personal correspondence, although this denial was not widely disseminated in the press. The sudden demand and the sense of urgency it created pushed the Pasteur Institute team to build up their serum production capacity very rapidly indeed. While there were only a dozen horses present at the Institute in September 1894, a little over three months later the Institute owned 136, and this growth was accompanied by the gradual transfer of all production to a single site at Garches, a village to the west of Paris. The huge personal and financial investments put into meeting the early demand meant that the Pasteur Institute ended up extracting over seven and a half thousand litres of blood from their horses in 1895, a volume that would not be matched again until four years later.[16] This rapid scaling-up was made possible thanks to the funds raised by the *Figaro*'s subscription. The initial target of 30,000 francs for the fund, would, it was predicted, enable the treatment of all the children of Paris.

15 Initially, Roux's idea seems to have been to use the press coverage to encourage a wealthy benefactor to fund a hospital dedicated to treating infectious disease, a project that would be realised subsequently under different circumstances.

16 These volumes are taken from loose notes inserted into the registers of the serum production centre at Garches now held in the archives of the Pasteur Institute (AIP).

After only a month, however, the appeal had already raised 240,000 francs, and showed no sign of losing steam. This money paid for the building work required to adapt the facilities at Garches to turn it into a dedicated serum production plant, leaving ample funds for Roux to invest elsewhere. The Institute also derived a considerable income from the sales of the serum, which, while being provided free to those unable to pay (subsidized by local and national government funds), was meant to be sold at 3 francs per dose (20 cc) to all those who could.[17]

Although the Pasteur Institute's production at Garches dominated the 'market' for diphtheria serum in France, there were a number of initiatives outside the capital that sought to make up for Roux's initial inability to supply the whole country. Thus, there were independent efforts to produce the serum in Nancy, Rouen, Le Havre, Toulouse, and elsewhere, mostly, it appears, with Roux's explicit, albeit reluctant support, as he was obliged to admit his incapacity to meet national demand, at least in the months following his triumphal presentation at Budapest.

As an example, I will cite the case of the *Institut sérothérapique de l'Est* founded in Nancy in 1895. As elsewhere, local newspapers in the East of France relayed the news of Roux's announcement in Budapest as well as the subscription launched by the *Figaro* in September 1894. The initial enthusiasm around the campaign gradually turned to frustration, however, as it became clear that the serum would not be distributed in any significant quantities beyond the Paris region in the immediate future. At the beginning of November, a group of representatives from the medical faculty of Nancy, headed by Pierre Parisot, a young but rising star of the faculty, went to visit the Pasteur Institute in Paris to find out more about how the serum was produced. At the same time, the money being collected by the local newspapers in Nancy as part of the subscription for the Pasteur Institute's diphtheria serum service stopped being paid to Paris and was instead diverted into a fund intended to pay for Nancy's own serum production. Following this change of tack, the size of the subscriptions rose dramatically, and on 13 November 1894, the *Institut sérothérapique de l'Est*

17 File 'Souscription pour la diffusion du vaccin du croup', dossier Roux, AIP.

was formed by a group of local notables.[18] The production of serum was to take place under the direction of Eugène Macé, Professor of Hygiene at the Medical Faculty, who had the immediate task of preparing two horses for this purpose. The official founding of the *Institut* had to wait until January 1895, with the construction of a dedicated building taking place between 1895 and 1897 and paid for by a very generous donation from the prominent French businessman, M. Osiris. One of the ironies of this story is that the money Osiris left to the Pasteur Institute in his will would finance the building of the Pasteur Hospital that Roux had initially hoped for in 1894.

The case of this provincial institute is typical in several respects. First, it mobilized local resources, diverting subscriptions initially intended for the Pasteur Institute, once it became clear that supplying the regions was not the priority in Paris. Second, the venture was linked to the local medical faculty, with Eugène Macé using the opportunity to set up not only a laboratory for conducting microbiological examinations and producing serum, but also courses in microbiology not provided by the medical faculty itself. Thus, as in other cities, the diphtheria serum provided the opportunity for forming a local 'Pasteur Institute'. In the case of Nancy, the Institute was physically and financially independent from the Faculty of Medicine (except for certain staff members), something that set it apart from most other large towns, although in 1899 it too was absorbed into the faculty. Third, local pride and patriotism combined with anti-German nationalism in asserting that the serum was French and should be produced locally, a sentiment that was expressed clearly in Nancy, which stood right next to the Franco–German border. Finally, like the Pasteur Institute, these institutions ran on a mixed economy of substantial charitable donations, which allowed the free distribution of serum, as well as the sale of a certain amount of serum, and an income derived from conducting the associated microbiological tests. Of course, we need to bear in mind that the production of such local centres, with only two or three horses in

18 'Institut sérothérapique de l'Est' file in Nancy's Municipal Archives.

their stables, could never rival the 'industrial' production that the Pasteur Institute mounted in Paris.[19]

Medical Legislation in Germany and Patent Law

After being presented at the congress in Budapest and at the regular meeting of the German Naturalists in Vienna at the end of September 1894, the serum enjoyed widespread scientific recognition in a short space of time all over Germany. The serum was not, however, just a scientific success, it also proved to be very profitable for those who produced it. Starting in August 1894, it was available in pharmacies in two units: five millilitres for five marks or ten millilitres for ten marks. With the margins these prices represented, Hoechst had by the end of 1895 made a profit of 707,000 Marks from sale of the serum.[20]

In August 1894, however, it was in principle possible for anyone in Germany to produce the serum. The research that led to the human use of serum was to be found in a series of articles dating from 1890 on, published by Emil Behring, Shibasaburo Kitasato, Erich Wernicke, Paul Ehrlich (1854–1915) and Hans Aronson in the *Deutsche Medizinische Wochenschrift*, the *Berliner Klinische Wochenschrift* or other medical periodicals. These articles were freely available, and were considered sufficient to allow a well-informed microbiologist to reconstruct the production process. Thus, there could be no patent covering the production and use of serum therapy, as all the relevant information had been published, and furthermore there was no protection of any particular trademark. As a member of the Institute for Infectious Diseases, Behring was not supposed to make any additional money out of his work over and above his salary, with his results being

19 Starting in the spring of 1895, the Pasteur Institute did supply the whole of France, as well as exporting serum abroad.

20 Throm, *Diphtherieserum*, Tab. IV.1.

freely available. Indeed, on joining, researchers signed over their rights to any discoveries to the institute.[21]

To make matters more complicated, Behring was a military doctor, who had been officially posted to the Institute, making him answerable to the Prussian army, his employer, as well as to Robert Koch, as director of the Institute. Thus, Behring had asked the surgeon-general of the Prussian Army, Alwin von Coler for permission to conduct his research on diphtheria serum, and paid for it with his own and Wernicke's money, making the legal situation particularly unclear.[22]

Neither Behring nor Hoechst had requested a patent for the diphtheria serum, both for the reasons explained above, and because no pharmaceuticals, chemicals or drugs could be patented in Germany. In 1891, the patent legislation that applied to chemicals was transferred onto pharmaceuticals, meaning that only processes and not products could be protected.[23] Indeed, only the trademark 'Behring's Diphtherie-Heilmittel dargestellt nach Behring-Ehrlich' was protected, meaning that anybody could, in principle, copy the product, and would just need to vary the production procedure to be safe. Thus, Merck in Darmstadt, Ruete & Enoch in Hamburg, as well as the Pasteur Institute in Stuttgart started to produce serum in 1895 without encountering any legal problems.

Due to this unclear legal situation and the profits to be made on the serum, the German medical administration started to worry about imitations of inferior quality as well as unscrupulous racketeers. Behind this

21 Emil Behring's contract concerning his work, results and discoveries at the Institute for Infectious Disease, 7 October 1891, BA Berlin, R 86/3801.

22 Behring eventually left the army to pursue a civilian career, landing himself in trouble with Robert Koch who considered his exploitation of serum therapy to be immoral, and who was jealous of Behring's business career, having failed to make money on his discredited cure for tuberculosis. See C. Gradmann, 'Ein Fehlschlag und seine Folgen. Robert Kochs Tuberkulin und die Gründung des Instituts für Infektionskrankheiten in Berlin 1891', in C. Gradmann and T. Schlich (eds), *Strategien der Kausalität. Konzepte der Krankheitsverursachung im 19. und 20. Jahrhundert* (Pfaffenweiler: Centaurus, 1999), pp. 29–52.

23 W. Wimmer, *'Wir haben fast immer war Neues' Gesundheitswesen und Innovation der Pharma-Industrie in Deutschland, 1880–1935* (Berlin: Duncker & Humblot, 1994), pp. 85–101.

concern there also lurked the spectre of tuberculin, the notoriously unsuccessful treatment for tuberculosis launched by Koch in Berlin in 1890 that had precipitated a high-profile public health fiasco.[24] Within the space of a few weeks, the unbounded optimism that had gripped the German population had turned to sour deception, with the administration failing to intervene effectively in the tuberculin affair. These fears and recent memories, combined with the novelty of serum therapy and lack of information concerning its long-term effects, pushed the medical officials in the Prussian Ministry for Cultural Affairs and in the Imperial Health Office – the highest medical advisory authority in the German Empire, answerable only to the Chancellor and the Office of Internal Affairs – to discuss regulating this medicine. At the end of October 1894 the extraordinary members of the Imperial Health Office were invited to meet and discuss the new serum therapy and the possibilities for implementing state control.

> The easy appraisal of Behring's diphtheria serum by the experts on the basis of its therapeutic success, and the rising demand for the diphtheria serum makes it necessary to discuss government measures concerning the production, distribution and use of this new therapy in order to protect public health.[25]

The conference took place in November 1894, bringing together medical officials from the Ministry for Cultural Affairs, representatives of the Federal States and the Imperial Health Office, and certain scientists from the Prussian Institute for Infectious Diseases, like Paul Ehrlich, Robert Koch and Emil Behring. At this conference, the participants discussed the new serum therapy and the need to protect the public against impure and/or ineffective serum.[26]

A series of meetings followed between November 1894 and February 1895 during which a system of state control was drafted. Significantly,

24 For a detailed description of Koch's development and use of tuberculin, and the criticism raised after it was seen not to work, see C. Gradmann, *Krankheit im Labor: Robert Koch und die medizinische Bakteriologie* (Gottingen: Wallstein, 2005), pp. 134–170.

25 Invitation of the extraordinary members of the Imperial Health Office to a conference on the new diphtheria serum, 3.11.1894, BA Berlin, R 86/1646.

26 See the minutes from 3 and 5 November 1894 in BA Berlin, R 86/1646.

starting in December, representatives of the pharmaceutical industry were also invited to participate in the meetings.[27] The decision was taken that the diphtheria serum could only be sold in pharmacies in accordance with the imperial decree from January 1890. Furthermore, the serum was only available on prescription, in accordance with a Federal resolution from July 1891 that regulated the commerce of potent and dangerous pharmaceuticals. A law to deal specifically with the production of serum was elaborated in the winter of 1894/1895 by members of the Imperial Health Office. Following a few weeks of consultation, the law was put into effect, and the diphtheria serum was included in the Pharmacopoeia germanica as 'serum antidiphthericum'.[28] In light of the dearth of empirical knowledge about the effects of the serum on human subjects, the sale of the product was to be accompanied by forms to be filled in by the treating physician that would allow the authorities to establish the safety and efficacy of the new serum therapy. The information gleaned from the returned completed forms was used by the *Deutsche Medicinische Wochenschrift* and the Imperial Health Office to prepare several bulletins, which disseminated the relevant statistics and made recommendations concerning the use of the serum.[29]

The most important element in the legislation was the state control of the production and distribution of the serum. Prior to the 1880s, the quality of the ingredients and the preparation of medicaments was the responsibility of the pharmacist who sold them. With the rise in mass-production by the pharmaceutical industry, however, it became increasingly difficult

27 *Idem*. For the context of the conference H. Zeiss and R. Bieling, *Emil von Behring. Gestalt und Werk* (Berlin: Bruno Schultz, 1941), pp. 153–157; A.C. Hüntelmann, *Gesundheitspolitik im Kaiserreich und in der Weimarer Republik. Das Reichsgesundheitsamt von 1876–1933* (Diss. Phil., University of Bremen, 2006), chap. 3.3.3.

28 Reichsgesetzblatt 1895, p. 1.

29 Minutes 3 and 5 November 1894 in BA Berlin, R 86/1646. The results of the statistics were published as 'Ergebnisse der Sammelforschung über das Diphtherieheilserum für die Zeit vom April 1895 bis März 1896' and sent to libraries within and outside the German Empire, BA Berlin, R 86/1646. A résumé of these results was published in *Arbeiten aus dem Kaiserlichen Gesundheitsamt*, 13 (1897): 254–292. Behring also published a summary of the collected data from more than 11 000 cases. E. Behring, *Die Statistik in der Heilserumfrage* (Marburg: N.G. Elwert'sche Verlagsbuchhandlung, 1895).

for the apothecary to analyse the ingredients of pre-packaged commercial specialities, meaning that he could no longer guarantee the quality of these medicines.[30] Furthermore, the efficacy and potency of the serum could only be determined by a trained expert with appropriate 'laboratory' facilities. The 'industrial' production of the serum pushed this process forward, with the industrial plant replacing the individual pharmacies as the site not only of production but also of quality control.[31]

Serum Legislation in France

When serum was first produced and distributed by the Pasteur Institute at the end of 1894 and beginning of 1895, there was no legislation covering this product in France. There were two possible approaches a jurist could take to the serum; one would be to consider it to be a medicament, in which case it would not be possible to apply for any patent, and would be covered by laws dating from the beginning of the nineteenth century that gave pharmacists a monopoly over the production and sale of medicines.[32] Alternatively, one could consider that it was not a medicament, in which case it would be subject to the laws governing any other similar commodity such as foodstuffs. The fact that the serum was not inscribed in the official pharmacopoeia,

30 In a context of industrial production it was easier to monitor a few producers than to inspect thousands of pharmacies.

31 J. Holsten, *Das Kaiserliche Gesundheitsamt und die Pharmazie. Dargestellt an der Entstehung des Deutschen Arzneibuches, fünfte Ausgabe* (Diss. Med., Free University Berlin, 1977); E. Hickel, *Arzneimittel-Standardisierung im 19. Jahrhundert in den Pharmakopöen Deutschlands, Frankreichs, Großbritanniens und der Vereinigten Staaten von Amerika* (Darmstadt: WVG, 1973); Wimmer, *Gesundheitswesen*, note 23.

32 Attempts by industrial interests to introduce patents for medicines in 1844 were successfully resisted by pharmacists. See J.-P. Gaudillière, 'Une marchandise pas comme les autres: historiographie du médicament et de l'industrie pharmaceutique en France au XXe siècle', in C. Bonah and A. Rasmussen (eds), *Histoire et médicament aux XIXe et XXe siècles* (Paris: Glyphe, 2005), pp. 115–158.

however, weakened the argument that it was a medicament, although many medical specialities produced by a thriving pharmaceutical industry were in a similar situation at this time.[33] For better or for worse, the new serum was introduced at a time when the French parliament was already debating a new law covering pharmacy, and the initial legislative response was to include an item in this new law that would cover the serum and other vaccination material as well – such as attenuated viruses or toxins.[34] Nevertheless, the sense of urgency felt by the members of both the National Assembly and the Senate pushed them into formulating a separate law to cover these products. The debate over this legislation in various forms took place in the Assembly and Senate in early 1895, and, after having been almost unanimously voted, the proposal was finally signed into law by the President, Félix Faure, on 25 April 1895.

The reporter to the National Assembly on an earlier draft of these regulations underlined the urgency of the situation, particularly in light of the paradoxical, if not straightforwardly illegal situation of the government with respect to the serum. An annual renewable subsidy of 80,000 francs had been voted for the Pasteur Institute to pay for the supply of free serum to the poor, and yet, as explained above, it was not clear that the Institute had the right to distribute the serum, in particular if it was considered a medicament. The method conceived for legalizing the existing situation was thus to introduce an approval system handled by the Ministry of the Interior and pass it quickly into law.

> Until the promulgation of the law on pharmacy, one can only deliver serum in France, whether free or for a fee, whose fabrication or provenance has been approved by the Ministry of the Interior, following the advice of the Academy of Medicine and the Consultative Committee on Public Health.[35]

33 For more on the history and functioning of this industry, see S. Chauveau, *L'Invention pharmaceutique: la pharmacie française entre l'Etat et la société au Xxe siècle* (Paris: Sanofi-Synthélabo, 1999).

34 *Journal Officiel* (12 avril 1895), documents parlementaires, chambre des députés, p. 313. Article 14 of the pharmacy law that was under debate in 1895 was the original legislative tool. *Journal Officiel, Documents Parlementaires, Chambre des députés* (10 janvier 1895), p. 2.

35 *Journal Officiel* (4 mars 1895), documents parlementaires, p. 247.

The initial idea was to insert the relevant clauses into a finance law, to get the situation settled quickly, and await the passage of the pharmacy law under discussion to lay the issue to rest. This solution was rejected on technical grounds, and so a specific law was drafted and debated as a matter of urgency in the National Assembly. During the drafting and negotiation process, the legislators vacillated between a system for approving the product – although avoiding the 'autocratic' excesses of the German government – and the producer, or just the producer. In the end it would only be the producer that needed official approval, which could be revoked at any moment. The instrument that would be used to approve producers, as suggested above, was an advisory committee to the Ministry of the Interior, the 'serotherapy commission'.

There are several noteworthy aspects of the law of 25 April, starting with its broad sweep. Covering 'attenuated viruses,' 'therapeutic serums', 'modified toxins', and similar products, as well as 'injectable substances of organic origin not chemically defined', the law was evidently intended to clear up the whole domain of opotherapy, the medical use of extracts of animal organs popularized by Brown-Séquard.[36] Thus, the introduction of the serum therapy, served as an excuse to take care of a range of other pharmaceutical grey areas at the same time.[37] Second, the law meant that, unlike other medicaments, this one was no longer the pharmacist's responsibility. In principle, the ultimate responsibility for any accident involving approved serum would lie with the Ministry of the Interior, although as this was never tested in court, one can only speculate concerning the responsibility of the producer, or even the pharmacist who sold the product. Third, there was no requirement that pharmacists be involved in the serum production, a novelty for the blossoming nineteenth-century French pharmaceutical industry.

36 For more on Brown-Séquard, see M. Borell, 'Brown-Séquard's organotherapy and its appearance in America at the end of the nineteenth century', *Bulletin of the History of Medicine*, 50 (1976): 309–320.

37 Article 1 of the law of 25 April 1895. The smallpox vaccine was explicitly excluded by article 2.

The debate over the law in the National Assembly was brief but heated, with opponents being caught short by its introduction via an emergency debate and vote. In this context, the main arguments offered against the law were the possible constraints it might put on scientific research, and the need to keep the state out of science. Indeed, the fiercest objection was against the state making the decision about who could and could not produce the serum. Notwithstanding, the law was approved by 516 votes to 2.

The commission was formed on 15 May 1895, and comprised sixteen members. Half were appointed as a result of their administrative position, and the other half were chosen by the Ministry of the Interior. Out of these remaining eight, half were members of the Academy of Medicine, the other half members of the Consultative Committee for Public Health in France. The four chosen from the Academy of Medicine were Nocard, Duclaux, Straus and Grancher, exhibiting a bias towards 'friends' of the Pasteur Institute that doubtless played a significant role in the functioning of the committee. The presence of 'interested' parties from the main serum producer in France can be interpreted in at least two ways. We can read it as a tacit acknowledgement that the Pasteur Institute was the official producer of French serum, or we can explain the coincidence in terms of the localization of relevant microbiological expertise around the Institute. The quasi-monopoly of French microbiology by the Pasteur Institute is an important point in the comparison between France and Germany to which we will return in the conclusion.

French serum production had now entered another grey legal zone, as from the 25 April 1895 until the first announcement of approved institutions, on 26 January 1896, the producers were technically engaging in an illegal activity. The first list of institutions approved for producing diphtheria serum named the Pasteur Institute in Paris and its subsidiary in Lille, as well as laboratories in Le Havre, Nancy, Lyon and Grenoble. None of these producers was a commercial enterprise, all being associated more or less intimately with a local medical faculty, as we saw in the case of Nancy above. Furthermore, every approved producer had some contact with the Paris Pasteur Institute, with directors such as Saturnin Arloing in Lyon and Eugène Macé in Nancy having spent time in Roux's laboratory learning how to produce the serum. It seems probable that this kind of personal contact was a necessary (if not sufficient) condition for having one's production approved, meaning that these local centres of production constituted part of a loosely-knit network of pastorian microbiologists spread out across France.

Finally, the French regulation of serum production left the choice of quality control entirely in the hands of the producers themselves. The Pasteur Institute had developed a version of the initial evaluation technique devised by Behring to fix the power of its serum. This involved injecting a guinea-pig with a certain volume of serum that represented a fixed proportion of the animal's weight, and following this with a fatal dose of diphtheria toxin. The survival of the test animal indicated the strength of the antitoxin. The Institute failed to evolve towards any more sophisticated technique, provoking the criticism of certain international observers.[38] Local production centres probably followed the lead of the Paris Institute, and once approved did not have to justify their choices to anyone. The situation in Germany, as we shall see, was quite different.

State Control of Serum Production and Distribution in Germany

Towards the end of 1894, the German authorities considered several different approaches to state regulation of the diphtheria serum. In this context, they raised the idea of founding an Imperial serum institute, viewing the Pasteur Institute's monopoly in France as an attractive model for serum production and distribution throughout the German empire. Nevertheless, Schering and Hoechst had already invested so heavily in serum therapy that if the state were to impose such an Imperial institute, it would then

38 See T. Madsen, 'Ueber Messung der Stärke des antidiphtheretischen Serums', *Zeitschrift für Hygiene und Infektionskrankheiten*, 24 (1897): 425–442.

have to find an acceptable way to compensate these companies for their investments and eventual loss of profits. This compensation would have been so costly by this point that the idea of an Imperial Serum Institute was quickly abandoned. In the ongoing discussions to put in place a system of state control, the medical officials took into account the fact that there were several producers in different locations, which argued in favour of a central monitoring organism.[39]

The final system governing the production of serum combined centralized and local elements, involving not only the local monitoring of production but also a centralized state institute for quality control. Thus, every production plant was permanently monitored by a medical officer, paid by the producer but answerable to the Prussian Ministry for Cultural Affairs via the district president. In addition, the serum was tested for purity as well as being evaluated and certified centrally at the Serological Institute, founded in February 1895. There were also strict regulations concerning the handling and packaging of the serum for dispatch, and the sale price was regulated, with special tariffs for social security insurance, welfare institutions and hospitals. Finally, the producers guaranteed the withdrawal of phials from pharmacies after two years, or immediately in the case of ineffective or impure serum. The system was implemented within a few months, and after 1 April 1895 only state-certified serum could be sold in Germany.[40]

In order to produce diphtheria serum, a company first had to apply for permission and prove their capability to do so. They also had to pay an 'entrance fee' of 1,000 marks to the Institute for Serological Research and

39 See the discussions on a meeting on 19 October 1894 in the Imperial Health Office, BA Berlin, R 86/1646; and the minutes of a meeting at the Prussian Ministry for Cultural Affairs on 24 October 1894, GStA PK, HA 1, Rep. 76 VIII B, No. 3747; see also Throm, *Diphtherieserum*, p. 71. Later on, however, the idea of a state-run institute for serological research and serum production was only raised by medical officials – as a bargaining tool – when difficulties emerged with the serum producers.

40 See the minutes of the meeting from 17.12.1894, 17.1.1895 and 1.2.1895 and the correspondence between the participants of the meetings in BA Berlin, R 86/1646; GStA PK, HA 1, Rep. 76 VIII B, No. 3747. For the industry side, see the Histocom archive of the Farbwerke Hoechst, Frankfurt, folder GL 18.1/3.

Serological Survey. Apart from this initial fee, the institute was financed by the regular fees for the test procedures. The producers then had to recruit their medical officer, mentioned above, another expense for the firm. Finally, the company would receive an initial inspection following the submission of their request to produce the serum to the Prussian Ministry for Cultural Affairs. The Ministry would make inquiries concerning the company's reputation with the district president, before organizing an inspection by the district veterinary and medical officials that would cover all aspects of production.[41] These stringent inspections were open to German and non-German companies alike, and permission to produce the serum could be granted only when all these formalities had been concluded with the Imperial Health Office and the Prussian Ministry for Cultural Affairs. Beyond these mandatory audits, there were also voluntary ones that would give the producer the right to claim state approval – 'Staatlich geprüft' – for their products. These optional tests were used not only to ensure that the quality of the serum was maintained but also as a marketing tool, passing on this message to the consumer.

At the production site, the medical officer had to be present while the horses were examined by a vet, and during the bleeding procedure. After the serum had been prepared, he then kept it under lock and key awaiting official approval. A sample from each batch was sent to the state-run control station for diphtheria serum, later on the Institute for Serological Research and Serological Survey, which tested it for purity and determined its immunizing power. The process used for assessing the quality of the serum was complex and sensitive and only those trained in the methods of bacteriology could carry it out.

At the Serological Institute, the samples were tested by two medical officials simultaneously and independently. The assay had to prove the harmlessness of the serum – it had to be free of all bacteria and other particles – and to determine its immunizing power. The Serological Institute based the

41 See for instance R. Otto, 'Die staatliche Prüfung der Heilsera', *Arbeiten aud dem Königlichen Institut für experimentelle Therapie zu Frankfurt a. M.*, vol. 2 (Jena 1906), chap. B and C.

evaluation of the efficacy of the serum on a standard serum and a standard toxin. The evaluation procedure was very tricky, involving the injection of very precise dilutions of the standard toxin and test serum, and an observation of the absence of any swelling at the injection site on a guinea pig. As one might imagine, the two medical officials sometimes failed to arrive at the same results. Likewise there were a number of exchanges between production sites and the Serological Institute after they had obtained different determinations. Nevertheless, despite the complexity of the procedure, the use of a standard serum and standard toxin meant that in principle the procedure was standardized and could be applied at any site.[42] After the official results of the evaluation process had been sent to the producer, the medical official could release the relevant batch of serum for transfer into phials and dispatch to the pharmacies. These phials were sealed and bore the mention 'Inspected by the State' on their labels.[43]

In the German case, every step of the process was carefully recorded in a system that was shared between state authorities and the serum producer, involving the exchange and inspection of ledgers. Each bleeding had its assigned operation number, with registers kept at the production site and reports made at the central Institute. Thus, the inclusion of this operation number on the label of the phial distributed for use made each dose of serum easily traceable. Overall, the German system was much more complex than the French one, but as the Danish bacteriologist Thorvald Madsen, later head of the Danish Serological Institute, pointed out, it was precise, and the dates could be compared by all those involved in the process, whether producer, government agent or consumer.

42 Regarding the evaluation procedure see the instructions in BA Berlin, R 86/1646; APEI, Dept. V, No. 1; Otto, 'Die staatliche Prüfung...', note 41..

43 If the serum was found to be impure, it was destroyed. In case of different determinations, the Serological Institute's value took precedence.

Conclusion: production and legislation in Germany and France

A first glance at the situation described above appears to confirm the idea that the administration procedures for the production and distribution of serum exemplified the stereotypes concerning the French and German political systems at the end of the nineteenth century. In Germany, there was an authoritarian bureaucratic centralized (and yet Federal) administration that oversaw every detail of production, while in France, a more relaxed, hands-off regulatory approach left serum production in the charge of a 'trusted' centralized (Parisian) institute that in certain respects was very close to being a state institution.[44] Indeed, the discovery of the serum treatment for diphtheria was of such capital importance for public health that one might have expected a country like France to create a nationalized dedicated serum institute with a monopoly over production. One can suppose that this would have been a realistic option, had the Pasteur Institute not existed. Even before taking institutional form in 1887, this group of scientists who federated around the emblematic, if not mythical figure of Louis Pasteur, effectively dominated all areas of microbiology in France, giving the Institute a 'technical monopoly' over the microbiological techniques needed to produce the diphtheria serum. This not only made the Pasteur Institute an obvious choice for an 'ersatz' state serum institute, but also made it difficult if not impossible for competitors – in particular private companies – to enter the field.[45] Nevertheless, as we have seen, regional serum production existed almost from the beginning. Local actors attempted to make up for the lack of serum coming from Paris, and often

44 For a description of the funding and functioning of the Pasteur Institute, see I. Löwy, 'On hybridizations, networks and new disciplines: the Pasteur Institute and the development of microbiology in France', *Studies in the History and Philosophy of Science*, 25 (1994): 655–688.

45 The fact that it was an 'obligatory point of passage' for serum production adds grist to Latour's pastorian analytical mill, B. Latour, *Microbes: guerre et paix* (Paris: Métaillié, 1984).

simultaneously exploited this opportunity for other ends as well. It should be noted, however, that none of these regional producers was a commercial venture, and all those involved seem to have had some personal connection to the Paris institute. In Germany, on the other hand, serum production and sale was a competitive field in which private companies invested with the aim of maximizing their profits, preferably by dominating the market. While this situation corresponds with an image of the pro-industrial orientation of the German Empire, we should bear in mind that the Imperial authorities seriously considered founding a single Imperial Institute with a monopoly over production of the serum.[46] In both countries, therefore, the legislation recognized the situation in place and sought to regulate it in a way that would avoid potential public health problems associated with the serum. By introducing a complex system of control both upstream and downstream, the German response suggests a relative lack of trust in the producers, while France's one-off (but revocable) authorization procedure seems to testify to a faith in the Pasteur Institute's ability both to assure the quality of its own production and to identify other reliable producers. Nevertheless, this vision needs to be refined, as industry involvement in the drafting of legislation in Germany suggests that it also represented an opportunity for an industry cartel, and maybe one single company to structure the serum market for its own ends. Thus, while an initial absence of competition in France was later effectively consecrated by the law (only approved producers could manufacture serum), it seems that the legislation in Germany, while apparently more favourable to free-market competition, ultimately served to force latecomers out of the market, and thereby restricted the initially fierce competition.

Behind this theme of competition lurks a thorny cultural issue that can once again easily lead us into exaggerated cultural stereotypes. Nevertheless, it does seem that serum production quickly took on the form of a philanthropic exercise in France, while remaining a business venture in Germany.

46 For more on the pro-industrial orientation of the German Empire, see T. Lenoir, 'Revolution from above: the role of the State in creating the German research system, 1810–1910', *The American Economic Review*, 88 (1998): 22–27.

Again, appearances can be deceptive, as price control was only enforced in Germany, where serum was offered to welfare organizations at a reduced price. Indeed, employees benefiting from national health insurance schemes also received the serum for free. In France, the vision of the Pasteur Institute as a charitable humanitarian organization obscured the fact that the Institute demanded payment from the *départements* or local government in return for the serum supplied for use by the indigent population free of charge. As in much of the history of the Institute, the distinction between gift and payment was never clearly defined.[47] Furthermore, people who could afford the serum were supposed to pay for it, and the Institute made money from exporting serum abroad.

Perhaps the most striking difference between France and Germany is the absence of private companies from the French landscape of serum production. Did this result from a French distaste for the involvement of business in medicine, or does it rather reflect the peculiar situation of technical dominance that the Pasteur Institute exercised over microbiology in France? The Institute had few connections with other industrial actors, and denser relationships only formed later.[48] Nevertheless, even outside Paris, local production was linked closely to academic faculties and not private enterprise, which seems to support the first explanation. The question as to why companies did not enter the fray in France remains open, however. Without the records of the serotherapy commission, it is impossible to say, for example, whether or not this possibility was actively blocked by the members of the Academy of Medicine.

The issue of centralization also becomes difficult to resolve when we look in detail. While one might expect centralized production in France and delocalized production in Germany, this was not the case. As in France, Germany had a dominant producer, and as in Germany, there were other producers in France besides the Pasteur Institute. Nevertheless, in Germany all the producers (large and small) competed on the national and even

47 See, Löwy, 'On Hybridizations'.

48 Ibid.

international market, whereas in France, producers outside Paris were essentially serving local demand.

Overall, many of the differences that might at first sight appear structural seem, when one looks closer, to be accidental. Thus the existence of the Pasteur Institute crucially shaped production and legislation in France, while the rapidity with which Hoechst and Schering invested in research and production of the serum similarly oriented German policy. Indeed, when Althoff considered setting up a central Imperial Institute for producing serum, his projections suggested that encouraging the existing competitive situation would result not only in better serum, but also in lower prices.

Finally, the creation of an independent quality control organism is an element in the German system that is strikingly absent in France. This seems to attest either to a greater concern for public health safety or to a lower level of trust in the producers on the part of the Germans or to both. On the French side, we can thus ask whether this lack of any independent mechanism for control resulted from the confidence the government had in the Pasteur Institute or whether they simply had neither the means nor the will to put a French equivalent of the German Serological Institute in place. On the German side, we can ask whether the multiplication of safety checks and control mechanisms reflects their recent unfortunate experiences with modern microbiological medicines or a genuinely greater concern for issues of public health. Whatever the ultimate explanation, however, a close examination of the systems that were developed in France and Germany around the production and control of diphtheria serum can teach us a great deal about the context for the introduction of industrially produced medicines and the orientation of public health policy in Europe at the beginning of the twentieth century.

FRANK HUISMAN

2 Struggling for the Market: strategies of Dutch pharmaceutical companies, 1880–1940[1]

It could easily be assumed that the pharmaceutical industry 'took off' only after the Second World War, when the introduction of drugs like penicillin, cortisone, streptomycin and chlorpromazine caused a true therapeutic revolution.[2] To be sure, the industry did witness an enormous growth in the post-war period, but its expansion was facilitated by structures that had come into place from the 1880s onwards.[3] It was a highly formative period, in which the dynamic relationship between many diverse producers, distributors and consumers resulted in the pharmaceutical infrastructure as we know it today. For a better understanding of the complexities of its formative period – that witnessed the introduction of new remedies, new advertising techniques, new distribution channels, new parties and changing loyalties, both economically and professionally – it is not enough to focus on the vicissitudes of successful, 'scientific' companies. For a proper understanding, the production, distribution and consumption of medicines need to be understood in their broader social and cultural context.

1 I would like to thank Prof. Ernst Homburg for his thoughtful comments on an earlier version of this article.

2 J. Le Fanu, *The Rise and Fall of Modern Medicine* (London: Abacus, 1999), esp. pp. 5–71.

3 This is also the position of Jonathan Liebenau, in *Medical Science and Medical Industry: the formation of the American pharmaceutical industry* (Houndmills etc: MacMillan Press, 1987), where he looks at the industry between 1890 and 1930, mainly from the point of view of successful 'elite actors' in the pharmaceutical field – like the other classic, by J.P. Swann, *Academic Scientists and the Pharmaceutical Industry: cooperative research in twentieth-century America* (Baltimore: the Johns Hopkins University Press, 1988).

I would like to suggest using the perspective of the marketplace for that purpose.[4]

I am not using the concept 'marketplace' in a limited economic sense, but rather in the broader sense in which it is used by the social and cultural historians of medicine.[5] It introduces the notion that medicine and health care are by no means static or self-contained entities; instead, they get their meaning from specific human actions in a particular context. By looking at health care as a (metaphorical) marketplace where the demand for pharmaceutical medical products and services meets their supply, one is forced to distance oneself from the rather linear view in which scientific innovation and government regulation create a pharmaceutical discipline and profession with clear identities and a market with clear boundaries. Consultation and consumption patterns result from the way disease is 'framed'.[6] Researchers, healers and patients negotiate ways of attaching

4 For a related approach, see Nancy Tomes, 'Merchants of Health: Medicine and Consumer Culture in the United States, 1900–1940', *The Journal of American History* (September 2001): 519–547. Tomes tries to transcend essentialism and finalism in medical history by developing the concept of 'consumer culture', which results in a more open, probabilistic perspective on twentieth-century health care.

5 Developed within the English historiography, the concept has mainly been used for the early modern era. See, for example, H.J. Cook, *The Decline of the Old Medical Regime in Stuart London* (Ithaca/London: Cornell University Press, 1986), esp. chapter 1; R. Porter, *Health for Sale: quackery in England 1660–1850* (Manchester: Manchester University Press, 1989); M.E. Fissel, *Patients, Power, and the Poor in Eighteenth-Century Bristol* (Cambridge: Cambridge University Press, 1991). On France, see L. Brockliss and C. Jones, *The Medical World of Early Modern France* (Oxford: Clarendon Press, 1997). On the Netherlands, see M. Gijswijt-Hofstra (ed.), *Geloven in genezen*, special issue of *Volkskundig bulletin*, 17 (1991); W. de Blécourt and G. van Vegchel (eds), *De medische markt*, special issue of *Focaal*, 21 (1993). For an elaboration of the concept for the modern era, see W. de Blécourt et al (eds), *De medische markt in Nederland, 1850–1950*, special issue of *Tijdschrift voor sociale geschiedenis*, 25 (1999).

6 C.E. Rosenberg and J. Golden (eds), *Framing Disease: studies in cultural history* (New Brunswick/New Jersey: Rutgers University Press, 1997). See also P. Berger and T. Luckmann, *The Social Construction of Reality: a treatise in the sociology of knowledge* (Garden City: Anchor Books, 1966); J. Lachmund and G. Stollberg (eds), *The Social Construction of Illness* (Stuttgart: Franz Steiner Verlag, 1992); L. Jordanova, 'The Social Construction of Medical Knowledge', in F. Huisman and J. Harley Warner (eds), *Locating*

meaning to a disease category, to diagnosis, to therapy, and to the way their results are to be evaluated. In this conceptualization process, many factors are in play: not just scientific ones, but religious, financial and political ones as well. Since supply and demand are shaped by many social, cultural and economic factors, one may conclude that the 'health care system' is the end result of the dynamic interaction between many ideas, interests and ideals. It could be qualified as a battlefield, with cultural authority and commercial success as the bounty.

On the market for medicines many factors need to be taken into account. Research questions include: to what extent did medical and pharmaceutical legislation structure the market? What was the role and influence of professional organizations of pharmacists and physicians? What role did pressure groups like the Society against Quackery play? How did the policies of purchasing companies and cartels influence patterns of consultation and consumption of Dutch citizens? Finally, we need to have a keen eye for the cultural dimension of the market. For example: what symbolic meanings were implicit in advertisements for patent remedies in the lay press? What was the attraction of secret remedies, denounced by pharmacists as quack remedies?

In their book on consumption, the anthropologist Mary Douglas and the economist Baron Isherwood consider production, technology and consumption as part of a cycle.[7] Consumption patterns are being shaped in an interactive way. Already in the early stages of the development of new products the manufacturer is anticipating on the needs and desires of the consumer. Therefore, the image of manufacturers offering their goods to a passive consumer is misleading. The consumer is active as well, sending signals to which the manufacturer in turn is responding. In short: the manufacturer is not just pro-actively shaping a market for his goods, but he is also – re-actively – meeting the needs of the consumer. When translated in terms of the market for medicines, this reads as follows: in order to sell

---

Medical History: the stories and their Meanings (Baltimore/London: the Johns Hopkins University Press, 2004), pp. 338–363.

7 M. Douglas and B. Isherwood, *The World of Goods: towards an anthropology of consumption* (London: Routledge, 2nd Edition, 1996).

their products, manufacturers had to create meaning and associate it with their product. This was by no means a creation out of the blue. In the case of medicines, they subtly played with dormant notions about health, illness and healing. Their advertising campaigns did not force behaviour on the patient in any way; nor could they force physicians to prescribe their products or indeed pharmacists to sell them.[8]

An important factor in the selling potential of a remedy was its 'cultural status'. Patients only entrusted their bodies to a healer they believed in or to a remedy in which they had put their hopes. Therefore, manufacturers of medicines realized how important it was to create a positive public image of themselves and their products. While some of them capitalized on a perceived need for exotic aesthetics among the public, others profiled themselves as being scientific. The designation 'scientific' represented an ethical, humanitarian and transparent public image; its opposite was considered to be unethical, commercial and secret. The dichotomy was part of a strategy of professional demarcation: whereas the 'ethical' camp claimed to produce 'ethical drugs', using academic knowledge for the common interest, its 'unethical' opponent was accused of being ignorant and selfish, producing over-the-counter or patent remedies.[9] As I hope to show, the construction of this antithesis was an important element of the corporate strategy of at least part of the early pharmaceutical industry, in an attempt to legitimize itself and obtain a market share in the process.

8 F. Huisman, 'Patiëntenbeelden in een moderniserende samenleving: Nederland, 1880–1920', in F.J. Meijman (ed.), *Geschiedenis van de medische publiekscommunicatie*, special issue of *Gewina. Tijdschrift voor de geschiedenis der geneeskunde, natuurwetenschappen, wiskunde en techniek*, 25 (2002): 210–225.

9 J. Liebenau, 'Ethical Business: the formation of the pharmaceutical industry in Britain, Germany, and the United States before 1914', *Business History*, 30 (1988): 116–129; idem, *Medical Science*, p. 137. On the related concept of 'trust', see R. Church, 'Trust, Burroughs Wellcome & Co. and the Foundation of a Modern Pharmaceutical Industry in Britain, 1880–1914', *Business History*, 48 (2006): 376–398. On the different meanings of 'science', ibidem, 6–7 and chapter 7. On the wider Dutch context, see K.E. Sluyterman, *Dutch Enterprise in the Twentieth Century: business strategies in a small open economy* (London: Routledge, 2005).

This paper cannot claim to be comprehensive about all actors on the pharmaceutical market, but it does hope to show the analytical potential of a market perspective. Using the Netherlands as my example, I would like to show the multidimensional character of the industry of the early twentieth century and the changing relationships between the industry and 'the field'. The story that results may be less straightforward than the traditional one, but it may do more justice to the complex character of the pharmaceutical industry, with its inherent tension between altruism and commercialism.[10]

Medicine Production between Legislation and the Market: the Netherlands

Traditionally, pharmacists had been the most important producers of medicines, operating on a local market. From the second half of the seventeenth century onwards, however, important competitors arrived on the scene.[11] In the field of production, chemists (*chymisten*) represented a growing threat to pharmacists. They established small chemical plants, mainly in the province of Holland, applying well-known methods of extraction and isolation in an industrial way, mass-producing cheap medicines for the

10 For a plea to open up the *black boxes* of 'science', 'technology' and 'industry', see the Introduction to E. Homburg et al (eds), *The Chemical Industry in Europe, 1850–1914: industrial growth, pollution, and professionalization* (Dordrecht: Kluwer Academic Publishers, 1998), in which an argument is developed against the classical historiography of chemistry, where these categories are considered as homogeneous and unproblematic.

11 E. Homburg, *Van beroep 'Chemiker'. De opkomst van de industriële chemicus en het polytechnische onderwijs in Duitsland (1790–1850)* (Delft: Delft University Press, 1993), pp. 77–82; W. Frijhoff, 'Ondernemerschap en cultuurpatroon: Van der Veen's elixer probatum van wondermiddel tot maagbitter', in W. Frijhoff and M. Hiemstra (eds), *Bewogen en bewegen. De historicus in het spanningsveld tussen economie en cultuur* (Tilburg: Gianotten, 1986), pp. 249-277; F. Huisman, 'Gezondheid te koop. Zelfmedicatie en medische advertenties in de Groninger en Ommelander Courant, 1743–1800', *Focaal*, 21 (1993): 90–130.

whole Dutch Republic. With the help of druggists (*drogisten*), booksellers and keepers of lodging houses they set up a distribution network, creating an outlet of national proportions.

Due to the general abolition of the guilds in 1798, corporate protection of pharmacists disappeared. At the same time, the druggists' trade was officially regulated in an instruction in 1818, which can be considered as a form of recognition. The druggists' trade was thriving, and it was to last almost 70 years until national legislation on the health care system ended this situation of professional insecurity for pharmacists. With his medical laws of 1865, the liberal minister J.R. Thorbecke had three goals in mind: first, he wanted to introduce a clear and formal distinction between medicine and pharmacy; secondly, he was keen on elevating pharmacy to academic levels and thirdly, he wanted to limit the druggists' trade to the wholesale of medicines.[12] The Law on Higher Education of 1876 stipulated that pharmacy and toxicology were to be taught in universities, while the Law on Physicians of 1878 limited the admission to the practical pharmacists' exam to those who possessed a degree in physics and theoretical pharmacy. Henceforth, the only possibility to become a pharmacist was to enroll in the pharmaceutical program of one of the four Dutch universities.[13]

Initially, pharmacists thought that they had acquired a monopoly for the preparation and delivery of medicines. Very soon however, it became clear that this was not the case – indeed, it had never been the intention of Thorbecke. The laws of 1865, 1876 and 1878 – that determined the legal boundaries of the Dutch health care system for most of the twentieth century – were meant to bring academic medical and pharmaceutical help within reach of every Dutch citizen.[14] But it was up to the patient to decide

12 F. Huisman, 'Pharmacists, Druggists and the Spirit of Thorbecke. The Shaping of Dutch Pharmacy, 1865–c.1920', in M. Gijswijt-Hofstra, G.M. van Heteren and E.M. Tansey (eds), *Biography of Remedies: drugs, medicines and contraceptives in Dutch and Anglo-American healing cultures* (Amsterdam: Rodopi, 2002), pp. 35–55.

13 M. Groen, *Het wetenschappelijk onderwijs in Nederland van 1815 tot 1980*, 9 vols. (Eindhoven: University of Technology, 1983–1987) vol. 1, pp. 132–134, vol. 2, pp. 54–56, and vol. 7, pp. 65–68.

14 Huisman, 'Pharmacists'.

if he wanted to use it: individual health care was no matter for the state. Every citizen was free to decide which healer to consult and what medicines to use. Thorbecke had helped to create a liberal political climate that offered excellent opportunities to druggists, who were the perfect outlet for the growing industrial production of medicines.

In addition, the druggists' trade was greatly facilitated by the unintended effects of two non-medical laws. The production, advertising and distribution by manufacturers other than pharmacists received a boost from the abolition of the tax on newspapers in 1869 and from the Law on Trademarks in 1880. Because the tax was abolished, newspapers could be produced more cheaply, which led to the establishment of many new ones. Because more people could afford a newspaper, more people could be reached, which made it more attractive for manufacturers to use newspapers to advertise their products directly to the public, bypassing traditional local producers like pharmacists. Thus, Holloway pills – to name but one example – began their march of victory on the Dutch market.[15] Advertising for medicines in the lay press was greatly facilitated when trademarks (being the name of a product and the logo of its producer) were legally protected and published in the *Dutch State Journal* (*Nederlandsche Staatscourant*). Indeed, one could argue that the protection of trademarks was a precondition for effective advertising. The law on trademarks was intended to enable consumers to differentiate between manufacturers and their products. It was not about the composition of a remedy or the formal qualification of its manufacturer, but rather about the protection of industrial rights.[16] From 1881 onwards, it was not those making medicines excluded from the

15 See, for example, *De Tijd*, 4 January 1870.

16 In this context, Tanner points to the fact that the course of industrial research was determined by market developments and opportunities rather than by the ambition to develop something new: J. Tanner, 'The Swiss Pharmaceutical Industry: the impact of industrial property rights and trust in the Laboratory, 1907–1939', in A.S. Travis et al (eds), *Determinants in the Evolution of the European Chemical Industry, 1900–1939* (Dordrecht: Kluwer Academic Publishers, 1998), pp. 257–271, on p. 268. A patent law was introduced in the Netherlands in 1912. E. Schiff, *Industrialization without National Patents: the Netherlands, 1869–1912, Switzerland, 1850–1907* (Princeton: Princeton University Press, 1971).

national pharmacopoeia who were liable for punishment (by a fine), but rather those who infringed the commercial rights of others.[17]

Many companies used the opportunity to protect themselves against competition by pharmacists and others. Immediately after the law had been put into effect, an enormous amount of mineral waters, laxative teas, elixirs vitae, salves, syrups, quinine products, bandages, mouth washes, capsules, cod-liver oil, soothing remedies and many different sorts of pills were registered, as can be seen from the entries in the *Dutch State Journal*.[18] Foreign companies were over-represented, especially those from France, England, Germany, Belgium and Switzerland. As a consequence of industrialization, the growth of advertising and the improvement of the infrastructure in the late nineteenth century, the Dutch market was flooded with new remedies, which were eagerly bought by the public.[19]

Concern about this situation led to the establishment of the Society against Quackery in 1880, an initiative of Dr G.W. Bruinsma and his brother dr Vitus Bruinsma, the former a practicing physician, the latter a teacher in chemistry.[20] Together, they edited the *Monthly against Quackery* (*Maandblad tegen de Kwakzalverij*). The Bruinsma brothers were very much convinced of the value of scientific research. At the same time, they established that the law did not offer much protection to those practitioners who shared their belief in science. Given the liberal political climate, they decided that public education was the only weapon left. The mission of the Society was to analyse the market and inform the public about it.

17 For examples of legal cases, see *Weekblad van het regt* 52 (1890), nr. 5893.2; ibidem 53 (1891), nr. 6073.2; ibidem 60 (1898), nr. 7186.2.

18 See the files for registration in the *Nederlandsche Staatscourant* from 1 February 1881 onwards.

19 *Pharmaceutisch weekblad*, 9 November 1889 and 11 April 1891. On the similar situation in Germany, see E. Hickel, 'Das kaiserliche Gesundheitsamt und die chemische Industrie im Zweiten Kaiserreich (1871–1914): Partner oder Kontrahenten?', in G. Mann and R. Winau (eds), *Medizin, Naturwissenschaft und Technik und das Zweite Kaiserreich* (Göttingen: Vandenhoeck und Ruprecht, 1977), pp. 64–86, on p. 74.

20 G. van Vegchel, *Medici contra kwakzalvers. De strijd tegen niet-orthodoxe geneeswijzen in Nederland in de 19e en 20e eeuw* (Amsterdam: Het Spinhuis, 1991).

Periodically, their analyses in the *Monthly* were collected in a book.[21] Over the years, they investigated and published on the composition of remedies, on the advertising strategies of their manufacturers, on the politics of newspapers and on the architecture of distribution networks.

The crusade of the Society against Quackery notwithstanding, pharmacists were relatively powerless in the face of the ever growing supply of industrially prepared medicines. As a profession, they felt threatened and superfluous.[22] Forced by the circumstances, pharmacists sold everything for which there seemed to be a demand. The Amsterdam professor Gunning established the fact that public esteem was very low, characterizing the pharmacist of his time as 'no more than a supplier of medicines'.[23] Members of the Dutch pharmaceutical elite – like the newly appointed professors of pharmacy and the executive committee of the Dutch Society for the Advancement of Pharmacy (NMP, a formal member of the Society against Quackery since 1888) – did everything they could to create a professional image that was ethical, humanitarian and scientific, but to little avail.[24] Although pharmacy had formally been elevated to academic levels by the medical laws of 1865, pharmacists suffered cut-throat competition from the emerging pharmaceutical industry, from druggists and from physicians with a pharmacy. In the field of the manufacturing of medicines pharmacists seemed on the verge of being pushed out of the market, because

21 V. Bruinsma and G.W. Bruinsma, *De kwakzalverij met geneesmiddelen en de middelen om haar te bestrijden* (Leeuwarden: Van Belkum, 1878) and idem, *De hedendaagsche kwakzalver. Een waarschuwing voor allen, die hun gezondheid en hun beurs op prijs stellen* (Leeuwarden: Van Belkum, 1880). Later on, this tradition of public education was continued by E.J. Abrahams. See E.J. Abrahams, *De kwakzalversmiddelen, hun inhoud en de gevaren die bij het gebruik dreigen, volgens analyses gedurende 35 jaar gemaakt voor de Vereeniging tegen de kwakzalverij* (s.l. 1916).

22 The same applied to their German colleagues: G. Huhle-Kreutzer, *Die Entwicklung arzneilicher Produktionsstätten aus Apothekenlaboratorien* (Stuttgart: Deutscher Apotheker Verlag, 1989), p. 288.

23 J.W. Gunning, *Een blik op de toekomst der pharmacie in Nederland* ('s-Gravenhage: Van Cleef, 1887), p. 8. See H.P. Wijsman, 'Over synthese van geneesmiddelen', *Chemisch weekblad*, 5 (1908): 317–326 on 317.

24 Huisman, 'Pharmacists'.

the industry could produce more, better (i.e. standardized) and cheaper, whereas in the field of distribution druggists were considered as good as pharmacists. On top of that, a major threat was developing from within the ranks of pharmacists: a company called Brocades & Stheeman.

Brocades: moving from pharmacy to industry

Unlike Organon – a Dutch company about which more will be said later – Brocades was not originally conceived as a concern for the large-scale production of medicines. Contrary to Organon, Brocades expanded very gradually, partly as a result of an extension of its product range, partly because of production increase. The history of the development of the company is interesting, because it gradually released itself from the ranks of pharmacists. With its large-scale and mechanized production methods, Brocades threatened the professional identity of pharmacists at its very roots. Therefore, it was forced to develop a new corporate image, of which 'science' and 'ethics' were the most important ingredients. The firm goes back to at least 1798.[25] We know that Willem Brocades was running a pharmacy shop in that year in the provincial town of Meppel, selling galenical preparations. Over the years, the range was extended with alkaloids like strychnine and quinine. Under his successor, Eisso Post Stheeman, the company was thriving, but its character as a pharmacy was left in tact. In 1880, Stheeman's son Sypko took over, marking an era of profound change. This change not only concerned the nature and size of the company (moving from galenics to synthetic medicines), but its relationship with pharmacists as well.

25 On Brocades see D.A. Wittop Koning, *N.V. Koninklijke pharmaceutische fabrieken v/h Brocades-Stheeman en Pharmacia, 1800–1950* (Amsterdam: KPF, s.a. [1950]); T. Rinsema, *De natuur voorbij. Het begin van de productie van synthetische geneesmiddelen* (Meppel: s.n., 2000), chapters 9 and 10.

In 1886 a pill machine was brought into use. Eight years later, Brocades & Stheeman – as the company was now called – acquired a steam engine. It enabled the large-scale manufacture of compressed medicines, increasing profits substantially. Using steam power, the company could produce 180.000 tablets and 150,000 pills a day. But mechanization was not the only reason for change. On top of that, product differentiation initiated a new era for Brocades & Stheeman. The company's price lists of the 1880s show 'the number of products almost doubling within a few years'. Apart from pharmaceuticals, the company sold products like vinegar essence, grape wine, dressing materials, mineral waters and soap. The expansion of production necessitated continuous building activity.[26] At the beginning of the twentieth century, Brocades & Stheeman was the largest pharmaceutical company in the Netherlands and it continued to grow because of a number of strategic takeovers and mergers after the First World War. By the middle of the twentieth century, the pharmacy shop in the rural town of Meppel had grown into an international company with correspondents in Europe, the Middle East, the East Indies, South Africa and South America.[27]

All of these developments – the large-scale production of familiar medicines, the introduction of new products and the commercial attitude of the company[28] – represented a serious threat to the traditional pharmacist. The high production levels of Brocades led to lower prices against which they could never compete. This was all the more threatening, because the market for compressed medicines expanded enormously at the end of the nineteenth century. The public considered them to be more hygienic, more compact, better balanced and less perishable than traditional pharmacists' preparations.[29] Moreover, the industrial distribution strategy

26 Rinsema, *De natuur voorbij*, 366.

27 I. van Goozen (ed.), *Van plattelandsapotheek tot Europees bedrijf: hoe Brocades Yamanouchi werd. Een beknopte bedrijfsgeschiedenis van 1798 tot 1995* (Leiderdorp: Yamanouchi Europe, s.a. [1995]), p. 20.

28 All representative trends of late nineteenth–century Dutch industry: Homburg et al, *Chemische techniek in Nederland in de negentiende eeuw* (Zutphen: Walburg pers, 1994), pp. 125–126.

29 J.E. Rombouts, 'De fabriek van gecomprimeerde geneesmiddelen van de firma Brocades en Stheeman te Meppel', *Eigen haard* (1902), 24–27 and 36–39 (11 and 18 January);

tended to bypass pharmacists by supplying druggists and physicians with a pharmacy.[30]

With its large-scale production methods, its product differentiation and its new distribution strategy, Brocades profoundly affected the professional identity of the pharmacist. In his inaugural address as Professor of pharmacy and toxicology, Johan Eykman expressed his worry that the large-scale production of medicines would push pharmacists out of the market, which he thought to be a highly undesirable development.[31] Earlier, R.J. Opwijrda, the editor-in-chief of the *Pharmaceutical Weekly* (*Pharmaceutisch weekblad*), had followed these developments on a more regular basis. Being a pharmacist himself, he was highly critical of the commercial character of the industrial production of medicines and opposed advertisements for them in the *Pharmaceutical Weekly* (that was to become the official journal of the Dutch Society for the Advancement of Pharmacy in 1902).[32] More than once, he explicitly referred to Brocades & Stheeman, expressing his doubts about the composition or the quality of their products.

Stheeman was very sensitive to such imputations, to which he always responded promptly.[33] He was keen on remaining in contact with pharmacists, because of their shared professional frame of reference but also because he did not want to alienate them as his customers. While conceding that the large-scale production of his company represented a profound break with the past, he appealed to the sentiment of a shared pharmacists' tradition and identity, which was both craftsmanlike and scientific: the defining characteristic of Brocades & Stheeman was its 'purely scientific orientation', that was embedded in its pharmacists' identity. In his responses he always stressed the fact that all his preparations were made according to the standards of the national pharmacopoeia. It was true that his company

T. de Ridder, 'N.V. Koninklijke Pharmaceutische Fabriek voorh. Brocades & Stheeman, Meppel', *Neerlands welvaart* (1918), 1–20 on 10.

30 Rinsema, *De natuur voorbij*, 363–365.

31 J.F. Eykman, *De roeping der pharmacie* (Amsterdam: s.n., 1897), p. 29.

32 *Pharmaceutisch weekblad*, 9 January 1886; 9 November 1889. Rinsema, *De natuur voorbij*, 349–357.

33 See for example *Pharmaceutisch weekblad*, 16 June 1878 and 24 October 1880.

was producing one million pills a day, but they were all of 'pharmacists' quality': the difference between his company and a pharmacy shop was only one in degree, not in kind. By a rhetorical *tour de force*, Stheeman tried to secure an optimal relationship with pharmacists: while suggesting an enormous gap between Brocades and the old pharmacy, he welcomed the modern pharmacist as a colleague. This is in line with the image the company created of itself in *Dutch Prosperity* (*Neerlands welvaart*), a journal geared to the promotion of Dutch products and Dutch industry. In a promotional article, the company was represented as being scientific, hygienic and modern, but also as having a pharmacists' identity.[34]

Thus, Brocades & Stheeman had detached itself skillfully from the world of the traditional pharmacist without cutting its ties from the modern profession. By stressing its identity as a pharmacists' company and by stretching the designation 'scientific' in such a way that it could also apply to the large-scale industrial production of medicines, the company succeeded in serving its commercial interests without endangering its relationship with pharmacists, from whose ranks it had emerged.

The Construction of an 'Ethical' Pharmaceutical Industry

Despite the idealist deontological calls made by the pharmaceutical elite, 'secret' or 'quack' remedies continued to be produced by companies like Brocades & Stheeman and distributed by pharmacists, simply because they were commercially attractive. The late nineteenth century had seen much protest against industrial medicines. Over time however, pharmacists realized that – given industrial developments and the liberal political climate –

34 T. de Ridder, 'N.V. Koninklijke Pharmaceutische Fabriek voorh. Brocades & Stheeman, Meppel', *Neerlands welvaart* (1918), 1–20.

they had to face reality.[35] The public was prepared to pay good money for *spécialités*, being products manufactured by the chemical industry and protected by the Law on Trademarks. At the same time, prices for pharmacists' preparations dwindled.[36] Over the years, the sale of industrial products by pharmacists had become accepted practice: the central committee of the NMP even decided to bargain for discounts with manufacturers.[37]

After pharmacists had pragmatically accepted industrially produced medicines, the issue that continued to concern many was that the majority was produced abroad, particularly in Germany. There was a growing feeling that the Netherlands should develop its own chemical and pharmaceutical industry. For the industry to blossom – it was argued by H.F. Tillema, a pharmacist who was appointed government adviser for the chemical and pharmaceutical industry – it was imperative for the chemist, the manufacturer and the merchant to cooperate in a harmonious way. Others argued for the establishment of a Dutch dyestuffs industry also involved in manufacturing medicines.[38] The First World War gave considerable impetus to the Dutch pharmaceutical industry. In spite of its neutrality, the Netherlands were isolated economically and cut off from imports, among them medicines. Initially, this seemed to have detrimental effects, but in retrospect it turned out to be a blessing in disguise.[39] The temporary cessation of competition from the German pharmaceutical industry, its most important rival, offered the Dutch industry unique opportunities

35 Some pharmacists continued to complain about the 'druggists' culture' among pharmacists (e.g., K. Scheringa, 'Specialités', *Pharmaceutisch weekblad*, 50 (1913): 188–189 and 288–289), but even they realized there was not much that could be done against it.

36 *Pharmaceutisch weekblad*, 50 (1913): 320.

37 W.P. Woutersen, 'Verkoop van specialités', *Pharmaceutisch weekblad*, 50 (1913): 160–162.

38 H.F. Tillema, 'Aan allen, die belang hebben bij de chem.-pharm. industrie in Nederland', *Chemisch weekblad*, 12 (1915): 12–13; C. Breuls, 'Ingezonden. Een eigen chemische Industrie', *Pharmaceutisch weekblad*, 53 (1916): 186–187; G.C.A. van Dorp, 'De toekomst van de Nederlandsche chemische industrie', *Chemisch weekblad*, 16 (1919): 265–269.

39 Sluyterman, *Dutch Enterprise*, pp. 75–91.

for development, like it did elsewhere in Europe.[40] In particular circles it was felt that in order to thrive, the Dutch chemical and pharmaceutical industry needed to establish itself on a scientific basis.

Thanks to the energetic efforts of people like G. Hondius Boldingh (industrialist and professor of pharmaceutical chemistry in Amsterdam) and H.R. Kruyt (professor of chemistry in Utrecht), a mutual rapprochement between academy and industry developed.[41] Both firmly believed that the Dutch industry could only be prevented from lapsing into its traditional, pre-war levels when scientific research would be intensified and improved. Scientists and industrialists should cooperate – as they had long been doing in Germany, where the industry had become very successful.[42] As chair of the Dutch Chemical Society (the NCV), Kruyt considered improving industry's appreciation for scientific research as his most important task. In 1918, Hondius Boldingh was co-founder of the Society of the Dutch Chemical Industry (VNCI) that had the explicit ambition of seeking contact with academic chemistry. Kruyt became a member of the board of the VNCI, while one of the industrials became a board member of the NCV. With the *Chemical Weekly* (*Chemisch weekblad*) becoming the joint organ, there were ties between the NCV and the VNCI at personal, organizational and financial levels. However, what was needed most was a change of mentality on both sides. Broadly speaking, scientists felt a certain degree of contempt for the industry, while industrialists and

40 G.J. Somsen, 'Selling Science: Dutch Debates on the Industrial Significance of University Chemistry, 1903–1932', in Travis et. al (eds), *Determinants in the Evolution*, pp. 143–168, on pp. 155–162; E. Homburg, 'De Eerste Wereldoorlog: samenwerking en concentratie binnen de Nederlandse chemische industrie', in J.W. Schot et al (eds), *Techniek in Nederland in de twintigste eeuw*, 7 vols (Zutphen: Walburg pers, 2000) vol. 2, pp. 317–331.

41 On Hondius Boldingh, see Homburg, 'De Eerste Wereldoorlog', pp. 322–329. On Kruyt, see G.J. Somsen, *"Wetenschappelijk onderzoek en algemeen belang". De chemie van H.R. Kruyt (1882–1959)* (Delft: Delft University Press, 1998), esp chapter 5. According to Kruyt the dichotomy between scientific use and social use was a false one: ibidem, pp. 171–172.

42 H.R. Kruyt, 'Samenwerking van wetenschap en industrie in Nederland', *Chemisch weekblad*, 15 (1918): 418–426. E. Homburg, *Speuren op de tast. Een historische kijk op industriële en universitaire research* (Maastricht: Universiteit Maastricht, 2003).

engineers reproached scientists for locking themselves up in an ivory tower. Hondius Boldingh and Kruyt felt that such attitudes were both unjustified and counter-productive, and decided to give a good example by acting as scientific advisors for several companies arguing for the establishment of industrial research laboratories.[43]

After the First World War, there was a new sense of potential for the Dutch industry.[44] In 1918, the *Chemical Weekly* proudly remarked that products that had to be imported before the war were now produced by Dutch companies and abundantly available.[45] This not only included chemical and pharmaceutical preparations, but tablet machines, distilling kettles, retorts, laboratory glassware and rubber products as well. All efforts were geared to the maturation of the Dutch industry, with Germany and the United States serving as the models to follow. 'Science' – both in reality and as part of the corporate image – was an important instrument for the new 'ethical' industry, in which pharmacists and chemists played an important part in daily management. Many of the chemical and pharmaceutical companies presenting themselves to the public at the Utrecht annual trade fair succeeded in merging science and commerce into a successful formula.[46] They included the N.V. Chemische Fabriek 'Naarden', the N.V. Koninklijke Pharmaceutische Fabriek v/h Brocades & Stheeman, the Amsterdamsche Chininefabriek and the N.V. Koninklijke Pharmaceutische Handelsvereeniging v/h d'Ailly & Zonen, Mastenbroek & Gallenkamp.

43 E. Homburg, A. Rip and J. Small, 'Chemici, hun kennis en de industrie', in Schot et al (eds), *Techniek in Nederland in de twintigste eeuw*, vol. 2, pp. 299–315, on pp. 310–311. On the rise of Dutch industrial research laboratories, see J.J. Hutter, 'Nederlandse laboratoria 1860–1940, een kwantitatief overzicht', *Tijdschrift voor de geschiedenis der geneeskunde, natuurwetenschappen, wiskunde en techniek*, 9 (1986): 150–174.

44 See, for example, W. Graadt van Roggen, 'De Nederlandsche jaarbeurs', in W.G. de Bas (ed.), *Gedenkboek 1898–1923. Uitgegeven ter gelegenheid van het zilveren regeeringsfeest van Hare Majesteit Koningin Wilhelmina der Nederlanden, op 6 september 1923* (Voorschoten: Fongers, 1923), pp. 907–916.

45 *Chemisch weekblad*, 15 (1918): 729–744, on 739–742.

46 For an early report on the presence of the chemical and pharmaceutical industry at the Utrecht annual fair, see *Chemisch weekblad*, 14 (1917): 291–297.

The gap that was created between 'ethical' and 'unethical' companies[47] as a consequence of the academization of the industry led to a renewed optimism among self-employed pharmacists. In a special issue of the *Pharmaceutical Weekly*, published on the occasion of its fiftieth anniversary jubilee, J.J. Hofman voiced this optimism. Hofman – who was a pharmacist, the owner of a pharmaceutical company in the Hague and a former board member of the Dutch Chemical Society (NCV) – saw a bright future for the profession, as long as pharmacists were prepared to turn their shops into places where patients could in their scientific expertise.[48] Much had changed since the days when Opwijrda, the editor of the *Pharmaceutical Weekly*, had criticized Brocades for its industrial production. During the war, the government advisor for the chemical and pharmaceutical industry, Tillema, argued that the Dutch pharmaceutical industry was in need of 'scientific management'.[49] Shortly after the war, this ideal was realized by the establishment of Organon.

Organon: cooperation between industry and science

There is an important difference to be noted between Organon (est. 1923) and Brocades & Stheeman. Whereas Brocades had grown organically, reinventing itself in the process, Organon conceptualized and organized itself as a company engaging in the large-scale and scientific production of

47 For the earliest Dutch examples, see *Pharmaceutisch weekblad*, 11 August 1894 (on the material and ethical danger of the plethora of new remedies); 'Vergadering der secties', ibidem, 21 August 1897 (a plea for the inclusion of pharmaceutical legislation and ethics in the curriculum); ibidem, 11 September 1897 (a call to Dutch pharmacists to take an ethical stance towards the industry); ibidem, 9 June 1900 (on the threat of secret remedies for the foundations of pharmaceutical ethics); ibidem, 6 July 1901 (a call for the establishment of 'Councils of Honour and Discipline' to guard and uphold pharmaceutical ethics).

48 J.J. Hofman, 'De maatschappelijke positie der apothekers voor 50 jaar en thans', *Pharmaceutisch weekblad*, 51 (1914): 325–335.

49 Tillema, 'Aan allen'.

medicines from the outset.[50] In a relatively short period of time the company developed not just into one of the most important manufacturers of hormone and vitamin preparations in the world, but into an expert in the field of endocrinology as well.[51] Organon sprang from the ambition of its founder, Saal van Zwanenberg, who was a butcher on an industrial scale, to put the offal produced by the family company to good use. Around 1920, Zwanenberg's slaughterhouses processed 2,000 pigs and 350 cattle on a daily basis. The offal was used as the raw material for oils, fats, soap, margarine, fertilizers, pig feed, glue, buttons and hides.[52] In 1923, insulin was added to this list. Insulin was even considered so promising that a private limited company was founded for its production.

In 1921 two Canadian researchers, Banting and Best, had succeeded in demonstrating an active substance in pancreas extract and isolating it before it was broken down.[53] This substance (called insulin later on) would prove to be very important in the battle against diabetes. When the Amsterdam pharmacologist, Ernst Laqueur, succeeded in producing the first insulin preparation, the industrialist Van Zwanenberg sensed an opportunity to put his offal to even greater use. He was keen on bringing scientifically prepared hormone preparations to the market without losing sight of the commercial interests of his company. The result was a close collaboration

50 For the statement of principles of the new company, see article 3 of the memorandum of association: archive Organon Oss, inv.nr. HA 53–55, 9 Juli 1923. On the company in general, see Marius Tausk, *Organon. De geschiedenis van een bijzondere Nederlandse onderneming* (Nijmegen: Dekker en Van de Vegt, 1978); Jeroen Verhoog, *75 jaar Organon 1923–1998* (Noordwijk: s.n., 1998). On Organon as a Jewish family firm, see Doreen Arnoldus, *Family, Family Firm and Strategy: six Dutch family firms in the food industry 1880–1970* (Amsterdam: Aksant, 2002).

51 Nelly Oudshoorn, *The Making of the Hormonal Body. A contextual history of the study of sex hormones 1923–1940* (Enschede: Alfa, 1991).

52 Tony de Ridder, 'N.V. Zwanenberg's slachterijen en fabrieken te Oss, Holland', *Neerlands welvaart* nr. 42 (1918), 8.

53 M. Bliss, *The Discovery of Insulin* (Chicago: University of Chicago Press, 1982); Idem, *Banting: a biography* (Toronto: McClelland and Stewart, 1985); C. Sinding, 'Making the unit of insulin: standards, clinical work and industry, 1920–1925', *Bulletin of the History of Medicine*, 76 (2002): 231–270.

between the butcher Zwanenberg (who supplied the raw materials) and the scientist Laqueur (who supplied the scientific expertise).

Organon is a good example of the kind of cooperation between science, technology, production and commerce that had been advocated by Kruyt.[54] Whereas in the past, 'science' had been used as a rather straightforward instrument of demarcation for the improvement of the corporate image and the optimization of profits, it developed into an indispensable element for industrial research and development departments. Between Zwanenberg and Laqueur, there was a perfectly symmetrical relationship, in which one could not do without the other. Laqueur was not only appointed scientific advisor of Organon, but became a full partner in the newly established company.[55] As one of its three directors, he received a third of all the profits. Also, Laqueur had the right to buy everything he needed for his laboratory and to publish all his scientific findings – both without prior consultation. He could stop the issue of preparations whenever and for as long as he thought necessary, and the company needed his permission for the appointment of all personnel related to medical preparations. After Van Zwanenberg had thus secured the loyalty and close collaboration of a well-known scientist, he signed agreements with the three most important Dutch pharmaceutical wholesalers.[56] In the agreements Organon committed itself to the exclusive delivery of insulin to these parties only. In return, the wholesalers promised not to buy or sell any insulin products

54 Kruyt, 'Samenwerking', 425. See also N. Oudshoorn, 'United We Stand: the Pharmaceutical Industry, Laboratory and Clinic in the Development of Sex Hormones into Scientific Drugs, 1920–1940', *Science, Technology and Human Values*, 18 (1993): 5–24. On 'entrepreneurial professors' after the First World War more generally, see P.J. Knegtmans, 'Onderwijs, wetenschap en particulier initiatief aan de UvA, 1920–1950', in J. Knegtmans and A.J. Kox (eds), *Tot nut en eer van de stad. Wetenschappelijk onderzoek aan de Universiteit van Amsterdam* (Amsterdam: Amsterdam University Press, 2000), pp. 79–105.

55 Archive Organon Oss, inv.nr. HA 53–54, July 1923.

56 These were: the Coöperatieve Apothekers Vereeniging De Onderlinge Pharmaceutische Groothandel in Utrecht (OPG), the N.V. Koninklijke Pharmaceutische Handelsvereeniging in Amsterdam (KPH) and the N.V. Koninklijke Pharmaceutische Fabriek in Meppel (KPF). See archive Organon Oss, inv.nr. HA 53–55, 1 July 1923 and inv.nr. HA 53–3, 23 August and 15 November 1923.

from others but Organon. With the agreements with Laqueur and the wholesalers, the new company not only succeeded in stamping its insulin products as scientific, but it also secured a market share that bordered on a monopoly.

The company expanded quickly: in 1928 net profits quadrupled as compared to the previous year, while they doubled the next year.[57] Organon's formula for success was a fortunate combination of a scientific foundation and a commercial organization. Because its production was science-based rather than just science-associated, Organon chose not to address itself directly to the public through the lay press (which would have been associated with quackery), but rather to physicians and pharmacists in the professional press. In 1928, 20,000 guilders – or 5.6 of the turnover – was spent on advertisements and on the free issue samples of insulin to physicians and pharmacists.[58] On top of that, it was decided to publish a scientific journal reporting on medicines developed by Organon. From 1931 onwards, the company published and distributed *The Hormone. Monthly for Endocrinology* (*Het hormoon. Maandblad voor de endocrinologie*). In an editorial note to the first issue, Marius Tausk – scientific director of the laboratory in Oss – vowed to be independent, 'without emotion or prejudice'.

To try and win over the general public, Organon used the instrument of philanthropy. Given both in money and in kind, it must have contributed to the creation of a positive public image of the company, and to the acceptance of its products. Among other things, Organon gave money to several hospitals in the Netherlands, to the Society of Patients Suffering from Tuberculosis, to the Fund for Deaf-and-Dumb Children in Amsterdam and to those who were 'persecuted for their race and faith'. Philanthropy in kind mainly included the free supply of insulin – via physicians – to the poor.[59] This worked both ways: on the one hand, Organon created a

57 Archive Organon Oss, inv.nrs. HA 83–84 (1926); HA 83–5 (1927–1929) and HA 79–3 (1930).8

58 Archive Organon Oss, inv.nr. HA 83–85, annual report 1928.

59 Archief Organon Oss, inv.nrs. HA 77–13 and HA 60–61. On average, Organon spent some 2000 guilders on 'philanthropy'.

profile for itself as a public benefactor, whereas on the other, the attention of physicians and pharmacists was drawn to its products.[60]

In a relatively short period of time, Organon had succeeded in becoming a commercially healthy company with a solid scientific reputation. While the company had initially used an external laboratory (of Laqueur in Amsterdam), later on it opted for its own lab nearby, with Tausk as its scientific director and editor of the journal *The Hormone*.[61] Laqueur and Tausk lectured on sex hormones and growth hormones.[62] From 1937 onwards, Organon organized courses for physicians, and in 1938 it even hosted an international gynaecological conference. In doing so, Organon made an important contribution to the recognition of endocrinology as an academic discipline. Shortly after the Second World War, in 1947, the Dutch Society for Endocrinology was founded. Nine years later Utrecht University established a chair for theoretical endocrinology, appointing Marius Tausk.[63] In 1955, Utrecht University awarded an honorary degree in medicine to the founder of Organon, the industrial butcher Saal van Zwanenberg.

60 In 1928 the company received a letter stating that in Groningen (in the north of the country), insulin could only be obtained with great difficulty: archive Organon Oss, inv. nr. HA 1 (24 December).

61 In Germany too the most successful pharmaceutical company (Bayer) used internal instead of external researchers, while research was organized on an institutional basis, in order to guarantee continuity: W. Wimmer, 'Innovation in the German Pharmaceutical Industry, 1880 to 1920', in Homburg et al (eds), *The Chemical Industry*, pp. 281–291.

62 See for example *Chemisch weekblad* 29 (1932), 314–316 and 390–391.

63 A. de Knecht-Van Eekelen, *Hoofdlijnen van het endocrinologisch onderzoek in Nederland na 1947* (Rotterdam: Erasmus Publishing, 1993); R.B. Welbourn, 'The Emergence of Endocrinology', *Gesnerus*, 49 (1992): 137–148. The inaugural lecture of Tausk: *Endocrinologie als methode* (Oss: s.n., 1956).

Ways of Structuring the Market

Some two decades after the First World War, much had changed in the Dutch pharmaceutical landscape. At the end of the 1930s, the use of chemical research was appreciated widely, so that manufacturers thought it wise to employ chemists and pharmacists. By 1938, some 50 per cent of all academic chemists worked in the industry.[64] Also, physicians and pharmacists increasingly considered chemical and pharmaceutical industrial expertise to be crucial or even indispensable for their professional practice.[65] From the perspective of pharmacists, the pharmaceutical industry had succeeded in reframing itself from a threat into an ally.

This did not mean, however that on the broader market 'quack' remedies had been replaced by 'scientific' ones.[66] It only meant that by the late 1920s, a few companies had released themselves from the rather variegated pharmaceutical landscape by profiling themselves as ethical and scientific, thus winning over the trust of pharmacists and physicians. During the 1920s, 1930s and 1940s, the Society against Quackery continued to point out to the readers of its *Monthly* the dubious or even dangerous nature of the products of unethical manufacturers. 'So-called laboratories' were mushrooming everywhere in the Netherlands. They were strongly backed up by advertisements in the lay press, while the authorities rarely acted against them. In 1920, the Society decided to launch a campaign against A. Mijnhardt in Zeist and L.J. Akker in Rotterdam, who were held up as examples because they were very active on the industrial 'unethical' market.[67] The grudge that the editors of the *Monthly* held against Mijnhardt and

64 Homburg et al, 'Chemici', p. 308.

65 G.L. Voerman, 'De werkzaamheid van het Rijksbureau voor onderzoek van handelswaren in verband met de chemie en de chemische industrie', *Chemisch weekblad*, 25 (1928): 384–385.

66 Good proof of this is the blossoming of a company like Roter: P.G. van Druenen, *Een halve eeuw Roter en de opkomst van de zelfmedicatie in Nederland 1938–1988* (Rotterdam: Waterstad, 1988).

67 *Maandblad tegen de kwakzalverij*, February 1920. See also ibidem, April 1920.

Akker was that they were completely lacking in scientific expertise, their only motive being maximizing profits to the detriment of the patient.

The objections physicians and pharmacists initially had against the entire industry were redirected to specific parts of it. No longer did they generalize about industrially manufactured medicines by repudiating all of them. On their part, some manufacturers – organized in the Association of Manufacturers of Medicines on a Scientific Basis (*Vereeniging van Fabrikanten van Geneesmiddelen en Artsenijen op Wetenschappelijken Grondslag*) – constructed a difference between ethical manufacturers (like Brocades and Organon) and unethical ones (like Mijnhardt and Akker), as they did between their products: *spécialités* and secret remedies.[68] *Spécialités* were defined as medicines that had been developed in close collaboration between chemists, clinicians and pharmacologists 'with the best intentions'. Examples included Adalin, Aspirin, Digaleen, Haematogeen, Iodipin, Lysoform and Norit: all of these were produced and offered by '*bona fide*' manufacturers, while their preparation and composition was disclosed in brochures and advertisements that were directed to experts only. At the request of the NMG and the NMP, a National Institute for Pharmaco-Therapeutic Research was established in Leiden in 1920, in an attempt to realize the ideal of scientific transparence. Its task was reminiscent of the mission of the Society against Quackery: to analyse medicines and report on its findings 'for the benefit of the nation'.[69] The proponents of the Institute argued that this ideal was not shared by the manufacturers of secret remedies, who were accused of keeping the composition of their products a secret to secure optimal profits for themselves, instead of serving the public good by making the results of research available to every-

68 'Geheim – of kwakzalversmiddelen', *Maandblad tegen de kwakzalverij*, February 1921; 'Specialité's', ibidem, March 1921 and 'Rapport van de Commissie tot het beramen van maatregelen, welke kunnen leiden tot beteugeling van den verkoop van kwakzalversmiddelen', ibidem, June 1921.

69 L. Lietaert Peerbolte, 'Volksgezondheid', in De Bas (ed.), *Gedenkboek*, pp. 584–595, on p. 590. On the occasion of his honorary promotion, Saal van Zwanenberg called for close cooperation between science and (bona fide) industry: *Ere-promotie dr. Saal van Zwanenberg* (s.l.: s.n., 1955), 16.

body.[70] Most secret remedies were to be considered as quack remedies, they held, meaning substances offered for many different disorders, without the manufacturer knowing the individual patient or the nature of his disease. From this it should be concluded that his intentions were not humanitarian, but rather commercial.

However, similar endeavours to rhetorically structure the market and deal a decisive blow to quack remedies were only partly successful. As we saw earlier, pharmacists and druggists continued to sell 'unethical' remedies, simply because they were in great demand. From their sheer perseverance on the market, we have to conclude that the success of a remedy and its producer is not just dependent on its intrinsic pharmacodynamic effects or its legal status but on its cultural image as well.[71] Whereas consumers were sometimes attracted to 'ethical', scientific remedies, at other times they preferred 'neat' products with appealing brand names and fancy packing. The Society against Quackery and the National Institute for Pharma-co-Therapeutic Research were forced to admit that they were unable to stop unethical manufacturers and medical laymen. Most products made by the food and the stimulant industries bordered on those made by the pharmaceutical industry. Many companies were moving into the over-the-counter market. At the Utrecht annual trade fair, they presented themselves to the public, 'ethical' and 'unethical' companies standing side by side.[72] Because the public was not always capable (or willing) to recognize the

70 Between 1921 and 1965, the Institute published 51 monographs on medicines. On secret remedies, see E. Ernst, *Das 'industrielle' Geheimmittel und seine Werbung* (Würzburg: Jal-Verlag, 1975).

71 'Therapeutics [...] involves emotions and personal relationships, and incorporates all of those cultural factors which determine belief, identity, and status': C.E. Rosenberg, 'The Therapeutic Revolution: Medicine, Meaning and Social Change in Nineteenth-Century America', in M.J. Vogel and C.E. Rosenberg (eds), *The Therapeutic Revolution: essays in the social history of American medicine* (Philadelphia: University of Pennsylvania Press, 1979), pp. 3–25, on p. 4. See '[...] the adoption of scientific techniques, or at least of a scientific *veneer* seems to have been crucial to success among the drug companies', in: Liebenau, *Medical Science*, p. 1 [my emphasis].

72 See, for example, *Chemisch weekblad*, 14 (1917): 291–297; ibidem 15 (1918): 354–361; ibidem 16 (1919): 280–287.

difference, the 'ethical' parties decided to use yet another strategy. Since coercion (by legislation and legal proceedings) and education (by the Society and the National Institute) had failed to structure the market, the field resorted to an economic instrument: the formation of cartels.[73] Pharmaceutical associations were founded in order to regulate a market that could not be regulated by law.[74]

Epilogue

When we want to understand the transformation of the pharmaceutical industry in the early twentieth century, several explanations may be given.[75] The first focuses on the profound changes that took place in medical therapy. The period witnessed a paradigm change from the use of traditional constitutional medicines (geared to the regulation of the humoral balance of the individual patient) to the use of specifics; 'magic bullets' to be used for specific diseases in every patient.[76] Another is the introduction of mechanization that facilitated large-scale, cheap and standardized

73 On pharmaceutical cartels in the Netherlands, see H.W. de Jong and R. de Lange, *A Study in the Evolution of Concentration in the Pharmaceutical Industry in the Netherlands* (s.l.: s.n., 1975); H.W. de Jong, 'Remedie of pijnstiller? De werking van combinatiebewegingen in de farmaceutische industrie', in Huisman and Vos (eds), *Farmacie*, pp. 46–58. For an early analysis of Dutch cartels, see F.M. Wibaut, *Trusts en kartellen* (Amsterdam: Soep, 1903).

74 On the Onderlinge Pharmaceutische Groothandel, a purchasing cooperation established by pharmacists, see Rein Vos et al, *OPG 100. De geschiedenis van een bijzondere apothekerscoperatie* (Utrecht: OPG, 1999). On the mammoth alliance between pharmacists and druggists in the Commissie uit Apothekers- en Drogisten-Organisaties and again in the Vereeniging tot behartiging van Pharmaceutische Handelsbelangen, see B. Kruithof, 'Wetgeving of marktordening? Apothekers en drogisten in het Interbellum', in Huisman and Vos (eds), *Farmacie*, pp. 34–45; Idem, *Het conflict tussen apothekers en drogisten. De professionalisering van twee beroepsgroepen tussen 1865 en 1932* (s.l.: s.n., 1995).

75 See Liebenau, *Medical Science*, pp. 5–6.

76 See Rosenberg, 'The Therapeutic Revolution'.

production of medicines, marginalizing traditional producers like pharmacists. In the above, I have focused on the role of representation and public image of a particular medicine and its producer.

Initially the new industry had been led by commercial motives only, defending itself by calling upon the liberal principle of the freedom of trade. Gradually, however, there was a diverging of opinion and strategy. Some pharmaceutical entrepreneurs realized that a 'scientific' profile need not stand in the way of commercial success. Quite the contrary: it had the potential of facilitating a new cooperation with pharmacists and physicians, calling on them as allies with a shared epistemological tradition. They introduced a distinction between ethical and unethical products; a distinction that was mirrored in the distinction between *spécialités* and secret remedies.[77] As it turned out, pharmacists, who had felt threatened in the beginning, were prepared to associate with the 'ethical' part of the industry, appreciating its complementary – and growing – pharmacological expertise. New alliances were shaped, and a new pharmacological landscape developed.[78]

I hope to have shown that by using the perspective of the medical marketplace, one is able to do justice to the many dimensions of the transformation of pharmaceutical production in the early twentieth century. The market perspective may enrich pharmaceutical historiography, because it does not limit itself to either 'scientific' or 'successful' companies.[79]

77 Among the group of 'ethical' firms two sub-groups can be distinguished: one that cherished its roots as a pharmacists' company (like Brocades), and one that had no ties with the pharmacy shop whatsoever, like Organon. For the German situation (where Hoechst and Bayer had a different profile than Merck, Schering and Beiersdorf), see Huhle-Kreutzer, *Die Entwicklung*.

78 After the Second World War, the ties between the 'ethical' industry and physicians became even closer, due to the introduction of the 'detail man', who had both a promotional and an educational function. On this phenomenon, see J.A. Greene, 'Attention to "Details": Etiquette and the Pharmaceutical Salesman in Postwar America', *Social Studies of Science*, 34 (2004): 271–292.

79 See for example H. Koopmans, *Vijftig jaar scheikundige nijverheid in Nederland* (Delft: Waltman, 1967); L. Reynders and F. van Winden, *De farmaceutische industrie in Nederland* (Amsterdam: SUA, 1976) and, for Germany, W. Wimmer, '*Wir haben fast immer was*

It could be argued that a similar one-dimensional view on the health care system – focusing on medical theory, on state regulation or on commercial success – is an impoverishment of historiography, because it excludes both historical dynamics and historiographical debate. The added value of the medical market perspective is that it does not selectively focus on medical progress or on growing professionalization. When using the market model, we are able to show how some groups decided to join forces and exclude others by constructing a dichotomy between ethical and unethical and between science and non-science. In the process, the narrative on a linear development of medicine and pharmacy is replaced by a story about a multidimensional structuring of the health market with a keen eye for the intrinsic tensions in the health care system between science, humanity and commerce – and that may very well be historiographical progress.

Neues': Gesundheitswesen und Innovationen der Pharma-Industrie in Deutschland, 1880–1935 (Berlin: Duncker & Humblot, 1994).

MARÍA JESÚS SANTESMASES[1]

3 Distributing Penicillin: the clinic, the hero and industrial production in Spain, 1943–1952

News about penicillin, its success in the treatment of war wounds, and its role in the therapy of infectious diseases during the Second World War, appears to have travelled fast. However, all data regarding the production, chemistry, and clinical trials of penicillin remained a war secret. Or so the story goes, according to the official documents produced by clinicians, scientists and government advisors in Spain.[2]

This paper describes and discusses the introduction of penicillin into Spain. I will begin by examining the public and clinical knowledge that was circulated about the drug, even before these official documents became available. I will argue that the heroic character penicillin acquired in Spain emerged in association with the ideology of the hero as saviour, which was attuned to the idea of national hero that Franco was constructing around himself, and his acclamation as such by his supporters, his successive government cabinets, and the wider public. The visit of Alexander Fleming to Barcelona and Madrid in 1948 illustrates this. The celebration of penicillin, and of Fleming himself, as heroes mirrored Spanish social life at a

1 I would like to thank Roberto Gallego for giving me access to personal files of his father, Antonio Gallego. I am grateful to archivist Elena Laruelo at the Archives of the Instituto Nacional de Industria (Madrid), to archivists of the Hemerotaca Nacional (Madrid), of the Archivo General de la Administración (Alcalá de Henares), and of the library of the Instituto Juan March de Estudios Sociales (Madrid). I also thank Pedro García-Barreno for his advice regarding Bustiza papers at the Real Academia de Ciencias (Madrid). I am indebted to Viviane Quirke for her comments, discussions and kind help in editing an earlier version of this article.

2 On penicillin reception in wartime, see the case of Germany in G. Shama and J. Reinarz, 'Allied intelligence reports on wartime German penicillin research and production', *Historical Studies on the Physical and Biological Sciences*, 32 (2002): 347–367.

time when penicillin was being distributed, albeit in scant quantities, and successfully used for the treatment of infections.

Although penicillin was greeted everywhere as a life saver, the singular political circumstances of Franco's dictatorship led to the particularities of its heroic reception in Spain. This specificity was also related to the extreme poverty that characterized the early years of Franco's regime. Rationing was in force throughout the 1940s, and the black market in basic foodstuffs and pharmaceuticals thrived. Corruption in local politics flourished in the formally centralized and bureaucratic 'New State'.[3] Out of poverty, illness, and the miseries of life, came the cure. In the devastated landscape of post-war Spain, the Spanish people may thus have become highly sensitized to the miraculous effects of penicillin, and better able to perceive its wonders as a new order in relation the role of science and the treatment of disease was being introduced.

In 1944, when penicillin was beginning to be mass-produced in the USA, not only for the treatment of war wounds, but also for the purpose of wider clinical trials, the Spanish Commission for Penicillin Distribution was set up with the objective of administering supplies of the drug. Its scarcity, together with its effectiveness in treating infections, made penicillin a desirable and expensive product. Its use preceded its manufacture almost everywhere except in the USA. Clinicians reported on its wondrous effects in Spain from 1944 onwards, and some medical researchers tried to produce it in the laboratory by means of surface culture.

I will then describe the early industrial production of penicillin in Spain. After several unsuccessful attempts by some industrialists and health officials to convince the industrial policy authorities of the importance – given the increasing number of successful treatments – of producing the antibiotic in Spain, finally in 1948 a government decree launched a public competition to build two factories, so that national requirements for the drug could be answered. In the immediate aftermath of the Second World

3 The historiography on the Franco regime is so vast that it cannot be quoted to its full extent here. For a recent review of the local politics of its early decade, see A. Cazorla-Sánchez, 'Dictatorship from below: local politics in the making of Francoist State, 1937–1945', *Journal of Modern History*, 71 (1999): 882–901; and idem, *Políticas de la Victoria. La consolidación del Nuevo Estado franquista (1938–1953)* (Madrid: Marcial Pons, 2000).

War, penicillin had been imported into Spain. Later on, when the decision was taken to start its production, according to the archival material available, the manufacturing process was imported from American firms without exploring any other option. Spanish penicillin manufacture was a state-promoted process, led by the industrial policy authorities, which kept any initiative towards industrial production as a government prerogative. Projects to create firms and factories to this aim had to be approved by the government, but the process could be into private hands, while the authorities retained permanent control over contracts and foreign currency expenses.

By the early 1950s, as penicillin was becoming cheaper in other countries, Spanish prices also began to decrease. However, industrial producers argued that they had not yet reached the point where a decrease in prices could be justified, for Spanish penicillin production had only begun in 1951, and they had not had the time to recoup their investments. By then, penicillin consumption was increasing, and antibiotics had achieved general recognition by physicians and the public alike, after almost a decade of scarcity and at least eight years on the black market. Then, in 1953, Spanish industry began to reduce imports of penicillin. Its rate of factory production was considered to cover national demand, while foreign overproduction was starting to threaten Spanish producers because of dumping practices. By then, streptomycin manufacture was also under way in Spain.

The Wonders of Penicillin: clinicians and the public

The large-scale production of penicillin began in the USA in May 1943 after promising clinical evaluations by the US Committee of Medical Research and with support from the War Production Board.[4] By the end of 1944, a

4 P. Neushul, 'Science, government and the mass production of penicillin', *Journal of the History of Medicine and Allied Sciences*, 48 (1993): 371–395; G.L. Hobby, *Penicillin. Meeting the Challenge* (New Haven-London: Yale University Press, 1985). See also R. Bud, *Penicillin: triumph and tragedy* (Oxford-New York: Oxford University Press, 2007).

National Committee for Penicillin had been created in Spain to ensure distribution of the drug in cases of necessity and to establish criteria according to which supplies should be approved or rejected.[5] Carlos Jiménez Díaz, a highly respected physician and clinical researcher, was appointed chairman of the Committee, which amongst its members counted his disciple Eduardo Ortiz de Landázuri. Jiménez Diaz was diagnosed as suffering from pneumonia during the summer of 1944, and was successfully treated with penicillin at the suggestion of his collaborators, colleagues and family.[6]

According to a report published in 1945, as a result of early contacts made by the *Comite Nacional del reparto de la penicilina en España*, a protocol describing which diseases should be treated with penicillin and which should not was received at the end of 1944. According to this report, it was Chester S. Keefer, head of the Committee on Chemotherapy of the US National Research Council, who had sent the protocol.[7] At that time penicillin production was increasing in the USA and civilian investigations of the uses of the antibiotic in the treatment of certain infectious diseases had begun. From July 1943 on, Keefer was in charge of all American supplies.[8] The Spanish committee for penicillin may well have been inspired by

5 The Commission could have been created in 1943, according to Guillermo Olagüe de Ros, *Del uso de la retórica en el discurso científico: a propósito de los programas de trabajo de Fidel Fernández Martínez (1890–1942) y Eduardo Ortiz de Landázuri (1910–1985)*, Discurso leído en el acto de recepción pública (Granada: Real Academia de Medicina y Cirugía de Granada, 2001), p. 33.

6 M. Jiménez Casado, *Doctor Jiménez Díaz. Vida y obra* (Madrid: Fundación Conchita Rábago de Jiménez Diaz, 1993), pp. 348–355.

7 E. Ortiz de Landázuri and J. Perianes – Clínica Médica Universitaria del Prof. Jiménez Díaz – 'Normas generales de la terapéutica penicilínica', *Revista Clínica Española*, XVII (1945): 361–372 (366–368). In this paper the committee was described as functioning from late 1944.

8 H.M. Marks, *The Progress of Experiment: science and therapeutic reform in the United States, 1900–1990* (Cambridge-New York: Cambridge University Press, 1997), pp. 106–107. On the US Chemotherapy Committee see D.P. Adams, 'War time bureaucracy and Penicillin allocation: The Committee on Chemotherapy and other agents', *Journal of the History of Medicine and Allied Sciences*, 44 (1989): 196–217; idem, *'The greatest good to the greatest number': penicillin rationing on the American home front, 1940–1945* (New York: Peter Lang, 1991).

the Penicillin Panel of the US National Research Council, but the shortage of the drug on the legal market in Spain contributed to enhancing its fame and the perception of penicillin as magical and heroic.

The medical perception of this new drug was constructed on the basis of its therapeutic properties. During the second half of the 1940s, penicillin was obtainable from the Committee or, more often, through the black market.[9] Fiction and non-fiction accounts of the post-war period in Spain told stories of buying penicillin in a well-known pub in Madrid, and in the fishing port of a seaside town, to cite but a few.[10]

Florencio Bustinza at the University of Madrid played an important part in early public information on penicillin. This British-born Spaniard, a pharmacist and natural scientist by training, began his career as a high school teacher and after the Civil War (1936–1939) was appointed professor at the Natural Sciences Faculty of the University of Madrid, in 1943. In 1944 he published his first article popularizing the discovery of penicillin in the widely distributed Madrid daily *Abc*.[11] He also published articles, two

9 The role of black market in Spanish industry and economy during the first decades of Franco regime should not be regarded as the main issue of this essay, it has been analysed by Spanish historians of economy and industry in the case of raw materials and basic food. See C. Barciela (ed.), *Autarquía y mercado negro: el fracaso económico del primer franquismo (1939–1959)* (Barcelona: Crítica; 2003); idem, 'Crecimiento y cambio en la agricultura española desde la Guerra Civil', in J. Nadal, A. Carreras and C. Sudriá (eds), *La economía española en el siglo XX. Una perspectiva histórica* (Barcelona: Ariel, 1987), pp. 258–279; and J. Catalán, *La economía española y la segunda guerra mundial* (Barcelona: Ariel, 1995). For an example on the availability of penicillin on the black market, see J. Chamorro, *Recuerdos de una vida* (Oñate: Diletante, 2002), memoir of a rural doctor in post-war Spain. I am grateful to Jon Arrizabalaga for drawing my attention to this text.

10 On the Spanish pub in Madrid, see, for example, J. Casado, *Doctor Jiménez Díaz*, p. 350. The pub owner was Jiménez Diaz's patient, according to Jiménez Casado.

11 F. Bustinza, 'El descubrimiento de la penicilina, momento estelar en la Historia de la Ciencia', *Abc*, 5 November 1944. On Bustinza's career see his *Currículum Vitae*, at the 'Expediente Florencio Bustinza 0687', Archivos de la Real Academia de Ciencias, Exactas, Físicas y Naturales, kept in the Royal Academy of Sciences library, Madrid. He was elected member of the Royal Academy of Sciences in 1943.

books and gave many talks on antibiotics during the second half of 1940s, the first of which in 1945 on the history of penicillin and antibiotics.[12]

Bustinza's trip to the USA from August 1945 until December 1946 took him to some of the main laboratories and facilities that had been involved in the early production of penicillin. Thus, he visited Robert D. Coghill's group, including Andrew J. Moyer and K.B. Raper, at the Northern Regional Research Laboratory of the US Department of Agriculture, Peoria, in which the efficient submerged culture of the mould had been developed. He also met Henry Welsh at the Food and Drug Administration laboratory for the control of penicillin in Washington, and Chester Keefer at the Evans Memorial Hospital in Boston, when Keefer was head of the Committee on Chemotherapeutics. Bustinza also met a number of other scientists involved in penicillin research, among them Leo Loewe at the Jewish Hospital in Brooklyn (New York) who used penicillin in bacterial endocarditis, and Harry Eagle at Johns Hopkins University (Baltimore) who applied it to the treatment of syphilis.

In addition Bustinza visited several penicillin factories, among them C.S.C. (Commercial Solvents Corporation) in Terre Haute (Indiana), Abbott's in North Chicago, Lederle's in Pearl River (New York), Squibb's in New Brunswick (New Jersey), Merck's in Rahway (New Jersey), and Wyeth's at West Chester (Pennsylvania). His American trip ended on 11 of December 1945 when he embarked for Europe in New York. He reached Southampton, in England, on the 17th, and on the 20th of December he visited Alexander Fleming at Saint Mary's Hospital in London. They met many times before Bustinza returned to Madrid in January. Bustinza's own nine year old daughter had some months before been saved by penicillin after conjunctivitis in the right lachrymal gland had caused a general inflammation of the right side of her face: 'a few intramuscular injections were enough' according to Bustinza's memoirs.[13]

12 See Bustinza *curriculum vitae*, dated January 1975, at the Real Academia de Ciencias.

13 On Bustinza's trip to the USA see F. Bustinza, *Diez años de amistad con Sir Alexander* (Madrid: M.A.S., 1961), pp. 39–41. On his stay in London, ibid. pp. 55–72; his daughter's infection and penicillin treatment are described on p. 72. On the NRRL, see P. Neushul, 'Science, government', pp. 374–381, and references therein. On the clinical

His report in an earlier book, *De Pasteur a Fleming* (1945), had contributed to public knowledge of the effects of penicillin in the treatment of infections, as well as to the perception of Fleming's work as a heroic achievement, and Bustinza remained an 'ardent devotee' of Fleming until he died, in 1982. He kept in personal contact with Fleming after his visit to Britain in 1945, and following Fleming's death published a book about his relationship with him. According to one of Fleming's biographers, Bustinza became 'one of Fleming's most ardent, and certainly eloquent, admirers'.[14]

By the time Bustinza published his first article in the Madrid daily *Abc*, many physicians were aware of the therapeutic success of penicillin. As soon as the early information on the beneficial effects of the new drug began to circulate, between late 1943 and mid-1944 a group of clinicians and chemical researchers in Barcelona tried repeatedly to develop a method for obtaining penicillin by culturing the mould *Penicillium notatum*. By growing a mould sample obtained from the Lister Institute in London, they found ways of extracting the antibiotic and testing it without access to literature on the methods available. Despite this, they reported success in testing penicillin in dogs and mice.[15]

In 1945, reviews were published in which the therapeutic usefulness of penicillin, as well as its limits, were described.[16] By that time, members

---

trials of penicillin in the treatment of syphilis, H.M. Marks *The Progress of Experiment*, chapter 4.

14 G. Macfarlane, *Alexander Fleming: the man and the myth* (Cambridge, MA: Harvard University Press, 1984), p. 226.

15 P. González Juan (de la Real Academia de Medicina de Barcelona), J. Suñer Pi (del Laboratorio Municipal), and F. González (médico asistente), 'Obtención de la penicilina y otras experiencias', *Medicina Clínica*, II (1944): 473–482. A. Esteve and A. Oriol Anguera, 'Aportaciones experimentales a la farmacognosia penicilínica', *Medicina Clínica*, III (1944): 116–128.

16 A. Ballabriga Aguado (Clínica pediátrica de la Universidad de Berna, 'Jenner Kinderhospital', Director: E. Glatzmann), 'Algunas consideraciones acerca de la penicilina y otras micoínas', *Medicina Clínica*, IV (1945): 152–156; R. Kutz Echave (San Sebastián), 'El empleo de la penicilina en la práctica médica', *Revista Clínica Española*, XIX (1945): 340–349.

of the Spanish Committee for Penicillin Supplies were already making contributions to professional knowledge on the type of infections for which penicillin treatment was recommended. Protocols for its medical use were becoming known, and the image of penicillin as a magical drug, wonderfully useful and efficient, was being widely publicized.[17]

Successful treatments at the hands of Spanish physicians were soon reported. A case of septicaemia caused by staphylococci was cured early in 1945, and within two days the haemoculture was negative. For this, 450,000 units of 'American penicillin' had been used after 'going through the necessary formalities'.[18] Such formalities might include either a request to the Committee, or contact with black-market traders. Between late 1944 and June 1945, more than a thousand patients were treated with penicillin by clinicians 'from all around Spain' who were collaborating with the Spanish Committee for Penicillin Supplies.[19]

In 1946, otorhinolaryngologist Mateo Jiménez Quesada reported sixty cases of successful treatment of throat, nasal and lower respiratory tract infections using penicillin. He treated sinusitis caused by the Preiffer bacillus and sent his results to Alexander Fleming, who, according to Jiménez Quesada's recollections, had stated that this bacillus was not sensitive to penicillin. By that time, Fleming was obtaining similarly good results against the same type of infection by administering the drug by inhalation. Thus, he suggested that 'by inhalation one can get in the sputum a concentration sufficient to kill [the bacillus]'.[20] Jiménez Quesada also travelled to the USA in 1945 to learn about progress made with antibiotic treatments. He visited hospitals and research laboratories, among them the Presbyterian Hospital Medical Center in New York and the Mayo Clinic in Rochester, where he

17 Ortiz de Landázuri and Peiranes, 'Normas generales', mentioned many treatments in Spain.

18 A. Crous Serdá, 'Septicemia estafilocócica tratada y curada con penicilina', *Medicina Clínica*, IV (1945): 415–419.

19 See Ortiz de Landázuri and Perianes, 'Normas generales', pp. 368–369.

20 Fleming to Jiménez Quesada, reproduced in M. Jimenéz Quesada, *De Fleming a Marañón. Anécdotas de mi archivo* (Madrid: Minaza, 1972), p. 140. See also Jiménez Quesada, *El tratamiento de las infecciones de garganta, nariz, oídos y vías bajas respiratorias con nebulizadores de antibióticos* (Madrid: Faure, 1947).

witnessed penicillin being used successfully to treat infections at a time when early antibiotics were only used for wounded allied soldiers.[21]

Fleming in Spain: propaganda for a scarce product

In May 1948 Alexander Fleming visited Spain as guest of the local governments of Barcelona and Madrid. The initiative came from the Barcelona Hospital of Infectious Diseases, Hospital del Mar, whose director, cardiologist Luis Trías de Bes, was an influential clinician.[22] On the day of Fleming's arrival in Barcelona, Trías de Bes published an article in the Barcelona daily *La Vanguardia*, calling attention to Fleming's achievements as an 'extraordinary benefactor of mankind' who was to be received with all the 'splendour and distinction [*señorío*] such an illustrious guest deserves'.[23] Fleming stayed from 26th of May until 14th of June, a long stay during which Spanish daily newspapers reinforced his fame as a scientific hero: 'in no other countries I visited have I been so kindly received as here' Fleming declared after his arrival in Barcelona.[24] Two days later, Fleming was on the front page of the widely distributed Catalonian daily, which

21 Jiménez Quesada, *De Fleming a Marañón*, pp. 132–135. Spanish university professors, including many clinicians, visited the USA from 1943 on.

22 One of Fleming's biographers, Gwyn Macfarlane, suggests that it was Bustinza's arrangements that made Fleming's tour of Spain a 'triumphal visit'. Macfarlane, *Alexander Fleming*, p. 227. However, Bustinza's reconstruction of the trip in his *Diez años de amistad con Sir Alexander Fleming*, the review of newspapers of those days, and the memory of this Catalonian Hospital suggest that Trías de Bes provided the main support for the trip, although while in Madrid Bustinza took a lead in the academic arrangements. Fleming was publicly acclaimed when visiting abroad at that time. On his visit to France in 1945, see Jean-Paul Gaudillière, *Inventer la biomedicine: la france, l'amerique et la production des savoirs du vivant* (Paris: la Découverte, 2002), pp. 46–47.

23 L. Trías de Bes, 'Alexander Fleming y su penicilina', *La Vanguardia*, 26 May 1948; reproduced and commemorated by *La Vanguardia digital* on 23 August 2002, as one of the 'Articles of the century', at http://www.lavanguardia.es, accessed 24 October 2002.

24 *La Vanguardia*, 27 May 1948, p. 9.

included a photograph of him walking along the popular Barcelona avenue of Las Ramblas, with the public surrounding him and a young woman giving him a bouquet of fresh flowers.[25] Fleming also joined in the Corpus Christi procession, a holy day that was widely and publicly celebrated in Spain at that time, when the State was officially catholic and religion was one of the social bases for Franco's dictatorship.[26]

Still in Barcelona, at the Hospital de Nuestra Señora del Mar for infectious diseases, 'the illustrious discoverer of Penicillin' gave a talk on septic wounds. On the same occasion, he inaugurated the hospital's new research department and the conference hall, in which he would give talks on penicillin and its uses. The room was full, as were the corridors and the old conference room next door. In the audience, a journalist recognised many well-known clinicians from Barcelona as well as public health authorities and faculty members of the Universidad de Barcelona medical school.[27] The talks were given in English, and were simultaneously translated by a physician sitting in a cabin in the hall, and the translation was communicated to the audience by way of loudspeakers. The correspondents mentioned the endless thunder of applause at the end of each talk, in which the properties of penicillin were lauded by the speaker. The newspapers published the detailed activities of each day of Fleming's visit in Barcelona, and also announced what he would be doing the next day. He gave two more talks at the Hospital for infectious diseases in Barcelona, on lysozyme, and on the use of penicillin. These talks were collected and published later on in the year in Barcelona, with a foreword by Trías de Bes.[28]

The main photograph on the front page of *La Vanguardia* emphasized the scientific expertise of the British hero. He wore a white coat, and was seated in front of the optical microscope of the Hospital, as if he were

25 *La Vanguardia*, 28 May 1948.

26 J. Tusell, *Franco y los católicos: la política interior española entre 1945 y 1957* (Madrid: Alianza, 1984).

27 *La Vanguardia*, 1 June 1948, p. 8.

28 Alexander Fleming, *Conferencias magistrales* (Barcelona-Madrid: Publicaciones del Hospital Municipal de Infecciosos-Diana Artes gráficas, 1948).

working in the hospital laboratory.[29] Fleming also attended bullfights in the Barcelona bullring, seated at the front of one of the best rows. His presence provoked a great public ovation, and the bullfighting teams joined in. This was something the columnist, who remarked that it lasted for more than three minutes, had never seen before.[30] Sharing all the features of Spanish public social life of the time with the authorities and distinguished clinicians from Barcelona, Fleming also attended a football match that opposed Spanish and Irish teams in the main stadium.[31]

Fleming went to Seville, where he was honoured by the Seville Academy of Medicine, which made him an honorary member. This time, the talk he gave had previously been translated into Spanish and was read at the event by the vice-president of the Academy, Cristóbal Pera. Fleming also visited Córdoba, as well as Jerez where he tasted the world-renowned sherry. Throughout his tour he was publicly acclaimed as the discoverer of a wonder drug.[32]

Fleming's Spanish tour ended in Madrid, where he received a tribute as a guest of honour, and as a 'peace messenger of medical science', thanked for his 'example and simple life as a laboratory worker', and for his work which 'benefited everyone equally'.[33] At Madrid airport, an airport porter joined the large group of political, medical and local authorities who welcomed Fleming, and handed him a letter of gratitude for penicillin, which had cured him from gangrene following surgery. On the same day, the Madrid government decided to name a street after Fleming. He gave a talk at the British Institute, and received many honours, invitations and public acclaim. Finally on the 12th June he was awarded doctor *honoris causa* at the University of Madrid in an academic session chaired by the rector and

29 *La Vanguardia*, 2nd June 1948. Similar pictures were also included in some other newspapers of the same date. See Barcelona weekly *Destino*, 22 May 1948.

30 E.P., 'Fiesta de los Toros', *La Vanguardia*, 1 June 1948.

31 *La Vanguardia*, 1 June 1948, p. 8.

32 A short chronicle of his trip to Sevilla, Córdoba, Jerez and Madrid can be found in F. Bustinza, *Diez años de amistad*, pp. 86–87. On Fleming's stay in Madrid see also Jiménez Quesada, *De Fleming a Marañón*, pp. 141–142.

33 *Abc*, 12 June 1948, p. 7.

many academic and education policy authorities, including the minister of National Education, José Ibáñez Martín.[34]

The chronicle of these events, some of them collected by Fleming's Spanish spokesman Florencio Bustinza, others detailed in newspapers, had a tone typical of the period: the heroic individual acclaimed at a time when groups were considered to be a threat to civil life by a highly controlling dictatorship. Public imagery included both the positive and dark sides of penicillin. Fleming even visited one of the famous pubs in Madrid where penicillin was sold 'de estraperlo', a popular term for the black market, on which this new drug was traded and sold.

The Economics of Penicillin

Encouraging the economical use of penicillin was an important aspect of its introduction into Spain. Members of the Spanish Penicillin Committee were concerned that the antibiotic should be used efficiently in order to avoid waste.[35] A 1944 report by a member of the Committee stressed the economic importance of using it in the smallest amounts possible.[36] When legally obtained, penicillin was scarce, and when illegally, it was expensive. This scarcity made access to it a distinct privilege.

In 1945, a full report on the methods of extraction and clinical trials of a pharmaceutical product from *Penicillium notatum* was submitted to the Pharmaceutical Products Section of the Dirección General de Sanidad (the public health authority). This was not a pure drug, but rather the concentrated extract from a mould culture that had shown activity against staphylococcal and pneumococcal infections. Among the successful trials

34 F. Bustinza, *Diez años de amistad*.

35 It should be remembered that in Oxford in 1941, penicillin was isolated from urine of patients who had already treated with it, according to Howard Florey's recollections. See Hobby, *Penicillin*, p. 67.

36 See Ortiz de Landázuri and Perianes, 'Normas generales'.

included in the report, there was the treatment of a four-month old girl suffering from a staphylococcal infection, and of a boy of two-and-a-half and a sixteen year-old sick with tonsillitis. The Spanish physicians who reported on its use recommended either topical or rectal administration.[37] As to the extraction process, it involved producing the mould by surface culture, followed by successive phases of extraction and concentration. The process and uses of the product fitted with Franco's autarchic industrial policy, which promoted autonomous national production, at a time, immediately after the Second World War, when the isolation of the regime was imposed by the industrial authorities as well as by the Allies. Political isolation and autarchy made national production a virtue out of necessity.

In the same year, 1945, penicillin was the subject of pharmacist Victor Villanueva's dissertation at the National Academy of Pharmacy. Villanueva was regarded as an influential expert in pharmaceuticals, as he was head of the pharmaceutical section of the Union of Chemical Industries, member of the Junta Técnica de Farmacobiología, of the Junta Interministerial de Farmacia (inter-ministerial committee for pharmacy), and advisor to the ministry of Industry. In his dissertation he stated that the fermentation process for penicillin was unknown, as was its chemical composition, and that such information was kept secret. He added that this valuable new drug was produced by fermentation of the mould *P. notatum*.[38]

Thus, by 1945, clinicians, scientists and the public in Spain were fully aware of the efficacy of penicillin in the treatment of infections, and this contributed to what Bud has called one of the 'few iconic tales' that perme-

37 'Penicilina-OM Tópica. Características y referencia clínica', in 'Memoria para el registro de la Penicilina-OM Tópica', Sociedad General Farmacéutica, Espulgas de Llobregrat, 'Extracto de penicillium notatum conteniendo 1,000 unidades Oxford de penicilina por ampolla'. This drug and its extraction procedure were approved for sale on May 1945. Archivo General de la Administración (Alcalá de Henares, Madrid), Ministerio de Gobernación, DG de Sanidad, Inspección General de Farmacia, Sección de registros Farmacéuticos, (8)8 44/18155, 6479.

38 V. Villanueva Vadillo, *Factores determinantes de un proyecto de industria químico farmacéutica nacional*, discurso leído en la toma de posesión como académico de número (Madrid: Real Academia de Farmacia, 1945).

ated public, medical and political consciousness in the post-war period.[39] The Spanish view of penicillin was linked to the public imagery of Franco's dictatorship, to the heroic overtones of the literature of the time, when censorship was imposed on every single text published, to the autarchy of his government during the 1940s, and, I would argue, to the concept of national hero. In Spain, physicians, patients and the public alike perceived penicillin to be such a hero. The historiography on corruption suggests that the illegal trade in penicillin might also be linked to 'business of power', when gasoline supplied for official use was being sold on the black market for example.[40] Spanish authorities regarded penicillin production in Spain as being necessary from about 1946, and by 1948 considered it to be of industrial interest. However, if Spanish-made penicillin reached the Spanish market in 1952, it was thanks to contracts with American firms.

The Spanish market, the Spanish public, and Franco's dictatorship were agents in constructing national medical life, and contributed to the inclusion of drug manufacture and medical practices in historical narratives.[41] Rather than being a national icon, as it became in the USA at the time of its mass-production and large-scale clinical trials involving government, or in France after the Second World War, during the early days of Franco's dictatorship penicillin functioned as a heroic medical and scientific object, as well as a State-controlled product. It saved lives but was also embedded in the regime's bureaucracy.[42]

39 R. Bud, 'Penicillin and the new Elizabethans', *British Journal of History of Science*, 31 (1998): 305–333 (305).

40 Cazorla-Sánchez, *Políticas de la Victoria*, p. 897.

41 On the role of penicillin in constructing national systems of research, see J.-P. Gaudillière and B. Gausmeier, 'Moulding national research systems: the introduction of penicillin to Germany and France', *Osiris*, 20 (2005): 180–202.

42 On the French and British cases, see also J. Liebenau, 'The British success with Penicillin', *Social Studies of Science*, 17 (1987): 69–86; Gaudillière, *Inventer la biomedicine*, chapter 4; V. Quirke 'Penicillin and post-war reconstruction in Britain and France: science and industry in the balance', *La lettre de la Maison Française d'Oxford*, 6 (1997): 87–98; idem, *Collaboration in the Pharmaceutical Industry: changing relationships in Britain and France, ca 1935–1965* (London/New York: Routledge, 2008), chs. 3 and 4.

Towards National Production

'Watching a fever decrease and the gravest symptoms of septicaemia, even those caused by a deadly staphylococcal infection, subside a few hours after beginning penicillin injections' was perceived as a 'miracle'. 'Those of us who have seen patients emerge out of insuperable pain can neither complain nor curse our times.'[43] Those words by the famous and distinguished Spanish clinician Gregorio Marañón in the foreword to Bustinza's first book echoed those by Bustinza himself, who described both the drug's wondrous effects and the curse of insufficient supply: 'What therapeutic results might be achieved when pure penicillin can be used as a medicine, at an affordable price, at the beginning of an infection and at suitable doses, without restrictions and the fear we have today of wasting it?'.[44] Marañón and Bustinza's views represent the Spanish *Zeitgeist* with regard to penicillin. Its therapeutic activity and scarcity were the bases upon which a legend was constructed. So unusual and unforeseen were its effects considered to be, that the telling of its story took on legendary form precisely because of its wide success.

Growing demand for penicillin, and its scarcity despite the notorious black-market trade of the drug, became so obvious that in September 1946, the Dirección General de Sanidad submitted a proposal to the industrial policy authorities in order to manufacture the drug in Spain. Private initiative could not develop without government approval at a time when the

43 'Acaso ninguno de esos prodigios pueda compararse con el de ver descender la fiebre y remitir los más graves síntomas de una septicemia, como la estafilocócica, que era inexorablemente mortal, a las pocas horas de empezar las inyecciones de la Penicilina. Los que hemos visto surgir todo esto del caos del dolor invencible, no nos podemos quejar ni maldecir nuestro tiempos'. Gregorio Marañón 1945, 'Prólogo', in Bustinza, *De Pasteur a Fleming* (my translation).

44 '¿Qué resultados terapéuticos se lograrán cuando se utilice como medicamento la penicilina pura, a un precio asequible a todas las fortunas, al comienzo de las infecciones y empleándola a dosis convenientes, pero sin restricciones y sin temor a la responsabilidad que hoy indudablemente tenemos de malgastarla?' Bustinza, *De Pasteur a Fleming* (my translation).

State controlled every type of industrial production. Penicillin consumption, the proposal argued, was so high that the official currency budget for importing penicillin – $5,000 per month at that time – was not enough. Instead of increasing currency expenditure for buying penicillin from abroad, it might be 'more logical' to promote the manufacture of the drug in Spain through an agreement between the industrial policy authorities and 'civil industry'.[45] The Director General of Health sent the proposal to the Minister of Industry in January 1947, and the Instituto Nacional de Industria, apparently acting as advisor for the Ministry of Industry 'discussed' the subject. In June, an application to build a penicillin factory was received by the Ministry of Industry.

This application for penicillin manufacture in Spain (apparently the first) had been submitted by Leopenicilina Española, a company from Bilbao. The firm asked for permission to establish a factory to produce penicillin under a Danish patent, that of Løvens Kemiske Fabrik, Copenhagen. As Leo was the pharmaceutical division of Løvens, the application seems to have been an attempt by the Danish firm Løvens to establish a penicillin plant in Spain. The accompanying memo stated that penicillin had achieved wide medical recognition in the treatment of many diseases, that its field of application was expected to expand, and therefore that sufficient quantities of the drug were needed.[46] As to the report, it expressed doubts about the

45 'Moción de 11 de septiembre de 1946', Ministerio de la Gobernación, Dirección General de Sanidad, copia con registro de entrada en el Instituto Nacional de Industria de 8 de enero de 1947, Archivo Histórico del Instituto Nacional de Industria (Madrid), legajo 270, documento 7 (1946–1947). The Instituto Nacional de Industria (INI) was created in 1941 to promote the Sate participation as entrepreneur and the State investments in the Spanish industrialization process after the Spanish Civil War. During the first decade of Franco's regime, INI was the main authority in industrial development, it created many industries and promoted State monopolies in several sectors. On the history of INI, see P. Martín-Aceña and F. Comín, *INI. 50 años de industrialización en España* (Madrid: Espasa Calpe, 1991). I have discussed the Spanish production of penicillin in M.J. Santesmases, *Antibióticos en la autarquía: banca privada, industria farmacéutica y cultura liberal en España* (Madrid: Fundación Empresa Pública, 1999).

46 'De Sección del Plan a Dirección Técnica. Asunto: Petición de la Sociedad Leopenicilina Española SA para implantar una fábrica de penicilina', Madrid, 1 de julio de 1947, Archivo Registro General del Instituto Nacional de Industria, División de Industria, Industrias

research capabilities of the Danish firm, compared to those of American firms such as Abbott Laboratories, Lilly, and Merck. The British Burroughs Wellcome & Co was also mentioned in this connection. As penicillin production was still undergoing improvements, it was hoped that those might yield both scientific and economic benefits. The report expressed concern about the high cost of royalties and the high price per 100,000 unit flasks of the penicillin that would be sold by the Spanish producers, and suggested that an investigation of the Danish patent and its cost be carried out. The report also stated that a representative from Abbott Laboratories had visited the Director General of Health, and recommended getting more information about Abbott's intentions concerning penicillin production in Spain. According to the available information, imports of penicillin from the USA were cheaper than the price proposed by Leopenicilina Española, and this should be carefully checked to avoid excessive costs, which would result in making Spanish-made penicillin more expensive than imported penicillin. The report ended by stressing the need to manufacture penicillin in Spain. Nevertheless, it recommended building contacts with other foreign producers and, more specifically, seeking clearer information about Abbot's intentions. It suggested approaching the health authorities to gain information on this issue. There seemed to be no hurry in establishing a factory, nor any interest in indigenous production. Buying in a process or a patent appeared to be taken for granted.

Both Leopenicilina Española's application and the report written about it for the industrial authorities reveal detailed technical knowledge about methods of production, about informal contacts established by pharmaceutical companies with health authorities, and about growing demand for penicillin, at a time when medical journals were beginning to report on successful treatments with the drug, which shared space in scientific publications with exciting results obtained with an even newer drug, streptomycin. Both also suggest that information about the efficiency of penicillin in the treatment of infections was already circulating widely, and that at

---

Varias, 'Leopenicilina Española SA', Legajo 200 documento 7 (1946–1947). Archivo del Instituto Nacional de Industria, Madrid.

least two foreign firms were trying to enter the Spanish market and begin penicillin manufacture in Spain under their own foreign patents. The historiography on Spanish industry shows that the industrial authorities were directly involved in heavy industry (power production, communications, and transport) as it was related to defence. In the first decade of Franco's regime, when autarchy was the official policy, the Instituto Nacional de Industria had among its duties not only the promotion of industrial production, but also the control of Spanish initiatives in all industrial sectors.[47] During this period, economic nationalism and the isolation of the dictatorship reinforced each other.[48] Private initiatives also developed, but only under the umbrella of the 'industry for national interests' regulated by the authorities through a decree that came into force at the end of the Spanish Civil War, and some of them were explicitly discouraged.[49] The State was also involved in different branches of the chemical industry, among them fertilizers, cellulose and artificial fibres, as well as in perfumes, aromatic and resin derivatives, in some instances after requests for help had been received from the industrial firms themselves.[50] Apparently the controlling State was not interested in pharmaceuticals.

47 Formulario de la gerencia a Presidencia, Petición de la Sociedad 'Leopenicilina Española SA' para instalar una fábrica de penicilina, 21.7.1947, Archivo del Instituto Nacional de Industria, Madrid, División de Industria, industrias varias, legajo 200, documento 7 (1946–1947).

48 J.L. García Delgado, 'La industrialización y el desarrollo económico de España durante el franquismo', in J. Nadal, A. Carreras and C. Sudriá (eds), *La economía española en el siglo XX. Una perspectiva histórica* (Barcelona: Ariel, 1987), pp. 164–189, and F. Guirao, *Spain and the Reconstruction of Western Europe 1945–57* (London: MacMillan, 1998).

49 This was precisely the case of early automobile production in Spain, see E. San Román, *Ejército e industria: el nacimiento del INI* (Barcelona: Crítica, 1999). On 'Industry of national interest', see the law of 24 October 1939.

50 P.M. Aceña and F. Comín, *INI. 50 años de industrialización en España* (Madrid: Espasa Calpe, 1991), pp. 251–259.

An Issue of National Interest: penicillin production

On 1 September 1948, the Spanish government took on the challenge of the growing demand for penicillin. On that day, the Ministry of Industry approved a decree entitled 'Penicillin: its manufacture is of national interest', which was published in October, launching an open public competition for the establishment of two factories.[51] By considering penicillin production to be of national interest, the authorities assumed control over its manufacture and included penicillin among the government's industrial priorities. The decree required competing firms to provide information about patents and technical advice, assumed to come from foreign companies, as the proposals were asked to specify any need for foreign currency. They were also required to give details about the origin of the technical material and devices requested in their proposals, as well as provide a detailed market study. The two firms that were selected and authorized to manufacture penicillin in Spain would need to be able to satisfy half the national requirements for penicillin and 'other antibiotics'.

Shortly after Franco's conquering army had entered Madrid in October 1939, an event that marked the end of the Spanish Civil War, a decree was published, giving special support to industries that were of particular relevance to national defence and the economy, such as a reduction of up to 50 per cent of taxes, a guarantee of at least 4 per cent benefits, and reduced customs duties.[52]

With the aim of establishing contacts with Merck & Co (Rahway, New Jersey), the young Professor of Physiology at the University of Madrid Medical School, Antonio Gallego, went to Rahway as representative of the industrial chemical and pharmaceutical group that belonged to the firm Banco Urquijo. His collaboration with Banco Urquijo had begun thanks

51 *Boletín Oficial del Estado*, 6 de octubre de 1948.

52 *Boletín Oficial del Estado*, 25 de octubre de 1939; 'Ley 24 de octubre de 1939 (Jefatura del Estado). Protección y fomento de la industria nacional a industrias que se declaren de interés nacional'.

to his brother José Luis Gallego, a director of the group. Antonio Gallego had been a postgraduate student of José María del Corral in Madrid, and a postdoctoral student of Rafael Lorente de Nó at the Rockefeller Institute of Medical Research in New York. Antonio Gallego's PhD dissertation in 1944 had been on the structure and function of the retina. This became his main research topic, carried out while collaborating with the antibiotics industry. His presence in New York, together with his close family relationship with the Urquijo industrial group, enabled him to become a representative of Banco Urquijo in the competition to create a penicillin factory in Spain.[53]

While he was in Rahway negotiating the production of penicillin under a Merck patent, in January 1949 Antonio Gallego signed an agreement, on behalf of the Banco Urquijo, with Merck's vice-president R.E. Gruber for the establishment of a penicillin factory in Spain.[54] The agreement included a commitment from Merck to assist, advise, and give detailed information about the fermentation process and technical requirements of a penicillin production plant 'if BU [Banco Urquijo] is granted such a concession'. Technical details, including all information necessary for the preparation of the bid to the Spanish government, were regarded in the agreement as 'strictly confidential', and could not be used except for the submission of the bid. The secrecy was absolute, and not a single document has been found containing such information. The letter of agreement mentioned that it was a 'deep fermentation method' to produce 'sterile crystalline sodium Penicillin G, sterile crystalline potassium Penicillin G, sterile crystalline procaine Penicillin and other kinds of penicillin regularly manufactured on the date BU is granted such concession by the Spanish government'. Merck would also supply suitable cultures of micro-organisms, as well as initial nutrients and a full description of the process. Merck and Banco

53 On Antonio Gallego, M.J. Santesmases, *Entre Cajal y Ochoa: ciencias biomédicas en la España de Franco* (Madrid: CSIC, 2001), ch. 4.

54 R.E. Gruber (vice-presidente of Merck & Co) to Juan Lladó (managing director of the Banco Urquijo), January 14 1949 (copy). Archivo Antonio Gallego, private collection preserved by his son Roberto Gallego, Alicante (Spain).

Urquijo negotiated a price for the agreement and a percentage of the sales as payments from Banco Urquijo to Merck.[55]

Approving Proposals, Establishing Factories

In August 1949, two decrees were published, in which two firms were approved for producing penicillin. A joint proposal from 'Consorcio Químico Español SA' and the Banco Urquijo was authorized to create a firm to establish a penicillin factory in Madrid, and a proposal from 'Industria Española de Antibióticos SA' was also approved to build a penicillin factory in León. Both firms would benefit from being declared of national interest, and therefore had to guarantee fifteen years' worth of the projected Spanish requirement for penicillin, with a minimum capacity of six million doses of one hundred thousand Oxford units per year. Contracts signed with foreign firms should be submitted to the industrial authorities for their approval, and imports would be controlled by the State, at a time when imports were severely restricted due to the lack of foreign currency.[56] Both firms were private firms whose existence was approved and controlled by the State. Their agreements with foreign companies suggest

55 'Agreement made and entered into as of the 14th of January, 1949 between Merck & Co., Inc and Banco Urquijo', contratos 37QC, cajas 5112 y 5113, IDD 1.04. Ministerio de Industria. Sección de Subsecretaría (Asistencia Técnica), Archivo General de la Administración, Alcalá de Henares, Madrid (in English in the original). I would like to thank Mar Cebrián for her guidance in consulting these files at the AGA.

56 *Boletín Oficial de Estado*, 11 de agosto 1949. Each project approvement by the ministry of Industry and Trade was separately published in succesive pages of the *Boletín*: 'Decreto de 17 de junio de 1949 por el que se resuelve el concurso convocado por el decreto de 1 de septiembre de 1948 para fabricar penicilina en España a favor de la proposición presentada conjuntamente por "Consorcio Químico Español SA y Banco Urquijo SA"'; and 'Decreto de 17 de junio de 1949 por el que se resuelve el concurso convocado por el Decerto de 1 de septiembre de 1948 para fabricar penicilina en España, a favor de la proposición presentada por "Industria Española de Antibióticos SA"'.

that nationalism and state controlled industry co-existed with imports of both the drug and its production process. From then on, buying patents was to be a usual feature of Spanish industrial life.[57]

A State share was not required for the proposals to be submitted. However, one of them, that of Industria Española de Antibióticos, did request a State share. The firm designed its project with the patents and technical support of the Schenley Laboratories, from the United States. For the creation of the firm itself, Schenley had agreed to participate in 10 per cent of the capital shares, reducing every year 25 per cent of the royalties that Antibióticos should pay for the 'cession' of manufacturing procedures until the value of 10 per cent of the capital was achieved. The firm Antibióticos was a consortium of well-known pharmaceutical laboratories including Abelló, Ibys, Instituto Llorente, Laboratorios Leti-Uquifa, and Zeltia, early producers of sera and vaccines before the Civil War. Schenley was, according to a report by Antibióticos, among the best penicillin producers 'in the world', and among the first involved in the early production of the drug. Schenley, the report stated, had recently installed a factory for 'Penicillin G' in France, Sofrapen (Societé Française de Pénicilline) in Paris, which only a few months after opening was already producing a greater quantity of penicillin than that anticipated in the project. The Instituto Nacional de Industria (INI), whose duty was precisely the promotion of big national industry, was considered by Antibióticos to be a 'best partner'.[58]

In November 1949, an INI report regarded partnership with one of the two firms approved to produce penicillin in Spain as not suitable. INI, the report argued, could be the partner of an enterprise only if it lacked financial capacity, which was not the case of these two firms. The possibil-

57 Mar Cebrián, 'Technological Imitation and Economic Growth during the Golden Age in Spain: 1959–1973' (Doctoral Thesis, Florence: European University Institute, 2004).

58 'Nuestro requerimiento al INI', undated but, according to the answering report (see below), ca. 1949; in 'Penicilina y Antibióticos', caja P-44, Archivo Altos Cargos, Joaquín Planell, vicepresidente, Archivo del Instituto Nacional de Industria, Madrid.

ity of the INI being a partner of both was also not advisable, according to the report.[59]

Meanwhile, in October 1949, the Banco Urquijo had created a new firm after the passing of a decree authorizing its project for a penicillin factory: the Compañía Española de Penicilinas y Antibióticos (Cepa). Thus two companies, Cepa and Industria Española de Antibioticos, were to share the production of penicillin in Spain, both with the use of the patents of US-based firms. Both would also later become involved in the production of other antibiotics. I deal with the case of Cepa on the basis of the archival resources that are available.

In January 1950 the construction of the Cepa bottling plant began. Factories were compelled by the decree that permitted their creation to set up a plant for bottling bulk imported penicillin first, while starting the construction of penicillin production plant, which might take even longer. In Cepa's case, seven months were needed for the construction of the bottling plant.[60]

Cepa's bottling plant for penicillin started production in Madrid in August 1950. Merck's vice-president J. Kerrigan came from Rahway (New Jersey) and joined the Spanish authorities for the inauguration, and publicly praised Cepa's technical facilities.[61] During 1951, Cepa bottled 12 million doses of 100,000 units, that is to say the maximum amount projected by

59 'De dirección técnica a gerencia: Solicitud de participación y colaboración en la Empresa Industria Española de Antibióticos SA', 25 de 11 de [19]49, legajo 051, documento nº 202 (1949). Archivo del Instituto Nacional de Industria, Madrid.

60 BOE 31 Agosto 1948, decree cit. artículo cuarto, para. i. According to Cepa annual reports, the aim of bottling in Spain imported penicillin was to save foreign currency, but the decree, by mentioning the possibility of bottling streptomycin, for which Spanish production no decision had been made as yet, rather suggests that by 1949 information about therapeutic effects of this new antibiotic was already widely known and authorities were once more expecting to import the product from abroad so as to bottle it in Spain as a first step toward production. See 'Memoria de la Compañía Española de Penicilinas y Antibióticos, 1952', p. 2, Fondos Banco Urquijo, Biblioteca del Instituto Juan March de Estudios Sociales, Fundación Juan March. Madrid (hereafter Fondos Banco Urquijo).

61 'Memoria de la Compañía Española de Penicilinas y Antibióticos, 1949–1950'; Fondos Banco Urquijo.

the decree.[62] In September 1951, Cepa's production of penicillin began, exactly, according to its 1952 annual report, two years after being granted manufacturing rights. By March 1952, Cepa was producing a higher output than that required by the decree. In June, monthly production reached one million doses of 100,000 units. This amount was 'insufficient to face market needs' as the bottling plant had been producing double the required number of doses per month during the whole year 1952, according to Cepa's annual report. This situation had already been foreseen, and for this reason, Cepa submitted a request to the authorities in November 1950 so as to get permission to double its penicillin production. By 1953, Cepa was producing 24 million doses of 100,000 units per year, on a working regime of 48 hours per week, and proposed doubling or even tripling this production rate by establishing supplementary work shifts and using reservoirs to increase fermentation capacity. This increased capacity enabled Cepa to stop importing penicillin as Spanish production, shared with Spanish company Antibióticos SA, was able to cover national demand for the drug. By that time, a Cepa streptomycin plant was under way, working at a 'satisfactory' rate.[63]

Concluding Remarks

In 1945, the authorities were worried about wasting the scarce and expensive amounts of imported penicillin, when its successful effects were widely known and demand was fast increasing. By 1952, however, demand seemed to stabilize and production costs began to decrease in neighbouring countries.

62 'Memoria de la Compañía Española de Penicilinas y Antibióticos, 1952', p. 1. Fondos Banco Urquijo. This amount was that being produced in January 1944 in the USA, see Hobby, *Penicillin,* p. 196.

63 'Memoria de la Compañía Española de Penicilinas y Antibióticos, 1952', pp. 3–4. Fondos Banco Urquijo.

In the meantime, in the USA early investments in production had already been amortized by the companies involved. In 1952 the Defence Production Administration requested American firms to over-produce antibiotics as a precaution against a new international, or world, war. The Korean War could have developed into a world war and this vision of a wartime future remodelled the industrial production and international distribution of antibiotics, when medical practices together with social confidence in the efficacy of antibiotics had become stabilized and medical uses and perceptions of infectious diseases had been transformed. Cepa's board of directors sensed a threat to its profits when the Spanish Director General of Health ordered a 20 per cent reduction in Spanish penicillin prices – the prices were under direct State control – suggesting that Spanish penicillin industry should follow its foreign counterparts. At that time, Spanish investments in penicillin production had not yet been amortized and the size of the Spanish market was considerably smaller than the US equivalent. The possibility that Spanish penicillin might have to compete with foreign and cheaper penicillin was a serious threat. Cepa 'respected' the order to reduce the price, notwithstanding the firm's 'delicate situation' in its early years of production, and considered that an evolution of Spanish prices 'in parallel' with foreign ones required 'time', a time comparable to that required in other countries to make penicillin production a cheap and profitable industrial process.[64]

A decade had passed between the early mass production of penicillin in the USA and its large-scale manufacture in Spain. During these ten years, Spain received penicillin as a wonder drug, a national committee to distribute it was created, treatment protocols were published, attempts were made to culture mould samples and extract the antibiotic from them, and finally, in the highly controlled industrial context of the time, two factories were approved by the Spanish government. Just as in the early 1940s penicillin had been imported, later in the decade so were the industrial processes for its manufacture in Spain. The need to import the whole technical procedure for its production was taken for granted. This included not only

64 Ibid.

the payment of royalties to US firms, but also the need to buy in many of the tools and devices required to install production facilities from abroad. Secrecy was crucial in the negotiations aimed at designing factories and setting them up, which helps to explain why not a single technical detail has been found in the archives of the industrial authorities. However, the annual reports of one of the firms involved, Cepa, suggest that by the late 1950s penicillin production had become a profitable business. Physicians and patients alike were waiting for Spanish-made penicillin, sold legally and in sufficient amounts to cure every infection. Spanish companies ultimately relied on the guarantees which were given in the government decree, and by which benefits were assured.[65]

According to the sources I have consulted, Spanish physicians travelled to the United States in the mid-1940s in order to learn about the therapeutic action of penicillin, but not about how to produce it. After some attempts to extract it by surface culture in some small laboratories, no further attempt was made in this direction. On the contrary, at a time when the period of autarchy was almost over, the interest of both private firms and industrial authorities converged towards the signing of agreements by which penicillin could be made in Spain under foreign patents. Hence, royalties were paid, and agreements were periodically renewed or, as happened with later antibiotics, new ones were signed, again with foreign firms.[66]

The early years of Franco's dictatorship should be considered as an agent in the process of making penicillin a hero and of celebrating Fleming as a particular hero in Spain. The heroic character of the new drug appeared to be associated with the idea of national hero that Franco created around his own persona. As the dictator was acclaimed by his supporters, and by a wider public in public events, so were both penicillin and its discoverer.

65 'Memoria de la Compañía Española de Penicilinas y Antibióticos, 1952', pp. 5–7. Fondos Banco Urquijo.

66 See Cebrián, *Technological Imitation*. For the case of Cepa, many other agreements were signed to produce many other new antibiotics, such as streptomycin, neomycin and many others. See AGA (8) 8 44/18155 'Compañia Española de Penicilinas y Antibióticos'.

Individual people and objects were celebrated while groups were regarded as a threat by the regime.

The Spanish public and Spanish clinicians were fully aware of its usefulness and penicillin became part of therapeutic reform, of the respect for laboratory science– reflected in the respect for the white-coated Fleming and the increasing involvement of clinicians in laboratory research – and access to it something that should be accessible to 'every pocket'.

Francoist autarchy and national protectionism laid the foundation for the subsequent industrial and technological dependency of Spain, when at the end of the period of autarchy, from the late 1950s onwards, a growing number of contracts with foreign firms for the production in Spain of industrial products became the basis upon which the *Wirtschaftswunder* of the 1960s was to be constructed.[67] The industrial production of penicillin is a good example of this: even before the end of the Second World War, its effects were known, growing demand due to its clinical success was evident by 1947, and some attempts were made at manufacturing the drug under the national autarchy which promoted indigenous initiatives for industrial and scientific production.

This case stresses the networks of scientific, technical and industrial knowledge, policy making and public concern that featured in the social, medical and political basis for the stabilization of the use of penicillin in Francoist Spain, their effects on medical practices and the role of a highly controlling industrial policy whose practices stabilized industrial dependency while contributing to economic development.

67 Cebrián, *Technological Imitation*; S. López and M. Cebrián, 'Economic Growth, Technology Transfer, and Convergence in Spain, 1960–1973', in J. Ljungberg and J.-P. Smits (eds), *Technology and Human Capital in Historical Perspective* (New York: Palgrave-Macmillan, 2004), pp. 120–144. For a general reflection on the lack of connections in Spain between industry and academic research, see S. López and M.J. Santesmases, 'La ciencia en España', in A. González Enciso and J.M. Matés Barco (coords), *Historia Económica de España* (Barcelona: Ariel 2006), pp. 891–918.

PART 2

Different Actors: scientists, doctors and patients

MARION A. HULVERSCHEIDT

4 The Scientist-Entrepreneur or Financing in Pharmaceutical Research: a portrait of the malariologist Werner Schulemann, 1888–1975

Introduction

Until the 1930s, German pharmacology was considered to be world leading, particularly in so far as industrial research was concerned.[1] This article contributes to ongoing discussions about the links between German universities and industry, and the degree to which they depended on, and benefited from each other, in this instance in the pharmacological field.[2] It focuses on the person of Werner Schulemann (1888–1975), who as a pharmacologist working at IG Farben in Wuppertal-Elberfeld is credited with the co-invention of the first synthetic anti-malarial, Plasmochin (pamaquin).[3] In 1937, he became Professor of Pharmacology and the Director of the Pharmacological Institute at Bonn University, where his contacts with industry remained excellent. This gave him an exceptional position among his academic colleagues who, although they admired him, envied his success and good relations with different pharmaceutical firms. Werner Schulemann is also a useful example in the ongoing debate on the extent to which science and scientists depend on political systems, or, on how much independence individual scientists are able to maintain.

1 W. Bartmann, *Zwischen Tradition und Forschritt: aus der Geschichte der Pharmabereiche von Bayer, Hoechst und Schering von 1935–1975* (Stuttgart: Steiner-Verlag, 2003), p. 2.

2 See for example P. Lundgreen, *Staatliche Forschung in Deutschland 1870–1980* (Frankfurt am Main-New York: Campus Verlag, 1986).

3 In English-speaking countries Plasmochin was marketed as Pamaquine or Plasmoquine.

Werner Schulemann, as a pharmacological researcher, a scientist, and an initiator of research projects, exemplifies the ways in which the spheres of academic research, politics, and industry are interconnected.

In 1942, after receiving large state subsidies from the DFG (Deutsche Forschungsgemeinschaft, German Research Fund), the RFR (Reichsforschungsrat, Reich Research Committee) and the REM (Reichserziehungsministerium, Reich Education Ministry), Schulemann wrote in a letter to Ferdinand Sauerbruch (1875–1951), subject branch leader for the subject division General Medicine in the RFR:

> [...] in Bonn, I have for the first time brought experimental therapy of infectious diseases to completion at the pharmacological institute of a German university, in research as well as in doctrine; this is something that hitherto had been exclusively in the hands of the industry.[4]

This quotation sheds light on Schulemann's special position as well as on university-based pharmacological research in Germany. Schulemann was no follower of any of the pharmacological schools; he had come from 'industry'. This is one of the reasons why he was an outsider among his professional peers.[5] Schulemann's declared goal was the training of pharmaceutical researchers for industry, as opposed to the implicit intention of his university collegues to train the next generation of pharmacologists, who would, like themselves, be ready to work within the academic world of research. As a professor, Schulemann was very interested in freeing university research from its image as ineffective, '[...] and given to redundant experiments with frogs and earthworms, and lots of reading', as he wrote in a letter to

4 Letter from Schulemann to Sauerbruch, dated 18 May 1942, Bundesarchiv Berlin (BA Berlin Berlin) R73 (Files of the DFG) No. 14576.

5 For more details on the pharmacological schools see J. Lindner and H. Lüllmann, *Pharmakologische Institute und Biographien ihrer Leiter: Zeittafeln zur Geschichte der Pharmakologie im deutschen Sprachraum von Anbeginn bis 1995* (Aulendorf: Editio-Cantor-Verlag, 1996) and A. Philippu, *Geschichte und Wirken der pharmakologischen, klinisch-pharmakologischen und toxikologischen Institute im deutschsprachigen Raum* (Innsbruck: Berenkamp Verlag, 2004).

a colleague.[6] Schulemann's opinion of his academic colleagues and their orientation towards basic research was not very high. He was much more interested in applicable research results, particularly in the development of new and more effective chemotherapeutic drugs.

He managed to acquire financial support for his institute, from state agencies as well as from industry, as he was an expert in formulating grant applications, with an acute sense of current scientific trends and political and financial interests, which he served very well, always managing to stick to his ultimate goal, the discovery of new anti-malarial medications.

After a short biographical sketch, this article will follow the different stages in Schulemann's work and its financial support in chronological order, thus opening up a new perspective on pharmaceutical research in twentieth-century Germany.

Pharmacological research in the field of infectious tropical diseases is often considered a prime example of the excellent quality of German research under the Weimar Republic. The synthesis of Germanin, the very first non-toxic therapeutic agent against sleeping sickness, had just been achieved in German laboratories, as well as that of the first synthetic anti-malarial, Plasmochin. But although these developments were 'German', they had been realized by industrially sponsored pharmaceutical laboratories, and not by state-supported university research. The development of remedies against tropical diseases was one of the focal points of Bayer's research labs in Wuppertal-Elberfeld, and Schulemann was directly involved in the invention of Plasmochin there.[7]

6 Letter from Schulemann to Wirth, dated 7 October 1939, BA Berlin R73 No. 14576.

7 H.-B. Dünschede, *Tropenmedizinische Forschung bei Bayer* (Düsseldorfer Arbeiten zur Geschichte der Medizin, 2 Düsseldorf: Trilsch, 1971).

A Biographical Sketch of Schulemann

Werner Schulemann was born in Breslau in 1888, and studied chemistry and medicine in Breslau and in Freiburg i. Br. He graduated as Doctor of Philosophy in chemistry in 1913, and received his PhD in medicine in 1914. During the First World War, he was Surgeon Major and head of the surgical department of a military reserve hospital.

In 1919, he joined Bayer, which in 1925 became part of IG Farben, where he would remain until 1936. After initially working in the pharmacological laboratory in Wuppertal-Elberfeld, he moved to the chemical research laboratory in 1923, becoming its Deputy Director in 1929. In 1932, he took on the position of Head of Pharmaceutical-Chemical Research at Bayer. From 1931 on, he also held a position as Honorary Professor at the Medical School in Düsseldorf. Together with Fritz Schönhöfer and August Wingler, his colleagues at Bayer, he synthesized Plasmochin, the first synthetic anti-malarial. He had not taken the habilitation examination; his scientific qualifications consisted in his doctorates in medicine and in chemistry, as well as the experience he gathered while working in the chemical industry. He was not affiliated to any of the pharmacological 'schools' which existed at the time in Germany.[8]

In 1937, he left Bayer to take up a position as Assistant Professor of Pharmacology at Bonn University. In 1938, he was appointed full Professor. Schulemann was a coveted candidate for membership of the National Socialist party, and Hitler himself signed his nomination certificate. He headed the Pharmacological Institute in Bonn until 1945. During the war, he also served as a consulting pharmacologist to the military forces, mainly

8 Lindner and Lüllmann, *Pharmakologische Institute* listed genealogies of pharmacologists. Particularly among experimental pharmacologists, the Buchheim-Schmiedeberg-school of thought was considered influential. But already under Hermann Albers and Carl Binz a center of modern pharmacological research had developed in Bonn. See K. Karzel, Geschichte der Pharmakologie und Toxikologie an der Universität Bonn, *Bonner Universitätsblätter* (1995): 69–80, p. 71.

working at the military malaria hospital in Bonn-Godesberg.[9] Between 1945 and 1950, he worked for Schering AG, in Braunschweig. After his denazification in 1950, he returned to Bonn to reclaim his chair and his position as a Director of the Institute. In 1956, he became Emeritus Professor, but still continued to head the Institute as a deputy until 1958. After retiring from this position, he remained active as a researcher and scientist. He died in 1975. Schulemann was married twice; his only son was born from his first marriage, which ended in divorce in 1929. His second wife died in 1955.[10]

During the National Socialist era, Schulemann received several important awards: in 1938 he received the Mary Kingsley Medal of Liverpool School of Tropical Medicine.[11] Two years later he received the Bernhard-Nocht-Medal of the Hamburg Institute for Tropical Medicine, and the Klebersberg-Medal of the University of Szeged in Hungary. In 1963 he was awarded with the Linnaeus medal by the Royal Swedish Academy of Science, though not for his research on the field of pharmacological malariology, but for his research and in-depth knowledge of Tibetan culture.

He was a member of several scientific associations, such as the Royal Society of Tropical Medicine and Hygiene. In 1969, he received the Schmiedeberg Insignia of the German Association for Pharmacology and

9 See Schulemann's reports to the Military Sanitary Inspection, in Bundesarchiv-Militärarchiv Freiburg (Military Archive Freiburg BA-MA) RH 12–23, No. 75, 83, 317; Schulemann consulted the internal expert at the HSI (Heeres-Sanitätsinspektion, Military Sanitary Inspection) on questions relating to malaria. He particularly influenced and was interested in the therapies at the special malaria hospital in Bonn-Godesberg. On the Military Sanitary Inspection and its role, see A. Neumann, *'Arzttum ist immer Soldatentum'. Die Heeressanitätsinspektion und das Amt ‚Chef des Wehrmachtssanitätswesens' im Zweiten Weltkrieg* (Düsseldorf: Schriften des Bundesarchivs, 64, 2005).

10 She suffered from rheumatism, as did Paul Martini (1889–1964), Professor in Ordinary for Internal Medicine in Bonn. Martini, who temporarily served as the Rector of Bonn university, strongly supported Schulemann's return to the Bonn pharmacological institute, not only for professional, but also for private reasons. Schulemann and Martini corresponded regularly: see Martini bequest in the Institute for History of Medicine Bonn and the letters of correspondence between Schulemann and Martini in the Universitätsarchiv Bonn (University Archive Bonn, UA Bonn) PA Schulemann.

11 Warrington Yorke, 'Presentation of Mary Kingsley Medals', *Annals of Tropical Medicine and Parasitology*, 32 (1938): 1–10.

Toxicology (Deutsche Gesellschaft für Pharmakologie und Toxikologie/ DGPT), in honour of his great scientific achievements. Between 1952 and 1957, he served as a board member of this society.[12]

Figure 4.1: Portrait of the Malariologist Werner Schulemann (1888–1975). Source: Günter Benz, Ralf Hahn, and Carsten Reinhardt, 100 Jahre Chemisch-wissenschaftliches Laboratorium der Bayer AG in Wuppertal-Elberfeld 1896–1996 (Leverkusen, 1996), p. 24.

12 Further biographical details to be found in: W. Mohr, 'Werner Schulemann zum 85. Geburtstag', *Zeitschrift für Tropenmedizin und Parasitologie*, 24 (1973): 129–130; L.J. Bruce-Chwatt, 'Obituary: Werner Schulemann', *Lancet*, 2 (1975): 85–86; R. Domenjoz, 'Nachruf auf Werner Schulemann', Jahrbuch der Rheinisch-Westfälischen Akademie der Wissenschaften (Opladen: Westdeutscher Verlag, 1977), pp. 30–32; R.O. Meisenzahl, 'In memoriam Professor Dr. Werner Schulemann', *Tribus Veröffentlichungen des Linden-Museums* 1977 (1977): 7–9; also BA Berlin personnel files (formerly BDC) Schulemann, Werner.

Schulemann's time at Bayer/IG Farben in Wuppertal-Elberfeld, 1919–1936

Initially, Schulemann had started his professional career in the chemical laboratory of the Bayer Company, developing varnishes and dyestuffs. As head of the Chemical Research Laboratory, Schulemann became involved in the development of anthelmintics, local anaesthetics, barbiturates, and analeptics. Compared with his university colleagues, he published very little.[13] But working in the research laboratory of one of the leading pharmaceutical companies in the world was not a bad thing, as far as money, career possibilities, and honours were concerned.[14]

While working for Bayer, Schulemann achieved his greatest scientific success, the invention of Plasmochin, which was rewarded by rapid promotions and pay rises. Together with Fritz Schönhöfer and August Wingler he developed this first synthetic anti-malarial, launched in 1927.[15] Schulemann stayed at the Bayer laboratories in Wuppertal-Elberfeld until 1937, becoming the Associate Director of the Pharmaceutical-Chemical Research Department there in 1929.

The time he spent in Wuppertal-Elberfeld may be considered Schulemann's most productive; he was able to concentrate exclusively on research, without having to spend time on the acquisition of funds, materials, and colleagues. In industry, everything was provided for research.

The organization of the German pharmaceutical industry changed in 1925, when Bayer AG, BASF AG and Hoechst AG fused to form the

13 In his publications he mainly concentrated on vital staining and chemotherapy with dyestuffs. Plasmochin, too, resulted from the introduction of a diethyl aminoethyl group into methylene blue. A list of Schulemann's publications up until and including 1941 is in his personal file BA Berlin, (formerly BDC) Schulemann, Werner.

14 G. Benz et al, '100 Jahre Chemisch-wissenschaftliches Laboratorium der Bayer AG in Wuppertal-Elberfeld 1896–1996', (Wuppertal: Bayer AG, 1996); Bartmann, 'Tradition und Fortschritt'.

15 In 1928, all three scientists together received the Emil Fischer Medal, an endowed award for German chemists; the fund had been set up by Carl Duisberg. F. Schönhöfer, '40 Jahre Plasmochin', Arzneimittelforschung 15 (1965): 1256–1258.

Interessengemeinschaft IG Farben, which cooperated with, and profited from, the Nazi political system.[16]

A Period of Transition, 1937–1938

In 1933, at the age of 45, Schulemann joined the SA. Before that he had not been politically active. Because of a personal dispute with Heinrich Hörlein, then Director of Research at Bayer, he was released from his position at the Pharmaceutical-Chemical Research laboratory in November 1936, but continued to receive a full salary.[17] He was offered, and turned down, a position at Farbwerke Höchst, also part of IG Farben. He also refused a professorship in Marburg, as he deemed the local working conditions there 'to be unsatisfactory'.[18] From October 1937, he headed the Pharmacological Institute in Bonn. Hermann Führer, who had previously held the chair, had died in 1937.[19] Why this institute attracted Schulemann more than the Marburg chair or a place in the laboratory at Höchst, remains unclear.

16 P. Hayes, *Industry and Ideology: IG Farben in the Nazi era* (Cambridge: Cambridge University Press, 2001).

17 Letter from Fritz TerMeer, board member (Vorstandsmitglied) of the IG Farben, to Schulemann dated 4 November 1936, Bayer Archive Leverkusen File 271/2.1 Quote: 'Ich glaube, wir waren uns bei unserer Besprechung ganz einig darüber, dass schwerwiegende Meinungsverschiedenheiten sachlicher und persönlicher Art ihr weiteres Zusammenarbeiten mit Herrn Professor Hörlein in der bisherigen Form nicht mehr möglich erscheinen lassen. Das kommt im geschäftlichen Leben öfter vor …' This and other citations were translated by the author.

18 Minutes of a meeting with Professor Dr. Schulemann in Leverkusen on 21 November 1936, Bayer Archive Leverkusen File 271/2.1 (Werner Schulemann); the transcription of a letter from 8 February 1937 is included in the documents relating to the new appointment to the pharmacological chair, Bonn in University Archive Bonn, appendix to a letter by the Rector of the University of Bonn, to the Minister of the Reich and Prussia for Science, Education and Common Education, dated 8 April 1937.

19 K. Karzel and M. Göthert, 'Institut für Pharmakologie und Toxikologie, Medizinische Fakultät der Rheinischen Friedrich-Wilhelms-Universität Bonn', in Geschichte und Wirken, pp. 135–145 (pp. 137–138).

The salary was the same as in Marburg, and both were lower than his salary at Bayer. Schulemann had agreed on 'a period of rest (Karenzzeit)' with Bayer AG/IG Farben, guaranteeing him 30,000 Reichsmark annually, the only conditions being that he continued to cooperate with the company and refrained from working for a competitor.[20] Independently of this agreement, during his 17 years at Bayer AG, he had also earned a sizeable pension.

In 1938 he was officially promoted to the status of full Professor of Pharmacology at Bonn University and obtained tenure as a civil servant, without ever having taken the habilitation.[21] Hitler personally signed the letter of appointment (Ernennungsurkunde).[22] Schulemann delivered his inaugural lecture wearing his SA uniform. He had joined the NSDAP on 1 May 1937, at a time when it was already evident that he would be promoted to a full professorship. His application for party membership as well as his wearing the SA-uniform at his inaugural lecture can be interpreted as the gestures of a political opportunist, mainly worried about his career, but not considering the long-term impact of his decisions. An alternative interpretation may be that the National Socialist regime had succeeded in convincing him, and this seems to have been true at least until the beginning of the Second World War,[23] as Schulemann openly criticized the pogroms carried out against the Jews in November 1938.[24] He was described as an SA 'personality', working energetically towards his goal of professionalizing

20 His basic salary as a tenured professor was 11,600 RM, in addition to a teaching fee of RM 6000 annually. The discrepancy between salaries paid by the industry and by the state was enormous.

21 His appointment was delayed, and in February Schulemann wrote a letter to the Dean of the Medical Faculty to counter abounding rumours that he had decided to return to IG Farben or to end his activities in Bonn because 'supposedly there is something wrong with my documents, as the endorsement by the Ministry of Culture is still not here', Schulemann to Siebke on 15 February 1938, UA Bonn MF-PA 351 Schulemann.

22 A transcription of the letter of appointment is stored in UA Bonn MF-PA 351 Schulemann.

23 R. Forsbach, *Die Medizinische Fakultät der Universität Bonn im 'Dritten Reich'* (München: Oldenbourg, 2006), pp. 102–103.

24 H.P. Höpfner, *Die Universität Bonn im Dritten Reich: akademische Biographien unter nationalsozialistischer Herrschaft* (Bonn: Bouvier, 1999), p. 284.

academic pharmaceutical research. A new building was planned for his institute. Although in the event the institute did not move due to wartime crises, it is clear evidence of Schulemann's success in obtaining funds to fulfill his own goals.

During his time with Bayer AG, Schulemann had established very good contacts with scientists in other countries. He was well liked and respected by his foreign colleagues. He himself often emphasized his links to Great Britain, to Warrington Yorke in Liverpool and Sir Henry Dale in London, as well as to France, Belgium, and Italy. In November 1938, when Germany had annexed Austria and Czechoslovakia after the Münchener Abkommen (Munich Agreement) Schulemann travelled to Liverpool, where he received the Mary Kingsley Medal.[25]

He intended to restructure the Bonn Pharmacological Institute, which had developed into an institution specializing in toxicology under his predecessor. His plan was to adapt it to his own preferred way of working and focus it around laboratory objects, in this instance compounds that were tested on laboratory animals, with the aim of laying the foundations for human therapeutic research at the institute. His approach was to move away from frogs and worms, and to use mammals as experimental animals. In 1937, he received RM 6000,00 from the budget of the *Vierjahresplan*, a four-year plan announced by Hitler in 1936 to streamline science and the economy for the war effort, for a new tropical medicine training and research facility. The money from the *Vierjahresplan* was distributed via the RFR and the DFG. The RFR had been founded in 1937 with the purpose of reorganizing German science from top to bottom, while the DFG functioned as the cashier's office of the RFR. Schulemann belonged to a group of people demanding the revision of the Versailles treaty vis-à-vis the loss of German colonies, and as a pharmacologist, he was meant to participate in the training of young medical doctors for the soon-to-be-recovered colonies.[26] Underpinning this training, was a complex policy intertwining imperialistic ideals and the so-called 'Kolonialschuldlüge'.

25 Yorke, 'Presentation'.

26 On the mentality of colonial revisionism amongst scientists and physicians in tropical medicine, see W.U. Eckart, *Medizin und Kolonialimperialismus: Deutschland 1884–1945* (Paderborn: Schöningh 1997).

As well as keeping in touch with his colleagues abroad, Schulemann cooperated with local clinicians. In the field of malaria research, he benefited more specifically from his contacts with psychiatrists. His collaboration with the director of the state mental asylum Düsseldorf-Grafenberg, Franz Sioli (1882–1949), was the most long lasting.[27] Sioli had been one of the first to set up an Anopheles-mosquito breeding station in his hospital, in order to treat progressive paralysis by inducing malaria via insect bites rather than blood transfusions.[28] In collaboration with Bayer AG, he tested Plasmochin on the paralytics treated with malaria. Even though medical ethical regulations banned such drug testing on the inmates of mental asylums, they did not consider this type of conduct to be problematic.[29]

Schulemann also cooperated with clinicians and scientists in Bonn. In a letter to the RFR, he mentioned that he had:

> … concluded an agreement on cooperation with Professor Pohlisch and Professor Reichensperger (psychiatry und zoology) … on the one hand, in order to be able to treat paralytics with malaria, and on the other hand, to conduct the necessary parasitological and entomological procedures.[30]

27 Schulemann to Breuer on 13 May 1938, page 5, BA Berlin R 73 No. 14575, 'The visit by Prof. Sioli has provided us with an opportunity to get insect breeding going without Dr. Jacobi having to go to Antwerp. So we can forego these foreign currencies.'

28 On Sioli see P. Kuhn *Die Geschichte des Rheinischen Landeskrankenhauses Düsseldorf-Grafenberg von 1918–45* (Düsseldorf: 1977); F. Sioli, A. Kentenich and O. Vollmer, 'Über die Zucht der Anopheles und die Verwendung der Anopheles zur Malariabehandlung der Paralytiker', *Zeitschrift für Neurologie* 155 (1936), pp. 783–797; F. Sioli, A. Kentenich and E. Boldt, 'Weitere Erfahrungen über die Zucht der Anopheles und ihre Verwendung in der Malariabehandlung der Paralytiker', *Archiv für Schiffs- und Tropenhygiene* 43 (1939), pp. 1–15.

29 In 1931, regulations concerning new therapies and the conduct of scientific experiments on humans were passed by the Reichsministry, strongly influenced by Julius Moses. W.U. Eckart and A. Reuland, 'First Principles: Julius Moses and medical esperimentation in the late Weimar Republic', in W.U. Eckart (ed.), *Man, Medicine and the State* (Stuttgart: Steiner Verlag, 2006), pp. 35–48. On the continuity between malaria experiments on paralytics and experiments on concentration camp inmates, see M.A. Hulverscheidt, 'German Malariology experiments with humans, supported by the DFG until 1945', in ibid, pp. 221–236.

30 Schulemann in a letter to the President of the RFR, 8 November 1937, p. 3, BA Berlin R 73 No. 14575.

The Beginning of the War, 1939

With the outbreak of the Second World War, the research funding policies of the German Reich changed dramatically. From then on it became imperative that all research projects be labelled 'war-important'. This war-importance was mainly attributed to short-term projects. It was not only the RFR, the subject branch leaders (Fachspartenleitern), and their staff who needed to adapt to the new application criteria, but also the scientists themselves, who showed a high degree of self-censorship, by terminating long-term research projects, and taking on new ones with a supposedly higher war-time priority.[31] Among them was Schulemann. Shortly after the German invasion of Poland, Schulemann's research projects on malaria and bacterial infections had been classified as only 'moderately' war-important, and he lost a large proportion of his state subsidies. Schulemann attempted to fight the decision, arguing that it had been extremely costly and time-consuming to set up the mosquito breeding station necessary for malaria research, and that his institute had carried on working, even though Bonn University had closed down.[32] He was partly successful, but, to his great regret, his experiments for the treatment of streptococcal infections were discontinued.

31 I. Kästner, 'Pharmakologische Forschung in Deutschland 1939–1945', in Medizin, Naturwissenschaft, Technik und Nationalsozialismus: Kontinuitäten und Diskontinuitäten ed. by C. Meinel (Stuttgart: Steiner Verlag, 1994), pp. 136–141 (p. 138).

32 Opinion by the expert consultant dated 13 September 1939, BA Berlin R 73 No. 14576. The consultant in question was Walther Straub (1874–1944), Munich, who was in charge of distributing the *Roche-funds*. Initially Straub evaluated only one of Schulemann's applications, on 'Comparison on the toxicity of benzene and its derivative homologues'. Straub wrote: 'If it has now been decreed that only war-important work should be supported, then we must say that Prof. Schulemann's work is at best indirectly war-important; and if further it is required that usable results should be available within a relatively short time, it must be said that Prof. Schulemann's work is purely scientific, that is to say that results will only be available in the remote future' (author's own translation). This expert opinion was preceded by a tendentially negative evaluation by the RFR-adviser Sergius Breuer, to Straub, in which Breuer had explained the exact wording of war-importance, as used by Straub. See Breuer to Straub on 12 September 1939, BA Berlin R 73 No. 14576.

After this first defeat, Schulemann contacted Wolfgang Wirth (1898–1996), at that time head of the Pharmacological-Toxicological Department at the Military Medical School of the Army Sanitary Inspection in Berlin, with a view to establishing a collaborative project on chemical warfare agents.[33] Even though before the war no work had been done in this field at the Bonn Institute, the necessary equipment was available (on loan from the DFG) and could easily be converted to this type of research. In addition, he argued, a well-trained assistant, who had so far busied himself with industrial toxicological tests, was already present on site.[34] Wirth responded positively to Schulemann's proposal.

Schulemann then suggested to Wirth that he should collaborate with the Director of the Bonn Medical Polyclinic, Professor Tiemann, in order to conduct experiments on the prevention of phosgene poisoning.[35] However, Wirth did not consider Schulemann's approach promising in terms of lowering the sensitivity to chemical warfare agents although he did describe the search for new chemotherapeutic drugs for the treatment of infectious diseases as 'undoubtedly war-important'.[36]

Nevertheless, Schulemann wanted to continue his research on chemical warfare agents. He offered to carry out comparative experiments on the development of phosgene-induced pulmonary oedemas, both with and without administration of pectins. Pectins and other similar substances were thought to alter blood-viscosity, thereby raising the level of sensitivity to phosgene. Schulemann, who painstakingly adhered to scientific protocol,

33 On Wirth see W. Woelk, 'Der Pharmakologe und Toxikologe Wolfgang Wirth (1898–1996) und die Giftgasforschung im Nationalsozialismus', in W. Woelk et al (eds), *Nach der Diktatur. Die Medizinische Akademie Düsseldorf vom Ende des Zweiten Weltkrieges bis in die 1960er Jahre* (Essen: Klartext-Verlag, 2003) pp. 269–287; on research on warfare-agents in Germany before and after the Second World War, see F. Schmaltz, 'Kampfstoff-Forschung im Nationalsozialismus: zur Kooperation von Kaiser-Wilhelm-Instituten, Militär und Industrie' (Göttingen: Wallstein Verlag, 2005).

34 Schulemann to the DFG on 14 September 1939, BA Berlin R 73 No. 14576.

35 Transcription of a letter written by Schulemann, to Wirth (Dear Mr. Wirth!) dated 2 October 1939, BA Berlin R 73 No. 14576.

36 Transcription of a letter written by Wirth, to Schulemann (Dear Mr. Schulemann), dated 13 October 1939, BA Berlin R 73 No. 14576.

and who normally insisted on testing every substance and method in his laboratory first, before applying them to animals, and finally to humans, departed from his principles and those of research in general, by accepting the tenets of 'war-importance'. He wrote:

> We have consciously decided to forego all theoretical preparations, but have immediately begun with practical applications on animals, in order to achieve useable results as quickly as possible.[37]

Thus, Schulemann used the positive attitude of the military towards his institute as an argument for drawing support from the RFR in order to continue to function.[38]

Research during the War: back to routine

From 1941 onwards, as infections from malaria on the Southern and South-eastern battlefields began causing epidemics among German troops, once again malaria research in Bonn was considered to be highly war-important, and started receiving financial support.[39] As a malariologist and pharmacologist, Schulemann was asked to provide expert advice to physicians at the special malaria field hospital in Bonn-Godesberg, hence coming into close contact with the Army Sanitary Inspection.[40] Under his guidance, the search continued for new combinations of agents for prophylactic treatment and, hopefully, also for the prevention of relapses. He initiated

37 Schulemann to Wirth on 13 October 1939, BA Berlin R73 No. 14576.

38 Schulemann to Breuer on 14 September 1939, Breuer to Wirth on 22 September 1939, with a request for an expert opinion, BA Berlin R73 No. 14576.

39 See M.A. Hulverscheidt 'German Malariology'; and W.U. Eckart and H Vondra, 'Malaria and World War II: German malaria experiments 1939–45', *Parassitologia*, 42 (2000): 53–58.

40 Schulemann to Breuer on 29 April 1942, BA Berlin R73 No. 14577, Report of Parrisius, Wehrkreis IV from 3 July 1942, BA-MA Freiburg RH 12–23 No. 85.

programmes in order to test new dosages of anti-malarials, new medications such as Sontochin and Certuna, and discover new methods for the early diagnosis of malaria infections or relapses. One of these methods was the investigation of bone marrow by sternal punction, a very painful diagnostic method.[41]

In 1942, Ferdinand Sauerbruch, Head of the Department for General and Classical Medicine at the RFR, criticized the fact that Schulemann was obviously working on two separate research topics at the same time, namely malaria research and the chemotherapy of infectious diseases. Sauerbruch demanded that Schulemann conclude one project before embarking on another. Schulemann rejected what he felt was a paternalistic attitude on the part of Sauerbruch. In his response to him, he presented his research topics as interconnected, and emphasized the support he received for his work by the Oberkommando of the Wehrmacht and the REM.[42] As it becomes apparent when reading these letters, Schulemann did not primarily consider himself to be a researcher confined to the laboratory, but rather a facilitator of research, someone who provided the surroundings, laboratories, personnel, and materials for top-quality research, and who acted as an intermediary with financial backers. He did not publish much himself during this period, but his contributions can be found in his assistants' publications.[43] In addition to research on the treatment of malaria and bacterial infections, after 1943–1944, he also searched for new insecticide substances to be used for delousing. Schulemann's research was integrated into the Department of Research into Contagion, headed by the RFR member and newly appointed Reich Epidemics Officer, Walter Schreiber. In addition to malaria and bacterial infections, in Bonn Schulemann was charged with delousing research and the search for preventive measures

41 See Hulverscheidt, 'German Malariology', p. 234.

42 Schulemann to Sauerbruch on 18 May 1942, BA Berlin R73 No. 14577.

43 W. Schulemann, 'Zur Pathologie der Malaria', *Deutsche Medizinische Wochenschrift*, 66 (1940): 253–256. In his own publication, which gives an overview of the current research goals in malariology, he quotes his assistants' publications. At the time it was not common to name the head of a department as the co-author of a scientific publication. He mentioned the publications of his assistants Kohlschütter, Zipf, and Jacobi.

against diseases transmitted by lice, such as epidemic typhus.[44] These orders came directly from the military, and were classified 'urgent'. Such a classification was at that time indispensable in order to receive any kind of funding and material from the 'martial-economic administration', which was part of the RFR.

The only documents still existing today are his applications for support and the orders received. There are no further papers containing his research results or any of his publications. During a bombing raid in October 1944, the Institute building was totally destroyed, and all documents and records were lost. After the end of the war, Schulemann was removed from his position by the British occupying forces, because of his SA and National Socialist party membership, and was banished from the city of Bonn. In November 1945, he wrote a detailed letter of protest against this decision, claiming that he had only been a passive member of the SA, and had worn his SA uniform during his inaugural lecture only because he did not own a gown or any other formal attire, which was mandatory on such occasions. He was also supported by his colleagues, one letter of support certifying that during his time as Dean of the Medical Faculty in 1942 he had '...courageously resisted party influence on research and doctrine'.[45] Schulemann himself stated that he had never been a member of the National Socialist party, i.e. had never received a membership book, whereas in a curriculum vitae, which he wrote in 1940, he said: 'I have been registered for party membership by the SA and expect to be accepted shortly'. A handwritten note was added later: 'membership received'.[46] Despite the willingness of his colleagues to support him by writing 'clean bills of health (Persilscheine)' and asserting his political opposition to the regime,[47] Schulemann had to

44 Schulemann to the DFG on 23 October 1943, BA Berlin R73 No. 14577. In this report he frankly mentioned his good contacts and cooperation with the military and the chemical industry.

45 Dean Redwitz to the Rector, dated 9 August 1945, and the evaluation by Paul Martini dated 27 August 1948, UA Bonn MF-PA 351 Schulemann.

46 Curriculum Vitae, Bl 102, UA Bonn MF-PA 351 Schulemann. In the BDC-files at the BA Berlin, however, no membership number may be found.

47 H.-P. Höpfner *Universität Bonn*, p. 88, pp. 283–285. As Dean of the Medical faculty, Schulemann had effected the removal of the Bonn Intelligence-Service Chief Müller from the dental clinic. Müller worked there as a dentist.

leave Bonn in December 1945, and payment of his salary ceased. During the denazification initiative led by the British Allied Forces, which aimed to rid German society, culture, press, economy, judiciary, politics, and science of any remnants of the Nazi regime, Schulemann was prosecuted as a 'Category IV', meaning a follower of the regime.[48] He wrote letters to the occupying forces and to the mayor, asking for the reasons for his dismissal, but never received an answer. He continued writing petitions, and persuaded several of his colleagues to do the same on his behalf.[49] He vigorously defended himself against the notion that he had been more an 'organizer' than a research scientist.[50] Before leaving, he began to set up a new provisional laboratory at the Villa Soennecken in Bonn Poppelsdorf, together with his former staff. He then moved to Braunschweig (Brunswick), taking on a position as a consultant for Schering AG, which ran a small branch operation there. This is also where his so-called de-nazification was completed.

He cut his ties with IG Farben after the end of the war. A major trial of the company was initiated, accusing it of holding a monopoly of chemicals as well as the production of Zyklon B.[51] Schering had not been part of IG Farben, so was not involved in this accusation. Nevertheless, there were fears that Schering might be divided, because it was based in Berlin. The location for its West-German subsidiary was to be Braunschweig.

48 For the process of denazifaction five categories of responsibility were identified: major offenders, offenders, lesser offenders, followers, and exonerated persons. But the categories differed depending on the allied zone in which they were applied, i.g. the Americans' decisions often differed from the British or French. Ultimately, the intention was the 're-education' of the German people.

49 Schulemann to the Rector of Bonn University, dated 24 November 1945, in German and in English, petition to the British occupying forces, UA Bonn MF-PA 351 Schulemann. Paul Martini sent a petition to the British occupying forces, dated 27 January 1946, to be found in Martini's bequest in the Institute for History of Medicine, Bonn. I am grateful to Peter Kuhnhenn for letting me view the correspondence between Martini and Schulemann.

50 Schulemann to the Rector of the Bonn university, dated 24 November 1945, in German and in English, petition to the British occupying forces, p. 4, UA Bonn MF-PA 351 Schulemann: 'But I must refer to these scientists in regard to the imputation against me, that I am to be considered more an "organiser" than a scientist.'

51 See F. Schmaltz, *Kampfstoff-Forschung*, F. Schneckenburger, *Die Geschichte des I.G. – Farben-Konzerns Bedeutung und Rolle eines Großunternehmens* (Köln: Pahl-Pugenstein Verlag, 1988); W. Bartmann, *Zwischen Tradition und Fortschritt*; P. Hayes, *Industry and Ideology*.

Figure 4.2: The Pharmacological Institute of Bonn University, former Villa Soennecken
Source: Heinz Schott (ed.), *Universitätskliniken und Medizinische Fakultät Bonn 1950–2000: Festschrift zum 50jährigen Jubiläum des Neuanfangs auf dem Venusberg* (Bonn: Schott, 2000), p. 120.

Biding his Time at Schering, at the Braunschweig Branch Establishment, 1945–1950

During the war, Schulemann had agreed to cooperate with the Berlin-based Schering AG.[52] Schering was not only active in the field of hormones, but

52 Board minutes, minutes of 22 January 1944, Scheringeanum, archive of the Schering AG SchA BO – 0381: 'During a meeting with Prof. Schulemann he was told that his

also did research on plant protection agents and insecticides. In 1945 it acquired the licence to produce Gesarol – DDT – from Ciba-Geigy. While the Allied Forces' conference in Yalta was taking place, Schering drew the conclusion that Berlin might no longer be a secure site, therefore the company quickly set up a subsidiary in the Western sector.[53] The Braunschweig location was opened in 1945, and existed until the end of the Berlin blockade in 1949. Schulemann's position was officially that of a freelance scientist, but his ties to Schering may be considered to have been very close. The company offered him shelter, work, and an income. His scientific interests at the time were the tissue hormone, histamine, and its antagonists, the anti-histamines.[54] Synthetic anti-histamines had first been developed in France by Rhône-Poulenc under German Occupation. Although French scientists had had to reveal their progress with these drugs to IG Farben as early as 1942, Schulemann received information about the anti-histamines via Schering, rather than IG Farben. The anti-histamines owned by Schering were tested and developed further in Schulemann's laboratory. Under his direction, anti-histamines were combined with acetyl salicylic acid for use in cold medications (Allercur). Schulemann's former assistant and the interim head of the Pharmacological Institute between 1946 and 1950, Hans Haas, had already experimented with histamines and anti-histamines, and the effects of certain irritants.[55] Hans-Friedrich Zipf, from the Department of Physiological Pharmacology in Berlin, had also cooperated with Schulemann in search of substances capable of increasing and stabilizing blood circulation.

---

cooperation is desired on a larger scale. This cooperation will therefore go beyond the field of malaria research and encompass general pharmacological concerns.'

53 On the history of the Schering Company see G.J. Wlasich, *Von der grünen Apotheke zum Weltunternehmen: historscher Jahresstrang* (Berlin: Schering 1996); C. Kobrak, *National Cultures and International Competition: the experience of the Schering AG* (Cambridge: Cambrige University Press, 2002). In these official histories of the Company, Schulemann is not mentioned personally, but the Braunschweig branch of the Company.

54 Synthetic anti-histamines are based on the same basic chemical structures as antimalarials, see V. Quirke, 'War and Change in the Pharmaceutical Industry: a comparative Study of Britain and France in the twentieth century', *Enterprises et Histoire*, 36 (2002): 64–83 (74).

55 A. Philippu, *Geschichte und Wirken*, p. 138.

Schulemann was therefore able to make good use of his relegation to Braunschweig, by intensifying his contacts with the Bonn Institute and Faculty, and by developing a second pharmacological 'leg' on which to base his scientific career. His financial situation also improved considerably.[56]

Re-Installation in Bonn and Rich Rewards, 1950–1975

In 1950, after completion of his denazification, Schulemann was allowed to return to his chair in Bonn.[57] He immediately set out to re-establish his institute. He took up residence in the attic flat of the Villa Soennecken, where the Pharmacological Institute was now housed.

He also returned to malaria research, but not exclusively.[58] Between 1946 and 1950, Gerhard Piekarski, Professor of Microbiology and Medical Parasitology, headed the Department for Parasitology, which was part of the Institute of Pharmacology. He subsequently became the first director of the only Institute of Parasitology to exist in a Medical School in Germany. Piekarski got along well with Schulemann. He also had good connections with the World Health Organisation (WHO), and on several occasions he too had applied to the DFG for funds. Using his research and his organizational skills, Schulemann gave his support to the foundation of the Institute of Parasitology.[59]

56 His resignation from his position in Braunschweig did not affect his relationship with Schering adversely: on leaving, he received a Ford Admiral car.

57 The denazification record of Schulemann is stored at at the Niedersächsischen Staatsarchiv, Landesarchiv Wolfenbüttel, Ndx. 92/1 No. 16260.

58 An application for material support for malaria research dated 5 June 1950 was shelved. DFG-Archive Bonn-Bad Godesberg, file card Schulemann, p. 1.

59 H.M. Seitz and F. Stader, 'Institut für Medizinische Parasitologie', in H. Schott (ed.), *Universitätskliniken und Medizinische Fakultät Bonn 1950–2000; Festschrift zum 50jährigen Jubiläum des Neuanfangs auf dem Venusberg* (Bonn, 2000), pp. 151–153.

He successfully created working models for testing potential therapeutic drugs in laboratory animals. Schulemann's malaria research was carried out on an avian model, and he continued to use this model even after colleagues had managed to develop a mammalian model using mice. Schulemann pursued his research using canary birds and chickens. The substances he synthesized were initially tested on these birds, before they were passed on to his colleagues, particularly in Tübingen and in Erlangen, for further testing on mice.[60] Schulemann had excellent connections not only with Schering and Bayer AG, but also with other pharmaceutical companies, such as Grünenthal GmbH in, Stolberg and Knoll AG in Ludwigshafen. The latter two companies were now the employers of his former assistants (Heinrich Mückter at Grünenthal, 1948–1954 and Hans Haas at Knoll, 1949–1971). Thanks to these numerous contacts, the Institute's financial situation was excellent: funds were received from Bonn University, the DFG, Bayer, and Schering, as well as other pharmaceutical companies. The research was also supported by the State of North-Rhine Westphalia and by the WHO.[61] According to his former students, Schulemann was an excellent organiser and fund raiser; however, these abilities were not matched by his scientific expertise. In the years that followed the war, he rarely published anything at all, even when he was requested to do so.

Schulemann himself was also well provided for. From 1956 on, after his 68th birthday, he received a pension from Bayer, amounting to 1,250 DM a month, in addition to his salary as an Emeritus Professor. Schering also gave him financial support, although the extent and amount are not known.[62]

In 1960, Schulemann applied for money from the DFG, for his work on malarial infections. He intended to test a group of new agents, which

60 On the canary bird model in malaria research, compare L.B. Slater, 'Malaria Birds: modeling infectious human diseases in animals', *Bulletin for the History of Medicine*, 79 (2005): 261–294.

61 The Ministry of Culture of North Rhine-Westphalia supported research on cancer ('Untersuchungen auf dem Gebiet der Carcinomtherapie') from 1967 on.

62 Access to the relevant records on file at the Schering-archives was denied for reasons of personal protection.

had so far been considered too poisonous for therapeutic use.[63] Schulemann was awarded 21,400 DM, even though one expert opinion stated that in his application he had referred to work done more than 30 years earlier, and expressed doubt as to whether an Emeritus Professor with only two technical assistants would be able to accomplish a task on which an entire Institute had spent decades. These doubts were dismissed, however. The head of the award committee strongly supported Schulemann's application. He wrote:

> It would be an unduly callous and ungrateful to deny these relatively modest means from a world-renowned researcher, if this is all he needs to complete his ongoing work.[64]

One year later, Schulemann's application for continued support was again reviewed, with a less favourable outcome: even though the experts did support further subsidies, they added the proviso that Schulemann be required to produce some kind of final report and publish his results. Schulemann was informed of these conditions and in January 1962 decided to turn down the amount offered.[65] He refused to be put under pressure, and in any case had managed to secure another source of funding for his research.

Together with his former colleague and co-inventor of Plasmochin, Fritz Schönhöfer,[66] Schulemann had contacted Bayer AG in 1961. These two former Bayer employees – now receiving pensions from the company – offered Bayer their services for testing unknown combinations of drugs and

63 File card Schulemann, Bl. 6, DFG-Archive Bonn-Bad Godesberg, Schu 8/27.

64 HA-Liste dated 22 July 1960, DFG-Archive Bonn-Bad Godesberg Schu 8/27.

65 final balance dated 24 April 1962, DFG-Archive Bonn-Bad Godesberg, Schu 8/30.

66 Biographical facts derive from an obituary: W. Schulemann, 'In Memoriam Professor Dr. Fritz Schönhöfer', *Arzneimittel-Forschung*, 15 (1965): 1256. In a letter to the author Fritz Schönhöfer's son said that his father had actually written his own obituary, F. Schönhöfer '40 Jahre'. Personal correspondence with Peter Schönhöfer, e-mail dated 13 December 2005.

their effectiveness against malaria.[67] It was the same project the DFG had agreed to fund. Schulemann's motives for contacting his previous employers once again may in part be found in his strong desire to complete what he considered as his lifetime's work. But monetary interests also played a decisive role. In July 1961, Bayer AG raised Schulemann's pension from 1,250 DM to 3,000 DM per month.

Contacts with the World Health Organisation (WHO)

Even though Schulemann had been given emeritus status in 1958, he actively pursued his scientific activities, as evidenced by his close contacts with Bayer AG. He conducted research projects, but actually published very little. Nevertheless, just as he had in the past obtained financial support from the DFG and Bayer AG, he now turned to the WHO.

In 1948, the WHO had been founded as the successor of the former Hygiene Department of the League of Nations. On 29 May 1951, the FRG obtained membership, but there had been contacts between German scientists and the WHO before this date; this was remarkably similar to the situation occurring between the wars, when Bernhard Nocht was appointed head of the Malaria Commission of the League of Nations, before Germany joined the League. In 1955, at the 8th General Meeting of the WHO in Mexico, the members decided on a programme for the world-wide eradication of malaria.[68] The FRG participated in this programme, making payments into a specially earmarked fund. The intention was to improve

67 Correspondence between Schulemann/Schönhöfer and the Bayer AG in the personnel file of Werner Schulemann in the Bayer Archive Leverkusen 271/2.1, separate file, entitled 'Überlassung Malaria-wirksamer Substanzen' (assignment of malaria-effective substances), letter Schulemann/Schönhöfer to Farbenfabriken Bayer AG, dated 29 July 1962, letter Dr. W. Salzer to Direktor Dr. Hansen dated 9 August 1962.

68 Y. Nishihara, 'The rise and fall of WHO malaria eradication campaign' (University of London BSc thesis?, 2006).

Germany's image and international scientific contacts.[69] German participation in this malaria programme had initially been the idea of German experts in tropical medicine, most significantly of Ernst Georg Nauck (1897–1967), Director of the Hamburg Institute for Tropical Diseases, who by 1952 had established good contacts with the WHO.[70] For research in the field of malariology the Hamburg Institute for Tropical Diseases received funding from the WHO; but Nauck was also responsible for calling the WHO's attention to Schulemann's work on chemotherapy.[71]

Since the mid-fifties, the WHO had unsuccessfully tried to set up a scientific research program in order to test the effectiveness of new antimalarial medication. It had attempted to persuade pharmaceutical manufacturers to conduct research programs in this area. This helps to explain why the WHO was very keen on cooperating with Schulemann. From 1962 on, he received financial support from the WHO for his chemotherapeutic research.[72] The WHO as well as Schulemann profited from this arrangement: the WHO was able to exploit Schulemann's malaria research, while Schulemann could continue to test the effectiveness of his 6-aminoquinoline series of compounds in avian malaria. Between 1962 and 1970, Schulemann received 34,800 US$ for his malaria research, which was enough to pay the salary of a medical technical assistant.

69 BA Koblenz B142 (Bundesgesundheitsamt, National Health Office) No. 3335 Malaria Eradication Program.

70 In June 1952, the WHO contacted the Hamburg Institute for Tropical Diseases, with the intention of creating cooperation in setting up training programs for students and young physicians and the conduct of courses. Staatsarchiv Hamburg 352–358/9 Bernhard-Nocht-Institut File 1–222 entitled WHO until 1964.

71 Nauck to Ms. Daelen in a letter dated 10 July 1962, BA Koblenz B142 No. 3335 Malaria-Eradication Program. In this letter he reports on a meeting in May 1962 with Dr. Alvarado and Dr. Göckel of the WHO: '... A special emphasis was put on the need to intensify research on chemotherapeutic fields. This suggestion was generally accepted. Professor Nauck informed the WHO of the scientific work done by Professor Schulemann, Bonn.'

72 Grant to the Pharmacological Institute-University of Bonn (Federal Republic of Germany), 1962–1975, WHO Archives Geneva Fonds 10: records of the Central Registry, WHO HQ M2/181/41.

Schulemann tested mainly two substances, RC 12 and B 505. He cooperated with Fritz Schönhöfer until Schönhöfer's death in 1965, and with Prof. Léon Schmidt in Davis, California, USA. In his Institute, Schulemann conducted the testing of substances and their effectiveness on avian malaria, while Davis in California tested the same substances on P. cynomolgi in monkeys. Schönhöfer's work on synthetic drugs was continued by Prof. Otto Dann (1914–1999) in Erlangen. Dann was the Director of the Erlangen Institute for Applied Chemistry (Institut für angewandte Chemie Erlangen) where he ran his own biological testing laboratory, financed by the Volkswagen-Foundation.[73] Schulemann tested the substances synthesized in Erlangen and at the Bayer labs in Bayer-Elberfeld using avian models; if the results seemed promising he sent the compounds on to Davis, who then tested them on monkeys. In addition, Davis developed another testing model, the Pl. berghei-malaria-infection in mice. This rodent model for malaria research actually rendered the avian model obsolete. Nevertheless, Schulemann continued to use the avian model, introducing improvements which allowed him to accelerate the tests.

In 1964, the head of the WHO-malaria program, Leonard Jan Bruce-Chwatt (1909–1991), approached Schulemann, asking him whether he could send a small amount of RC 12 to Prof. Ciuca in Bukarest, who planned to try the substance on neurosyphilitic patients, who were experiencing malaria relapses.[74] Schulemann was pleased that Ciuca wanted to test a substance, which he, Schulemann, had developed. In 1966, the WHO received a report on the tests involving RC12, with a commentary by Dr. Weniger of the WHO Division Malaria Eradication. Schulemann expressed his opinion on the report and on the commentary: in his view it would be

73 On Otto Dann's scientific career before 1945, see F. Schmaltz, *Kampfstoff-Forschung*, pp. 430, 497–503.

74 Grant to the Pharmacological Institute-University of Bonn (Republic Federal of Germany), 1962–1975, WHO Archives/Fonds 10: Records of the Central Registry, WHO HQ M2/181/41 For his 'Comment on the Romanian Trial with RC 12 in Sporozoite Induced Malaria tertiana, VT Strain and Malaria Quartana' Schulemann presented the preliminary events leading to his experiments on humans, '...who intends to try it out on relapsing cases of P. malariae in neurosyphilitic patients.'

a mistake to publish the report, as the Romanian reseachers had not asked for a second expert opinion. Nor was he happy with the experimentees selected, concluding that:

> A sound decision whether RC12 is an advance or not may only be made after performing the intended clinical investigation on volunteers in the USA.[75]

Cooperation with the WHO led to wider international collaborations, which seemed always to seek the path of least resistance. Clinical trials were run wherever resistance to such trials was minimal, and where 'human material' was readily available: advanced syphilitics in Romania, and so-called 'volunteers' (prison inmates) in the US. Schulemann used all the opportunities offered to him by international cooperation. By passing on the substances to be tested to Romanian doctors, he was co-responsible for these experiments. During the 1930s, Ciuca had conducted research on malaria, though at that time he was interested mainly in immunity against malaria.[76] Even though the tests on volunteers in the US were disappointing, Schulemann did not give up his hope of having found a potent remedy in RC 12, and in 1970 attempted to combine it with Chloroquine as a radical cure against Pl. Vivax malaria. As the production of RC 12 was extremely inexpensive and simple, Schulemann declared himself able 'to arrange the supply of almost any quantity required for a field trial, should we propose a research worker interested in carrying out a study in a malaria-infected area'.[77] Schulemann insisted on the potential of his chemotherapeutic substances, and also on testing them on humans.

75 Comment on the Roumanian trials with RC 12 in Sporocoite Induced Malaria tertiana, VT Strain and Malaria quartana, VS Strain, dated 1 August 1966, Grant to the Pharmacological Institute-University of Bonn (Republic Federal of Germany), 1962–1975, WHO Archives/Fonds 10: Records of the Central Registry, WHO HQ M2/181/41.

76 M. Ciuca et al, 'Immunity in Malaria', *Transactions of the Royal Society of Tropical Medicine and Hygiene*, 27 (1934): 619–622.

77 Memorandum to Dr. G. Sambasivan, Director of Malaria Eradication Division on a discussion with research workers during a visit to Bonn, Nijmege, Moscow and Warsaw, dated 28 December 1970, Grant to the Pharmacological Institute-University of Bonn (Federal Republic of Germany), 1962–1975, WHO Archives/Fonds 10 : Records of the Central Registry, WHO HQ M2/181/41.

Conclusion

What is so special about Schulemann? He seems to have been the perfect opportunist, with a talent for organization and for raising funds for his research projects. He made excellent use of his early success and his international reputation as the inventor of Plasmochin. By contrast, neither Fritz Schönhöfer nor Fritz Wingler acquired the same fame as Schulemann for the discovery of Plasmochin.

Schulemann was a member of the SA and of the Nazi party, but during the National Socialist era he changed his opinion about its political leaders and their goals. After the end of the Second World War, he was forced to leave Bonn by its Lord Mayor, moreover he was forbidden to enter his Institute. He was denazified in 1949, thereby receiving permission to return to his chair in Bonn. But he was not totally rehabilitated within the scientific community. In his professional association, the DGPT, for example, he remained a member of the board, but there is little information about him in the association's archive. He received most of his awards under National Socialism, and was only rarely mentioned in his later years. He was also never asked by the DFG to provide an expert opinion, a request which could have been interpreted as a recognition of his denazification. Schulemann did receive funds from the DFG, but was never involved in their allocation. Within his professional association, he held the position of a quiet 'senior partner'. It was only during the 1960s that he was referred to as a 'much respected member of the [WHO's] Expert Malaria Panel'.[78] Objections against him and his achievements seem to have been greater at the national rather than the international level.

He managed to transform his removal from Bonn and his subsequent denazification into a personal success story, by cooperating with another large pharmaceutical company, Schering AG. And he was expert at reading the writing on the wall: during the 1950s, when the entire world was enthusiastic about the possibility of exterminating malaria with DDT, he

78 L.J. Bruce-Chwatt, 'Obituary', p. 86.

turned his attention to anti-histamines and put his energy behind rebuilding the Institute of Pharmacology in Bonn. When the first cases of resistance against DDT appeared, he took hold of the quest for new anti-malarials, returning to the substances he had synthesized during the 1920s and testing them for their toxicity against malaria. He used new methods and made new alliances, with researchers, and with funders, signing a research contract with the WHO in the 1960s. Schulemann's research projects did not always win support because they were innovative, but because he knew how to ask the right questions, in the right place, and at the right time.

Schulemann was attracted to the prestige offered by the position of academic director, and to the financial rewards from working with industry, and was exceptional in his success at securing both. A possible interpretation of the DFG's approval for funding in 1960–1961, is that rules for the allocation of research funds always include a certain element of flexibility, which can be used to sponsor well-known and established scientists, who for a long time have failed to achieve any results. By contrast, the cooperation with the WHO was a win-win situation, as until 1962, it had unsuccessfully tried to establish a research group on anti-malarials, and Schulemann had been searching for new sources of funding and scientific collaborations. For eight years, the WHO supported him and his research as an Emeritus Professor, including his testing in Rumania of a compound, which was ethically questionable at best.

The view this article provides on Werner Schulemann is a very limited one. The objective was to outline the permeable boundaries between industrial and university research, and the often marginal role of state-subsidized research. The relationship between science and politics has a major influence on the direction of research and the fate of individual projects. Werner Schulemann is an exceptional example, as there is no contemporary who had similarly close ties to industry. His colleagues tended to remain in their intellectual world of basic research (Grundlagenforschung), rejecting any ties with industry and commerce, whilst envying the rewards such ties might bring. It would be interesting to research this further at an international level, one of many opportunities clinical pharmacology and its history might offer scholars.

CAY-RÜDIGER PRÜLL

5 Scientists, Doctors and Drug Development: the history of Raymond P. Ahlquist's receptor theory, 1948–1988[1]

This paper deals with the influence of medical expertise on drug development. There are basically two groups whose work in this area is important for modern medical history. Firstly, there are the medical scientists who work in laboratories, where they carry out basic research – in this case in the field of pharmacology. Secondly, there are the medical students, who go on to treat patients after graduating, approach pharmaceutical problems from a clinical point of view, and perform clinical trials. The main purpose of this paper is therefore to study the different factors which influence the work of members of both groups: the factors which affect the individual researcher; those linked to the scientific community and

1 The analysis of Raymond Ahlquist and his receptor concept is one outcome of the Wellcome Trust-funded research project 'The Impact of the Receptor Concept on 20th Century Pharmacology' (grant number 061819), which has been undertaken by Andreas-Holger Maehle, Robert Halliwell, and myself. I wish to thank both my colleagues as well as Hans-Georg Hofer, Institute for the History of Medicine, Bonn, and Viviane Quirke, Centre for Health, Medicine and Society, Oxford Brookes University, for their advice and suggestions, and the Wellcome Trust for its support. Also, I am grateful to Lowell Greenbaum, former head of Department of Pharmacology at the Medical School of Georgia, Marilee S. Creelan, Head of the Collection Services, Robert B. Greenblatt Library, Medical School of Georgia, Augusta, and furthermore to Richard E. White, Lois T. Ellison, Armand M. Karow, Jerry J. Buccafusco (see the following footnotes) for their support with material and information on Raymond P. Ahlquist. Many thanks also to Sir James Black about additional information on the development of the receptor concept and the β-blockers. Last but not least many thanks also to Danny Knapp, Gudrun Pischinger, for checking the language and style.

the wider historical context; and those related to the interplay between basic and clinical research.

Consequently, this paper contributes to modern trends in science studies, which, in my view, provide the most valuable explanations of how scientists work.[2] Knowledge gained from the investigation of nature is not 'objective'. It is influenced, amongst other things, by the world view, mentality and social position of the researcher. Some historians of science even assume that knowledge is, in this way, 'socially constructed', as discoveries depend on the cultural setting in which they occur. This does not necessarily mean that valid facts about nature do not exist, but rather that the academic origins of the researcher, his social roots and connections, as well as his mentality, influence his view(s) of the natural world, his research questions, and therefore affect his results. This means that science is a particular way of considering phenomena, of collecting and sorting them, and giving them meaning. Science is part of a national culture or subculture, but is also a culture in itself. Scientific knowledge and practice can therefore be viewed as local and contingent phenomena, which depend on the time and place in which they develop.[3]

Consequently, the main focus of this paper is not on experimental settings and – as it is sometimes presumed – their decisive role in the conduct of science.[4] On the contrary, I will focus on the people conducting the science. The scientists' standing in their respective private and academic realms is key to understanding the history of scientific developments in

2 See R.C. Olby et al. (eds), *Companion to the History of Science* (London/New York: Routledge, 1996); M. Hagner (ed.), *Ansichten der Wissenschaftsgeschichte* (Frankfurt/M.: Fischer, 2001).

3 On the constructivist approach in the history of science see: J. Golinski, *Making Natural Knowledge: constructivism and the history of science* (Cambridge: Cambridge University Press, 1998); Thomas Schlich, 'Wissenschaft: Die Herstellung wissenschaftlicher Fakten als Thema der Geschichtsforschung', in N. Paul and T. Schlich (eds), *Medizingeschichte: Aufgaben, Probleme, Perspektiven* (Frankfurt/New York: Campus, 1998), pp. 107–129, esp. pp. 114–116, 119–123; T. Lenoir, *Instituting Science: the cultural production of scientific disciplines* (Stanford: Stanford University Press, 1997).

4 See H.-J. Rheinberger and M. Hagner (eds), *Die Experimentalisierung des Lebens. Experimentalsysteme in den biologischen Wissenschaften 1850/1950* (Berlin: Akademie Verlag, 1993).

general, and, as in our case, the history of scientific concepts in particular.[5] In the following pages, the social context of drug research will be analysed through a case study. This case study tells the story of the American pharmacologist Raymond P. Ahlquist (1914–1983) and his theory of the effect of specific substances, catecholamines, on cells, published in 1948. With this new theory, Ahlquist contributed to a breakthrough in the chemical understanding of drug binding, the so-called 'receptor concept'. Modern biomedicine relies heavily upon this theory, which challenged twentieth-century medicine to apply basic research to medical practice in an efficient way. Ahlquist's story is therefore that of the fate of the theory and its inventor's tireless efforts to gain recognition for his work. Hence, this paper contributes not only to the history of the receptor concept, and to the history of pharmaceuticals, but also more widely to the history of science.

Before turning to Ahlquist, I will define what 'receptors' are, and give an introduction to the history of the receptor concept. Then, I will concentrate on the development of Ahlquist's theory, and will deal with the publication and the reaction of the scientific community to his approach. I will analyse the reasons for the late acceptance and ultimate breakthrough of Ahlquist's dual adrenoceptor theory. Finally, I will summarize my argument and draw some conclusions.

The 'Receptors' and their History up to 1948

The receptor concept is one of the fundamental theories of twentieth-century scientific medicine. It answers the question of how information is transmitted to the cell. Today, receptors are described as 'small, discrete area(s) on the cell membrane or within the cell with which molecules or molecular complexes (e.g. hormones, drugs, and other chemical messengers)

5 C.-R. Prüll, 'Disziplinen: Entwicklungsmöglichkeiten der Medizingeschichte als Disziplinen- und Wissenschaftsgeschichte', in Paul and Schlich (eds), *Medizingeschichte*, pp. 216–242.

interact'.[6] The origin of the receptor concept dates back to the last third of the nineteenth century, and is chiefly linked to two names: Paul Ehrlich (1854–1915), and John Newport Langley (1852–1925). Neither of them was a member of the community of pharmacologists, which became institutionalized around 1850 with the creation of university chairs and research institutes, primarily in the German speaking lands.[7] Moreover, they approached the receptor concept from different backgrounds and via significantly different routes. Both mentioned the idea for the first time in 1878, and both returned to it only several years later – at the turn of the century. The two scientists developed their initial thoughts on receptors independently from each other.[8]

John Newport Langley, Professor of physiology in Cambridge, was predominantly concerned with the investigation of the functions of the autonomic nervous system, i.e. of those nerve fibres that regulate, without any conscious influence, the vegetative functions of the body, such as blood pressure and respiration. From the late 1880s onwards, he combined animal experimentation, chemical analysis, and a careful histological examina-

6 G. MacPherson (ed.), *Black's Medical Dictionary* (London: A&C Black, 39th Edition, 1999, repr. 2001), p. 463.

7 H.-H. Eulner, *Die Entwicklung der medizinischen Spezialfächer an den Universitäten des deutschen Sprachgebietes* (Stuttgart: Enke, 1970), pp. 112–138. For a history of medical disciplines see also: G. Weisz, *Divide and Conquer: a comparative history of medical specialization* (Oxford: Oxford University Press, 2006).

8 For current views on the history of the receptor concept see: A.-H. Maehle, C.-R. Prüll, and R.F. Halliwell, 'The Emergence of the Drug Receptor Theory', *Nature Reviews. Drug Discovery*, 1 (2002): 637–41; C.-R. Prüll, A.-H. Maehle, R.F. Halliwell, 'Drugs and Cells – Pioneering the Concept of Receptors', *Pharmacy in History*, 45 (2003): 18–30; A.-H. Maehle, 'Historical Foundations of the Receptor Concept in Pharmacology', *Gesnerus*, 61 (2004): 57–76; A.M. Silverstein, *Paul Ehrlich's Receptor Immunology: the magnificent obsession* (San Diego, London etc.: Academic Press, 2002); J. Parascandola and R. Jasensky, 'Origins of the Receptor Theory of Drug Action', *Bulletin of the History of Medicine*, 48 (1974): 199–220; J. Parascandola, 'The Development of Receptor Theory', in M.J. Parnham and J. Bruinvels (eds), *Discoveries in Pharmacology*, vol. 3, *Pharmacological Methods, Receptors & Chemotherapy* (Amsterdam/New York/Oxford: Elsevier Science Publishers, 1986), pp. 129–156; idem, 'The Theoretical Basis of Paul Ehrlich's Chemotherapy', *Journal of the History of Medicine and Allied Sciences*, 36 (1981): 19–43.

tion of specimens in order to investigate the autonomic nervous system. In the course of this work, Langley addressed the question of whether the substances administered to the experimental animals influenced the tissue of the end organ directly or only via the effect on the nerves and their endings. By 1905, Langley had become convinced that the administered drugs influenced the cells directly, and he speculated about the existence of a 'receptive substance' of the cell that mediated the effect of a drug. He elaborated his ideas about 'receptive substances' in the following years, despite heavy criticism.[9]

Paul Ehrlich, the Berlin bacteriologist and immunologist, was keen to examine the relations between bacterial toxins and antitoxins, as part of a programme to combat infectious diseases. He was the first to develop a logical and consistent receptor concept.[10] Thus, in 1897, while working on immunological problems, he developed his chemical 'side-chain theory'. Certain 'side-chains' of the cell were able to bind to certain bacterial toxins. Because the occupied side-chains would then become unable to fulfil their physiological functions, the cell would compensate by producing a surplus of additional side-chains. These side-chains would be released into the blood stream, where they would act as antibodies or antitoxins.[11] Ehrlich elaborated this chemical theory further during the following years.

9 J.N. Langley, 'On the Reaction of Cells and of Nerve-endings to certain Poisons, chiefly as regards the Reaction of Striated Muscle to Nicotine and to Curari', *Journal of Physiology*, 33 (1905): 374–413, esp. 399; A.-H. Maehle, '"Receptive Substances": John Newport Langley (1852–1925) and his path to a receptor theory of drug action', *Medical History*, 48 (2004): 153–174; J.N. Langley, *The Autonomic Nervous System*, Part I (Cambridge: Heffer & Sons, 1921), p. 44.

10 For the best account of Ehrlich's life so far, see C.E. Dolman, 'Paul Ehrlich', in *Dictionary of Scientific Biography*, vol. 3 (New York: Charles Scribner's Sons, 10th Edition, 1981), pp. 295–305. Dolman also gives an overview of the literature on Ehrlich up to 1980.

11 P. Ehrlich, 'Die Wertbemessung des Diphtherieheilserums und deren theoretische Grundlagen' (Klinisches Jahrbuch, 1897), in F. Himmelweit (ed.), *The Collected Papers of Paul Ehrlich in four Volumes including a complete Bibliography*, vol. II, *Immunology and Cancer Research* (London/New York/ Paris: Pergamon, 1957), pp. 86–106, Engl. transl.: ibid., pp. 107–125. For a contemporary description of Ehrlich's side-chain theory, see L. Aschoff, *Ehrlich's Seitenkettentheorie und ihre Anwendung auf die künstlichen Immunisierungsprozesse* (Jena: Fischer, 1902), pp. 1–25. See also A.M. Silverstein, *A*

In 1900, he introduced the term 'receptor' to describe the functions of side-chains, and turned his attention towards studying their character.[12] In 1908 his influential work in immunology led him to being awarded the Nobel Prize together with Elie Metchnikoff (1845–1916). The 'receptors' formed the theoretical basis for his subsequent work on Salvarsan, the first chemotherapeutic substance for the treatment of syphilis, introduced in 1910. It was this discovery that made Ehrlich one of the most prominent figures of twentieth-century medicine.[13]

Ehrlich's and Langley's ideas shed new light upon the role of chemical affinity and specificity in drug effects, thus opening the door for analysing the mechanisms of drug action in greater detail. Both their approaches laid the groundwork for the detection of the specific causes of diseases and, in the end, for improvements in therapy. Such improvements were, in fact, what people expected from scientific medicine after 1900. However, Ehrlich and Langley's receptor concept was not immediately accepted by the pharmacological community. The most important representatives of the discipline promoted competing theories, as well as alternative research strands.[14] Nevertheless, the receptor concept survived, thanks to the development of quantitative theories of drug action.

The quantitative approach received clear impetus in the 1930s with Alfred Joseph Clark's (1885–1941) work on drug binding.[15] After his early

---

History of Immunology (San Diego, New York etc.: Academic Press Inc., 1989), pp. 64–66, 94–99.

12 P. Ehrlich and J. Morgenroth, 'Über Haemolysine. Dritte Mittheilung' (Berliner Klinische Wochenschrift, 1900), in Himmelweit, *The Collected Papers*, vol. II, pp. 196–204, Engl. transl. pp. 205–212.

13 The development of Ehrlich's receptor theory was the combined result of his social biography, his personality and his scientific career. For more details on the subject, see: C.-R. Prüll, 'Part of a Scientific Master Plan? Paul Ehrlich (1854–1915) and the Origins of his Receptor Concept', *Medical History*, 47 (2003): 332–356. For a traditional explanatory approach, see Silverstein, *Paul Ehrlich's Receptor Immunology*.

14 C.-R. Prüll, 'Caught between the Old and the New – Walther Straub (1874–1944), the Question of Drug Receptors and the Rise of Modern Pharmacology', *Bulletin of the History of Medicine*, 80 (2006): 465–489.

15 A.-H. Maehle, 'The Quantification and Differentiation of the Drug Receptor Theory. c. 1910–1960', *Annals of Science*, 62 (2005): 479–500.

death, colleagues including John Henry Gaddum (1900–1965), Everhardus Jacobus Ariëns (1908–2002), and Robert P. Stephenson (1925–2004), pursued the question of the mechanism of drug binding to receptors. Consequently, in the late 1940s, receptor theory was strengthened by a new group of pharmacologists who devoted their research to receptors. But their work was hampered by the fact that there was still no clear correlation between the chemical structure of drugs and their specific biological effect, and that receptors remained unspecified. Therefore, most research in pharmacology continued to be devoted to traditional approaches, which were deeply rooted in the educational and research programmes of the field. These traditional approaches explained drug binding as a process depending, above all, on physical laws. Furthermore, within pharmacology the work on receptors competed with the research programme on transmitter substances. After the Second World War, Sir Henry Hallett Dale (1875–1968), with his eminent reputation as a Nobel Prize winner, dominated research policy in the UK and internationally, and was as an influential spokesman for this second group of researchers.[16]

Consequently, those promoting the receptor concept were still in a defensive position when it received fresh impetus from investigations undertaken after 1945. The Second World War against the Third Reich and Japan had led to major new scientific and technical developments. Although these were often considered as a threat and a danger to civilization, war-related developments did encourage further investigations in the civil sector. Areas of rapid development included electronics, computers, and other scientific instruments, such as manometers for measuring differences in pressure, many of which proved to be useful in medicine. This bolstered the natural and medical sciences, and generated trust in scientific results and values. Science, despite all its ambiguities, was seen as the key to solving people's problems, and medicine as a mainly unquestioned institution for dealing with all matters related to health and disease.[17] The

16 See C.-R. Prüll, A.-H. Maehle, and R.F. Halliwell, *A Short History of the Drug Receptor Concept* (Houndmills/Basingstoke/New York: Palgrave-Macmillan, 2009).

17 R. Cooter and J. Pickstone, 'Introduction', in R. Cooter and J. Pickstone (eds), *Medicine in the 20th Century* (Amsterdam: Harwood Academic Publishers, 2000), pp. xiii–xix.

pharmacologist Raymond P. Ahlquist (1914–1983) benefited from these developments, and gave a decisive impulse to the research on receptors.

Raymond P. Ahlquist and the Development of the 'Dual Adrenoceptor Theory'

Ahlquist was a child of Swedish immigrants and was born in Missoula, Montana (USA) in 1914. He studied pharmacy at the University of Washington in Seattle. It was there that he received his Bachelor of Science degree in 1935. Under the supervision of the pharmacologist J.M. Dille, he obtained his Master's degree in 1937, and his Doctoral degree in 1940. Between 1940 and 1944, he was Assistant Professor of pharmacology and pharmacognosy at South Dakota State College in Brookings, USA. In 1944, he started work as Assistant Professor of pharmacology at the Medical College of Georgia in Augusta. In 1948, he succeeded Robert A. Woodbury (1904–2001), Professor and Chairman of the department of pharmacology since 1943. Ahlquist held this position until 1963, and then again between 1970 and 1977. In 1977, he became holder of the Charbonnier Professorship. This short description of his life shows that Ahlquist, although interested in the application of his work to medical practice, remained a medical scientist throughout his career. He lived in Augusta until his death in 1983.[18]

During his time at South Dakota State College, Ahlquist became increasingly involved in the research on the functional aspects of the autonomic

18 G.O. Carrier and M. Shlafer, 'Raymond P. Ahlquist. 26 July 1914 – 15 April 1983', *Trends in Pharmacological Sciences*, 5 (1984): 41–44, esp. 41; R.C. Little, 'Raymond P. Ahlquist (1914–1983)', *Clinical Cardiology*, 10 (1987): 583–584, esp. 583; J.H.R. Sutherland, 'History. Department of Pharmacology and Toxicology. School of Medicine. Medical College of Georgia', 1988, typewritten manuscript, p. 10, courtesy of Lois T. Ellison, Medical Historian in Residence, Augusta; J.H.R. Sutherland, G.O. Carrier, L.M. Greenbaum, 'Obituary Raymond P. Ahlquist, July 26, 1914 – April 15, 1983', manuscript, in: *Dean's Office Newsletter*, MCG (Medical College of Georgia), 8 (1983), No.15, July 29, one page, in: Papers of Lowell Greenbaum, Augusta.

nervous system. He was interested in sympathomimetic drugs – substances with an adrenaline-like effect. In the course of his experiments, Ahlquist encountered the crucial problem of trying to understand how drugs bind to cells, at a time when the concept of receptors was not yet well defined. Unlike now, when scientists have much more effective knowledge about, for example, receptor qualities and binding sites, for Ahlquist and his contemporaries the functions and effects of receptors were still uncertain. And the idea of the existence of a multitude of receptors, which had been formulated by Paul Ehrlich, remained highly speculative. Worryingly, different studies had shown that sympathomimetic substances could elicit diverse and contradictory responses in the cardiovascular system of research animals. This made it impossible to assign a particular pharmacological profile to a specific amine and, therefore, to a specific chemical structure. Especially troubling was the discovery of the 'reversal' effects of these substances. For instance, although Octin, a sympathomimetic amine, caused a 'typical' sympathomimetic rise in the heart rate, it also caused a depressor response. Moreover, these substances showed a different pharmacological profile in different tissues and organs.[19]

After arriving at the Medical College of Georgia in Augusta (1944), Ahlquist provided the answer to this problem. The starting point were studies he performed together with Robert A. Woodbury (at the time head of the small pharmacology department). During his collaboration with Ahlquist, Woodbury had tried above all to find a drug to treat the symptoms of dysmenorrhoea. Ahlquist's own contribution showed his early interest in investigating the cardiovascular system with specific substances, and it was the search for a sympathomimetic substance producing a relaxation of the uterine muscle that finally enabled Ahlquist to make his successful contribution to receptor research.[20]

19 G.O. Carrier, 'Evolution of the dual adrenergic receptor concept: key to past mysteries and modern therapy', in Parnham and Bruinvels (eds), *Discoveries in Pharmacology*, vol. 3, pp. 203–221, esp. pp. 208/209.

20 R.P. Ahlquist, R.A. Huggins, and R.A. Woodbury, 'The Pharmacology of Benzyl-Imidazoline (Priscol)', *Journal of Pharmacology and Experimental Therapy*, 89 (1947): 271–288; R.P. Ahlquist and R.A. Woodbury, 'Influence of drugs, and uterine activity upon uterine blood flow', *Federation Proceedings*, 6 (1947): 305. Adrenaline itself could not be used because of its cardiovascular side-effects, see: Carrier, 'Evolution', p. 209.

This research work was very much aided by new experimental tools developed during the war by the physiologist William F. Hamilton (1893–1964). Since 1934, Hamilton had been the head of the physiology department of the Medical College of Georgia, which became much strengthened in this period. He was also in charge of the Pharmacology Department until 1942. Hamilton's influence on Ahlquist cannot be overestimated. One source of support for Ahlquist's research was Hamilton's innovative engagement in developing instruments that enabled a more accurate measurement of blood flow and blood pressure. These instruments were the product of Hamilton's 'pioneering work on the relations between pressure and flow of blood in the cardiovascular system',[21] which reached its peak while at the Medical College. At a time when there was no financial support for research, to obtain additional funding, between 1934 and 1946 Hamilton sold his new invention, the Hamilton metal membrane manometer, to other laboratories, including Woodbury's. Ahlquist therefore obtained the instrument, which made possible very sensitive measurements of blood pressure via the deflection of a small copper membrane, which was optically recorded. The instrument also enabled investigators to sort out and examine the specific impact of the various parameters that influence blood pressure. These were mainly cardiac functions and disturbances that originate in the vessels themselves. Other gadgets developed during the Second World War included electromanometers. Their action depended on the deflection of a metal membrane, which was recorded electrically. Electromanometers were the product of the aircraft industry, and were introduced into routine laboratory practice after 1945.[22]

Ahlquist's work therefore depended not only on his own innovative brainwork, but also on local support, such as William F. Hamilton's.

21 C.H. Baker, D.L. Davis, L.T. Ellison, C.E.H. Schoenborn, R.C. Little, and W.H. Waugh, 'William F. Hamilton', *The Physiologist*, 27 (1984): 64–66, esp. p. 65. During the war years (1939–1945), as organizer of the Society, Hamilton also maintained contacts among members of the American Physiological Society. See P. Dow, 'William F. Hamilton', *The Physiologist*, 8 (1965): 95–96, esp. p. 96; W.F. Hamilton, 'Personal Recollections as to the Affairs of the Society from 1921 – to Date', *The Physiologist*, 7 (1964): 38–41, esp. 38.

22 Baker et al., 'William F. Hamilton', pp. 64–66; Carrier, 'Evolution', pp. 205/206; Carrier and Shlafer, 'Ahlquist', pp. 41–42.

Ahlquist attempted to determine the exact pharmacological profile of various sympathomimetic drugs, concentrating mainly on six substances: Noradrenaline, Cobefrine, Adrenaline, Levo-Adrenaline, Methyladrenaline and Isoprenaline. Their activity was tested on 23 different tissues of research animals, namely cats, dogs, rats and rabbits. Five organ systems were tested: the cardiovascular system (renal, mesenteric and femoral vascular beds, isolated and intact heart);[23] the intestine (isolated and intact ileum); the uterus (isolated and intact uterus); the urethra (intact urethra); and the eye (intact dilator pupillae and nictitating membrane).[24] The new tools described above allowed Ahlquist to investigate the heart and blood vessels independently, and to distinguish between the effects of the different substances in question. The blood flow and the vasomotor resistance (mechanical resistance of vessels reducing the speed of blood flow) of the vascular beds was measured and recorded using modern flowmeters, which were introduced into the arterial inflow or the venous outflow. For investigations on isolated hearts, Ahlquist used an optical lever system, which was made out of a former bombsight.[25] For recording the general arterial pressure, Ahlquist did not only use the older, mercury manometers, but also a new one designed by Hamilton and, in the case of several dogs, a high-

23 'Isolated' means removed from the animal body, 'intact' means left within the animal body.

24 R.P. Ahlquist, 'A Study of the Adrenotropic Receptors', *American Journal of Physiology*, 153 (1948): 586–600, here esp. 587–595. The number of 23 different tissue results from the fact that Ahlquist did not use every animal species for every organ system. See also J.D. Robinson, *Mechanisms of Synaptic Transmission: bridging the gaps (1890–1900)* (Oxford: Oxford University Press, 2001), p. 153. The number of tissues differs from other documentation and Table I of Ahlquist's paper, which summarizes the results, is reduced to a simplified version without any comment; Carrier, 'Evolution', p. 211. Carrier reduces the number of investigated substances to five, leaving out Levo-adrenaline, presumably for didactic purposes.

25 A bombsight is a device in an aircraft for aiming bombs. The most well-known is the 'Norden bombsight', which was developed during the Second World War in the USA, allowing the consideration of various factors influencing the dropping of the bomb into the ground target as e.g. the altitude of the aircraft. See: http://en.wikipedia.org/wiki/Bomb_sight; S.H. Ross, *Strategic Bombing by the United States in World War II: the myths and the facts* (Jefferson/NC: McFarland & Co., 2002).

frequency Hamilton manometer. The Hamilton manometer was also used for the investigation of the intestine and the uterus. Experimental results on the eye and urethra were recorded with a string and lever system or a drop counter. In investigations of intact animals, the substance in question was injected into a vein or an artery. In experiments on isolated organs, it was applied directly or the organs were perfused.[26]

Ahlquist summarized the results of his experiments in a table, in which he ranked the effect of all substances on each organ system in terms of potency and function. Potency meant the strength of effect in relation to the function of the organ, e.g. vasoconstriction or vasodilation, i.e. excitatory or inhibitory effect. His analysis had the remarkable result that the effect of the six amines examined could be reduced to two orders of potency for specific functions. The first order was:

1. Levo-adrenaline, 2. Adrenaline, 3. Noradrenaline, 4. Cobefrine, 5. Methyladrenaline, 6. Isoprenaline

With decreasing potency these substances induced vasoconstriction, excitation of the uterus, nictitating membrane, dilator pupillae and urethra in all animals. They also had an inhibitory effect on the intestine.

The second order was:

1. Isoprenaline, 2. Levo-Adrenaline, 3. Methyladrenaline, 4. Adrenaline, 5. Cobefrine, 6. Noradrenaline

With decreasing potency these substances induced vasodilation and myocardial excitation in all animals. They also had an inhibitory effect on the contraction of the uterus.[27]

26 Ahlquist, 'A Study of the Adrenotropic Receptors', pp. 588, 591–595; Carrier, Shlafer, 'Ahlquist', p. 43.

27 Ahlquist, 'A Study of the Adrenotropic Receptors', pp. 590, 596.

Ahlquist concluded that there must be two different kinds of adrenergic receptors, which he termed α and β receptors. The α-receptors would be related to the first order of amines and therewith to 'most of the excitatory functions (vasoconstriction, and stimulation of the uterus, nictitating membrane, urethra and dilator pupillae) and one important inhibitory function (intestinal relaxation)'. The β-receptors would be related to 'most of the inhibitory functions (vasodilation, and inhibition of the uterine and bronchial musculature) and one excitatory function (myocardial stimulation)'.[28] With this 'dual adrenoceptor theory', Ahlquist did not simply add a new receptor theory to those already existing, he brought the concept of receptors to a new level. While Clark, Gaddum, Ariens, Stephenson, and others concentrated their investigations on the mode of action of 'the' receptor, Ahlquist, to a certain extent, went back to Langley and Ehrlich's theories. Both of them had speculated that there were different kinds of receptors. The latter had even speculated about a microcosm of receptors. Ahlquist somehow confirmed their speculations, opening the door to the idea that there were different types of receptors, thus paving the way for our modern understanding of an always-changing multitude of receptors.

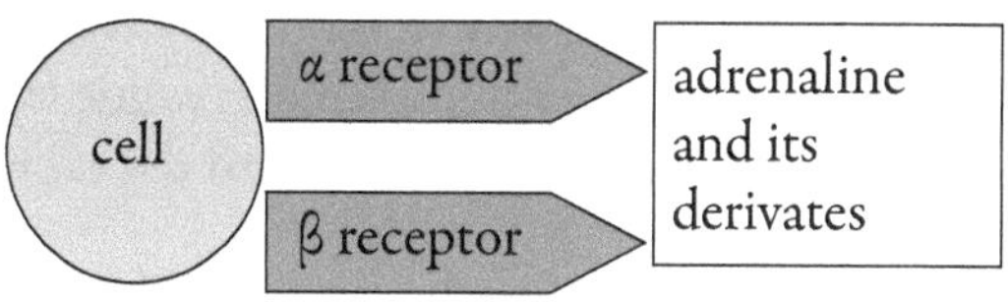

Figure 5.1: The dual adrenoreceptor concept of Raymond P. Ahlquist.
© Prüll, (2009)

28 Ibid., p. 599. For Ahlquist's dual adrenergic receptor concept see also: Carrier, 'Evolution', pp. 211–217.

The Publication of and the Reaction to Ahlquist's Theory

Ahlquist published his results in 1948 – but only with difficulty. The problem was not only that he had ventured into the hotly debated field of receptors, but also that he had launched an attack on the well-known and widely accepted theory of Walter Bradford Cannon (1871–1945) and his collaborator Arturo Rosenblueth (1900–1970) from Harvard Medical School. These two researchers had explained all the ambiguities regarding the effect of adrenergic substances in terms of a 'baroque theory', which they had deduced from their investigations with adrenaline. According to them, two mediator substances in the cells, namely the inhibitory 'sympathin I' and the excitatory 'sympathin E', were responsible for the divergent pharmacological profiles of adrenaline and related substances, i.e. one receptor could release two different 'sympathins'.[29] Ahlquist launched his attack – in a style true to his personality – clearly and plainly. Of course, criticism was unavoidable, as it was an implicit result of his findings. There were two types of receptors, each of them having a specific pharmacological profile, but both having basically excitatory as well as inhibitory functions. Also, according to Ahlquist, epinephrine (adrenaline) was the most active substance on both receptor types, having, therefore, inhibitory as well as excitatory properties. This meant that it was impossible to argue for two mediator substances E and I, one being 'excitatory' and the other 'inhibitory'.

29 J.W. Black, 'Ahlquist and the Development of beta-adrenoceptor Antagonists', *Postgraduate Medical Journal*, 52, Suppl 4 (1976): 11–13, esp. 12; H.H. Dale, 'Walter Bradford Cannon. 1871–1945', *Obituary Notices of Fellows of the Royal Society*, Vol. V, 1945–1948 (London: Morrison & Gibb Ltd., 1954), pp. 407–423; W.B. Cannon and A. Rosenblueth, 'Studies on Conditions of Activity in Endocrine Organs, XXIX. Sympathin E and Sympathin I', *American Journal of Physiology*, 104 (1933): 557–574; W.B. Cannon and A. Rosenblueth, *Autonomic Neuro-Effector Systems* (New York: The Macmillan Company, 1937); E.L. Wolfe, C.A. Barger and S. Benison, *Walter B. Cannon, Science and Society* (Boston, Cambridge/Mass.: Boston Medical Library in the Francis A. Countway Library of Medicine. Distributed by the Harvard University Press, 2002).

Furthermore, it was the receptor type, not the mediator itself, which determined the effect of the substance.[30]

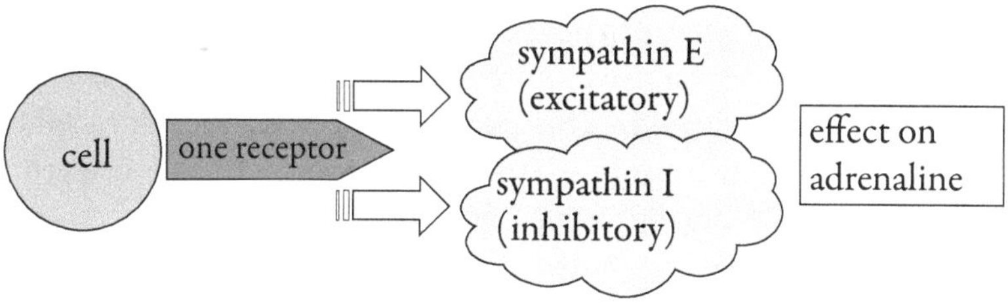

Figure 5.2: The drug binding theory of Walter Bradford Cannon and Arturo Rosenblueth (Harvard Medical School).
© Prüll, (2009)

The consequence of Ahlquist's implicit criticism of Cannon and Rosenblueth's theory was that his paper was rejected by the *Journal of Pharmacology and Experimental Therapeutics*. The argument of the editor of the journal was that Ahlquist's work condemned the theory of the two sympathins, which at the time was considered as a 'law' of physiology.[31] It was therefore his friend and colleague Hamilton, as editor of the *American Journal of Physiology*, who enabled its publication.[32] Hence, once again, Ahlquist's social contacts in Augusta proved most important for his work

30 Ahlquist, 'A Study of the Adrenotropic Receptors', pp. 586/587, 596–599. In the course of my investigations on Ahlquist's receptor theory during my research stay in Augusta in Summer 2003, two of Ahlquist's laboratory books turned up. The first one was entitled 'Sympathomimetic Amines. Book I. Compound and Reference Numbers. R.P. Ahlquist 1945'. It contains only formulas, reference numbers with company names and literature-hints, but no remarks or descriptions of his experiments. The same is true for the second book, entitled 'Sympathomimetic Amines. Book II', which contains only an evaluation of the literature about the topic from the years 1945/1946. Both books may have been saved by Richard E. White in 1983, when Ahlquist's room was emptied after his death. His papers have been destroyed.

31 Ahlquist, 'A Study of the Adrenotropic Receptors', p. 586; Carrier, 'Evolution', p. 217.

32 Baker et al., 'William F. Hamilton, p. 65/66.

– in this instance for the dissemination of the first decisive description of his new drug-binding theory. Hamilton's move was well thought through, however. He was keen to support experimental work in physiology and pharmacology at the Medical College in Georgia. Furthermore, he encouraged the collaboration between physiologists and pharmacologists. The staff of both departments spent their lunchtimes together, and discussed their findings and theories frankly. This created an open-minded atmosphere, which almost certainly contributed to the development of Ahlquist's theories. Hamilton had helped to equip Ahlquist's laboratory with his newest experimental gadgets. He promoted original and critical work in pharmacology in general, and Ahlquist's innovative research in particular. Together, the two friends 'turned out many fine postdoctoral trainees'.[33] Hamilton also developed close contacts with the clinicians in Georgia, and it might be that he could see the practical implications of Ahlquist's work. Based on this, Hamilton's support for the publication of Ahlquist's decisive paper in 1948 seems to be only one element in a consistent strategy.[34]

The next steps, namely the public acceptance of Ahlquist's theory, the realization of the full impact of his ideas, and finally the award of honours in recognition of this impact, were beset with difficulties. In the following years, there was almost no reaction to his 1948 paper. It was brought into the scientific arena only ten years later, in 1958. Its important practical implications were considered only in the 1960s, and Ahlquist never received what many of his friends and colleagues expected for him: the Nobel Prize in Physiology or Medicine. It is tempting to explain the late breakthrough of the dual adrenoceptor concept mainly by the scientific disputes that arose around the theory that fitted best the knowledge of

33 *Speech of Raymond P. Ahlquist (draft) on occasion of the receipt of the Memorial Award in Therapeutics in 1973*, typewritten manuscript, 6 pp., p. 3, in: Folder on Raymond P. Ahlquist, Department of Pharmacology and Toxicology, Medical College Augusta.

34 Concerning the relationship between Ahlquist and Hamilton, see: Ibid., p. 3; Carrier, 'Evolution', p. 217; Carrier and Shlafer, 'Ahlquist', p. 42; Interview with Lois T. Ellison, M.D., Medical Historian in Residence, Provost Emeritus, Professor Emeritus of Medicine and Surgery, formerly director of the cardiopulmonary lab of the Medical College, Georgia, Augusta, on September 9, 2003, Transcript part 1, pp. 1/2, 6/7.

drug binding.[35] But if we look in greater detail at the post-1948 period, at the measures taken by Ahlquist and what happened to him, the story becomes more complicated.

Although Ahlquist received no immediate positive feedback on his paper, he did not stop working on amines, especially adrenaline (epinephrine), after 1948. During the next ten years he experimented with adrenergic blocking agents to examine in greater detail the pharmacological profile of the α- and β-receptors. This work was done primarily with his first postdoctoral trainee, Bernhard Levy.[36] It is important to note that, in the course of his further work on the sympathetic nervous system, Ahlquist also helped to characterize β-blocking agents, which would become more and more important in the decades to follow.[37] α-blockers were well known,[38] but there was not much information about β-blocking agents, and this led to investigations being carried out by other research groups as well.

One of these groups worked in the Lilly Research Laboratories in Indianapolis, USA. Eli Lilly had already sponsored Woodbury and Ahlquist's research in Augusta, and the company laboratories stuck to the same research area as they began looking for a substance with

35 Carrier, 'Evolution', pp. 218–220; R.G. Shanks, 'The discovery of beta adrenoceptor blocking drugs', in Parnham and Bruinvels, *Discoveries*, Vol. 2, *Haemodynamics, Hormones & Inflammation* (Amsterdam: Elsevier Science Publishers B.V., 1984), pp. 37–72, esp. pp. 39/40.

36 See B. Levy and R.P. Ahlquist, 'Inhibition of the Adrenergic Depressor Response', *Journal of Pharmacology and Experimental Therapeutics*, 121 (1957): 414–420; *Speech of Ahlquist (draft) on occasion of the receipt of the Memorial Award in Therapeutics in 1973*, p. 4. In 1962, Levy became Professor of Pharmacology at the University of Texas, Galveston. See Sutherland, 'History. Department of Pharmacology and Toxicology', p. 2.

37 See the short account on Ahlquist's work on the adrenergic system and especially on β-blockers after 1948 in: R.P. Ahlquist, 'Historical Perspective. Classification of Adrenoreceptors', *Journal of Autonomic Pharmacology*, 1 (1980), 101–106, esp. 103–104. In 1961, Ahlquist together with Levy even developed a multi-organ testing system so study adrenoreceptor blocking substances, see ibid., p. 103.

38 R.P. Ahlquist, Draft of a speech, no date, typewritten manuscript, pp. 11, p. 7, in: Folder on Raymond P. Ahlquist, Department of Pharmacology and Toxicology, Medical College Augusta.

bronchodilator activity.[39] This fact tells us much about the therapeutic value attributed to research on substances influencing the autonomic nervous system. The Eli Lilly Company had been founded in 1876 by Colonel Eli Lilly, a veteran of the US Civil War, with the aim of improving the quality of already existing drugs in order to create new ones later on – an aim that was still being pursued by the company in the twentieth century. Eli Lilly were clever enough to back the right horses when producing the 'world's first commercially available insulin product' in 1923, shortly after the substance had been isolated by Frederick Grant Banting (1891–1941) and Charles Best (1899–1978), and by developing 'a method to mass-produce penicillin' in 1940.[40]

In the 1950s, in response to the growing importance of cardiovascular diseases, Eli Lilly set up a huge cardiovascular research programme. Catecholamines were among the substances that were studied. One of the compounds, which was tested by the Eli Lilly Research Group, namely dichloroisoprenaline (DCI, compound No. 20522) had been synthesized by J. Mills in the Organic Chemistry Division of the Company. In the course of experiments, the group showed that DCI had the ability to prevent the inhibitory actions of adrenaline, including broncho- and vasodilation. The results were published by C.E. Powell together with Irwin Slater in 1958. The two authors remarked that the results of their experiments had 'special relevance when considered in terms of drug-receptor theory', but they did not relate their findings to Ahlquist's dual adrenoreceptor concept, and they did not cite their colleague from Augusta.[41] The reason for this is that

39 Ahlquist, Huggins, and Woodbury, 'Pharmacology', p. 287; Ahlquist and Woodbury, 'Influence of Drugs', p. 305; Shanks, 'Discovery', pp. 40–41.

40 See: *Eli Lilly and Company History*: http://www.lilly.ca/searchable/about/history2.htm; *Milestones in Medical Research*: http://www.lilly.com/about/milestones.html. Banting and John James Richard Macleod (1876–1935) got the Nobel-Prize for the discovery of insuline in 1923.

41 C.E. Powell and I.H. Slater, 'Blocking of Inhibitory Adrenergic Receptors by a Dichloro Analog of Isoproterenol', *Journal of Pharmacology and Experimental Therapeutics*, 122 (1958): 480–488, esp. 480, 486 (quotation). On DCI and Eli Lilly's cardiovascular research programme, see E.J. Kahn, *All in a Century: the first 100 Years of Eli Lilly and Company* (West Cornwall/Ct.: Eli Lilly, 1975), pp. 15, 114/115.

they saw their experiments as belonging to the research tradition created by Clark, and followed by Gaddum, Ariens, and Stephenson. This tradition focussed on the quality of the substance-receptor attachment, the competition between substances in their search for a receptor site, and the strength of the substances when binding to receptors. But this research line did not consider the possibility of the existence of different receptors.[42]

It was another researcher who related Powell and Slater's blocking agent of adrenergic sites to Ahlquist's dual adrenoceptor concept. He was Neil C. Moran (1924–1997), who at the time was head of the department of pharmacology (Division of Basic Health Sciences) at Emory University in Atlanta, Georgia, only 120 miles away from Augusta. He had heard a presentation by Irvine Slater at the meeting of the Federation of American Societies for Experimental Biology in Chicago, Illinois in 1957. In this lecture, Slater presented the results of his research on DCI, on which the paper published together with Powell one year later was based. Moran was a keen rival of Ahlquist's and knew his work and theories very well. He was struck by the idea that Slater and Powell's results fitted Ahlquist's theory, and that the heart played a central role within the range of activity of DCI. He even thought that DCI might be a β-receptor blocking substance. He therefore asked Slater for a sample of DCI, which he received two weeks later.[43] Together with his colleague Majorie E. Perkins, he then investigated in detail the effect of the substance. They used anaesthetized dogs and isolated rabbit hearts to examine the influence of DCI on the effect of different amines. The main result was that DCI blocked exactly those effects Ahlquist had assigned to the second order of sympathetic substances. It was 'a selective blockade of most inhibitory functions' and

42 Powell and Slater, 'Blocking of Inhibitory Adrenergic Receptors', pp. 480–488; Shanks, 'Discovery', p. 41.

43 This information comes from Moran himself, given in a personal communication to Shanks in 1982. Although given in hindsight, his description fits perfectly well other facts and information on the story of Ahlquist's theory. See Shanks, 'Discovery', pp. 41–42. Powell and Slater, 'Blocking of Inhibitory Adrenergic Receptors', p. 480; Interview with Armand M. Karow, PhD, who was Assistant Professor since 1968 and Associate Professor at the Pharmacology Department since 1971, then President of Xytex Corporation, on September 9, 2003, Augusta, typewritten transcript.

'cardiac positive inotropic and chronotropic effects'. DCI caused a 'specific cardiac adrenergic blockade' and, in the view of Moran and Perkins, this supported the 'postulate of Ahlquist (1948) that the adrenotropic inhibitory receptors and the cardiac chronotropic and inotropic adrenergic receptors are functionally identical, i.e. that both are *beta* type receptors.' DCI was explicitly described as a β-blocking substance. In this way, Moran and his colleague related their research on blocking agents of the sympathetic system to Ahlquist's theory, thus helping to introduce it into experimental pharmacology as an explanatory tool for the study of adrenergic receptors. With a clear feeling that Ahlquist had made an important contribution to research on the sympathetic system, Moran enhanced his own reputation by association with Ahlquist's work. Like the Eli Lilly team, Moran and Perkins submitted their paper six months later to the *Journal of Pharmacology and Experimental Therapy*, where it was also published in 1958.[44]

Powell and Slater quickly adopted Moran's view. Only one year later, in 1959, they published a contribution to a symposium on catecholamines, this time with Slater as first named author. Their topic was, again, the effect of DCI, but this time they explicitly pointed out that the substance would be a useful tool for separating α- from β-receptors. Moreover, quoting the papers of Ahlquist and of Moran/Perkins, they now postulated that their data supported Ahlquist's concept of α- and β-receptors.[45]

In later publications, Moran kept up this image of himself as Ahlquist-discoverer and consistent supporter. In 1961, again with Majorie E. Perkins, Moran published his work on α-receptors in the *Journal of Pharmacology and Experimental Therapeutics*. This was a reaction to some reports that α-receptor blocking agents – similar to the β-blocking ones – would also block sympathetic inotropic cardiac action. Moran and Perkins saw their work as an extension of their experiments performed in 1958. They used the same experimental settings but now with the aim of investigating the

44 N.C. Moran and M.E. Perkins, 'Adrenergic Blockade of the Mammalian Heart by a Dichloro Analogue of Isoproterenol', *Journal of Pharmacology and Experimental Therapy*, 124 (1958): 223–237. All quotations are on pp. 234–235.

45 I.H. Slater and C.E. Powell, 'Some Aspects of Blockade of Inhibitory Adrenergic Receptors or Adrenoceptive Sites', *Pharmacological Revue*, 11 (1959): 462–463.

impact of some α-blocking agents (phenoxybenzamine, phentolamine, chlo-rpromazine, and dihydroergotamine) on positive inotropic cardiac response. They found no evidence of selective cardiac adrenergic properties.[46] The paper was a defence of Ahlquist's theory and the contribution they had made by helping to disseminate it within the scientific community.

Reasons for the Late Acceptance of Ahlquist's Concept

The scientific community of pharmacologists

Looking at the network of pharmacologists after Ahlquist had published his theory, one can argue that Ahlquist's theory was not accepted at once for a variety of different reasons, which can be summarized in three major points.

First, Ahlquist competed with other research groups, which were also devoted to the investigation of the sympathetic system – sometimes with the same sponsors. Therefore, it was difficult for any new concept to gain acceptance. Powell and Slater accepted Ahlquist's approach only after strong support had been shown towards it by Moran and Perkins. The latter supported Ahlquist because his theories were the basis for their own investigations. Furthermore, there still was competition between the well-known and dominant theory of Cannon and Rosenblueth, and the idea that adrenaline is the substance that increases blood pressure and cardiac activity, but not vasodilation. Therefore, Ahlquist's differentiated theory of the adrenergic system was hard to accept and understand.

Second, there were inherent problems with Ahlquist's theory itself. One of the points Ahlquist had made was later viewed as incorrect: it was

46 N.C. Moran and M.E. Perkins, 'An Evaluation of Adrenergic Blockade of the Mammalian Heart', *Journal of Pharmacology and Experimental Therapeutics*, 133 (1961): 192–201, esp. 192–194, 199–201.

not adrenaline, but rather noradrenaline, which was the main mediator of the sympathetic system. This was identified by Ulf von Euler (1905–1983) in 1946, i.e. at about the same time as Ahlquist's discovery.[47] Much more problematic were the technical limitations which existed in Ahlquist's time, which led to methodological problems, and to the inability to give direct evidence for the existence of receptors. Therefore Moran and Perkins, as well as Powell and Slater, were reluctant to admit Ahlquist's theory as the decisive event in the understanding of the sympathetic system. For example, in their 1958 paper, Moran and Perkins pointed out that the 'structural requirements for cardiac adrenergic blocking drugs are poorly defined as yet.' In this sense, Ahlquist's concept was little more than 'a useful classification of adrenergic receptors'. Later, in their 1961 paper, they called their proposal to classify adrenergic blocking drugs according to Ahlquist's terminology of adrenotropic receptors a mere 'suggestion'.[48]

Third, from 1958 until the early 1960s, support for Ahlquist came mainly from theoretical pharmacology. Moran and Perkins, as well as Powell and Slater, investigated DCI with the aim of fostering research on the sympathetic system as part of a basic research programme in pharmacology. They did not intend to promote therapy in a direct way, nor, as Slater commented later in hindsight: 'We had few ideas for clinical utility and were not in a position to do detailed cardiovascular studies'.[49] Even Robin Shanks (*1934), who worked at the department of pharmacology at Queen's University Belfast and who, together with A.D.M. Greenfield (*1917) and W.E. Glover, was the first to demonstrate the effect of a β-blocking drug (in this case DCI) in man, did not think about any therapeutic application. This meant that Ahlquist's theory had only a limited impact on medicine, and it was only thanks to an initiative from clinical practice that this situ-

47 Shanks, 'Discovery', p. 39.

48 Moran and Perkins, 'Adrenergic Blockade of the Mammalian Heart', pp. 234/235; Moran and Perkins, 'An Evaluation of Adrenergic Blockade', p. 200.

49 Personal communication of Shanks with Slater in 1982, quoted from Shanks, 'Discovery', p. 41.

ation changed.[50] Finally, one has also to consider that Ahlquist's theory received no support from physiologists, that there was only a small group of pharmacologists working on receptors, and they were investigating the relationship between dose and response on the basis of A.J. Clarke's work.[51]

Ahlquist's character and his own attitude towards the 'dual adrenoceptor concept'

As well as the scientific rivalries that shaped the debates on the dual adrenoceptor concept, Alhquist's personality and attitude influenced the fate of his own theory. Often, he was brutally frank in his remarks, especially if these were on scientific matters. His open criticism of Cannon and Rosenblueth's theory had already shown glimpses of this particular side of his nature. This tendency was soon well known among his colleagues in Augusta. Although he could be socially charming, Ahlquist said what he thought, without caring much about the result. Even at his funeral in 1983, mention was made of this aspect of this character. In contrast to Hamilton, who was polite, personable, and obliging, Ahlquist was brusque and difficult to handle.[52] He could be 'a very private person' and 'at times abrasive'.[53] His

50 Ibid., pp. 44–47. Carrier's statement, that the dual adrenoceptor concept was 'unanimously accepted' in the 1960s, is therefore misleading. See Carrier, 'Evolution', p. 219; *Speech of Raymond P. Ahlquist (draft) on occasion of the receipt of the Memorial Award in Therapeutics in 1973*, p. 4.

51 Interview with Sir James Black: 'Learning by Doing', *Molecular Interventions*, 4 (2004): 139–142, esp. 141.

52 Interview with Lois T. Ellison, September 9, 2003, Transcript part 1, p. 3; Transcript part 2, p. 3; Interview with Richard E. White PhD, former student of Ahlquist, now Associate Professor of the Department of Pharmacology and Toxicology at Medical College of Georgia, Augusta, on September 12, 2003, p. 8. See also Little, 'Ahlquist', p. 584.

53 Sutherland, Carrier, and Greenbaum, 'Obituary Raymond P. Ahlquist'.

brutal frankness was not confined to his students. He was also very open in his criticisms when talking to his colleagues.[54]

Ahlquist's style of communication seems to have affected the diffusion and reputation of his receptor theory. Of course, Ahlquist's work was acknowledged and he was awarded many prizes. In 1974, he received the Hunter Memorial Award of the American Society for Clinical Pharmacology and Therapeutics. In 1976, together with Sir James Black (*1924), he received the Albert Lasker Clinical Medical Research Award, and in the same year the Ciba Award for Hypertension. Ahlquist's merits were pointed out at several international symposia on β-blockers. In 1980, he was honoured with an 'International Symposium on Adrenergic Receptor Pharmacology'. This was the first annual meeting of the 'South Eastern Pharmacology Society' (SEPS), which had been founded by Lowell Greenbaum.[55] But Ahlquist failed to receive such important honours as the Abel Award, one of the most important prizes in American pharmacology, and he also failed to get the most important award in science in general, the Nobel Prize. Everyone in Georgia expected him to receive it, but Ahlquist 'was a complex man who had love/hate relationships with some which no doubt at the international

54 C. Deriso, 'Ahlquist laid foundation for Nobel', *Beeper* (The Medical College of Georgia), 10 (1988): No. 5 (Nov. 2, 1988), 1; Interview with Jerry Buccafusco, since 1979 Assistant of Lowell Greenbaum (Ahlquist's successor on the chair of Pharmacology), now director of the Alzheimer's Research Center, Professor of Pharmacology and Toxicology, Professor of Psychiatry and Health Behavior, Medical College of Georgia, Augusta, September 11, 2003.

55 'MCG Professor to receive Hunter memorial citation', *The Augusta Chronicle*, (Dec. 27, 1973), in: Special Collections, Greenblatt Library, MCG, Augusta; 'Dr. Raymond Ahlquist, noted researcher, dies', *Beeper* (The Medical College of Georgia), 4 (1983): No. 16 (April 20, 1983), 1; 'Symposium given for Dr. Ahlquist', *Beeper* (The Medical College of Georgia), 1 (1980): No. 43 (Nov. 12, 1980), in: Special Col. Ahlquist, Greenblatt Library, MCG, Augusta; Sutherland, 'History. Department of Pharmacology and Toxicology', p. 5.

level cost him the Nobel Prize.'[56] Ahlquist himself believed that one of the Nobel officials was 'a personal enemy' of his.[57]

Diplomacy was not one of Ahlquist's strengths, and there are enough hints that this also had an impact on his career in Augusta. During his time as Dean of the Medical School, he made sometimes rigid decisions about the funding and research activity of the School, which earned him not only friends, but enemies as well. In 1977, he was sued for not granting tenure to an individual researcher in his department. He even lost the chair of the department, and was banished to an isolated building of the MCG campus without any laboratory facilities. Becoming 'Charbonnier Professor' was therefore at once an honour and a punishment for Ahlquist. Five years later, in 1982, his successor Lowell Greenbaum, who in 1979 came from Columbia University to Augusta, offered him an office and a laboratory in the department and therefore a chance to return. Ahlquist happily agreed, but unfortunately, he died shortly before moving to the new facilities.[58]

Another feature of Ahlquist's personality was his scepticism, and also a certain pessimism.[59] He did not restrict himself to grumbling about those honours that had been withheld from him. He was very critical of his own work and publications as well. In his view, he was 'too ambitious' and expected 'perfection' from his collaborators, but he was also self-

56 Information from Professor Lowell Greenbaum, successor of Ahlquist as head of the Department of Pharmacology and Toxicology at Medical School of Georgia, Augusta, in 1979, given on January 26, 2003; Personal comments of Lowell Greenbaum, September 7, 2003.

57 Quotation taken from: Judy Taylor, 'Nobel odds called good by Ahlquist', *Augusta Chronicle*, 1976, press cutting, material of Lowell Greenbaum, Augusta.

58 Interview with Armand M. Karow, September 9, 2003, p. 6; Information from Lowell Greenbaum, January 26, 2003; curriculum vitae of Raymond P. Ahlquist, typewritten manuscript, pp. 7, in: material of Lowell Greenbaum, Augusta; personal communication with Lowell Greenbaum, September 7 and 9, 2003, Augusta; Interview with Jerry Buccafusco, Augusta, September 11, 2003; 'Recognition justly due', *Augusta Chronicle*, editorial (March 23, 1977). Acting Chairman of the Department of Pharmacology between 1977 and 1979 was J. Malcolm Kling. See also Sutherland, 'History. Department of Pharmacology and Toxicology', pp. 5–6.

59 Personal communication with Gloria Greenbaum, Augusta, Sept. 9, 2003; Interview with Lois T. Ellison, September 9, 2003, part I, p. 3.

demanding.[60] Ahlquist was extremely scrupulous and cautious in promoting his dual-adrenoceptor concept, above all in the early stages of his research. Even in his famous paper of 1948, he described his theory as simply a 'useful' tool for further investigations.[61] His reluctance to boast about a new theory was a part of his research philosophy, and not only a reaction to early criticism of his concept.[62] After 1948, Ahlquist did not hesitate to point out that his dual-adrenoceptor concept was 'an abstract concept', and was 'totally unnecessary from a physiological point of view'. In the strictest sense, α- and β-receptors did not exist, and naming them 'alpha' and 'beta' was a formal decision. Ahlquist viewed the rapid development of receptor subtypes with suspicion. He debated the question with A.M. Lands from the department of pharmacology at the Sterling-Winthrop Research Institute in Rensselaer, New York, who, in 1967, together with some colleagues, had distinguished between β1- and β2-receptors. Ahlquist believed that 'if there are too many receptors something is obviously wrong'. In the 1970s, he even postulated that there was only one adrenergic receptor. This receptor acted in response to epinephrine, and its α- and β-ness was determined by the biological environment. An example of this for Ahlquist was the frog heart, whose receptor-types depended on the temperature: a warm heart would have β-, and a cold heart, α-receptors.[63] Ahlquist told

60 'Profile (R.P. Ahlquist)', *Intercom*, 5 (May 9, 1975), in: Special Collections, Greenblatt Library, MCG, Augusta.

61 Ahlquist, 'A Study of the Adrenotropic Receptors', p. 598.

62 See also R.P. Ahlquist, 'The Adrenotropic Receptor-Detector', *Arch. Int. Pharmacodyn.*, 139 (1962): 38–41, esp. 40.

63 R.P. Ahlquist, 'Adrenergic Receptors: a personal and practical view', *Perspectives in Biology and Medicine* (1973), 119–121, esp. 121 (here also the first quotation); *Speech of Ahlquist (draft) on occasion of the receipt of the Memorial Award in Therapeutics in 1973*, pp. 3, 5 (second quotation on p. 5); *Ahlquist, Draft of a speech, no date*, pp. 6, 9–11 (third quotation on p. 10); 'Lasker Laurels for Medical Researchers', *Medical World News* (Nov. 29, 1976), 48–50, esp. 48, in: Folder on Raymond P. Ahlquist, Department of Pharmacology and Toxicology, Medical College Augusta; Interview with Jerry Buccafusco, September 11, 2003; Shanks, 'Discovery', p. 65; Robinson, *Mechanisms of synaptic transmission*, p. 154/155.

story as if he had developed his concept by chance, without any foreseeable outcome.[64]

This very sceptical or critical attitude towards his own concept also manifested itself in his teaching of pharmacology, which became more important for him than research in the late 1960s.[65] Although he was proud of his theory, he did not present it to the students as being a decisive step in the research on the sympathetic system. On the contrary, he saw the concept predominantly as a 'teaching aid', and assumed that the concept had had 'for a number of years, ... primarily an educational value, simplifying the teaching of sympathetic nervous system function to medical and graduate students.' At the beginning of the 1980s, Ahlquist confessed that he had never believed in different alpha- and beta-receptors, but that he had found it the easiest way to help medical students to conceptualize the sympathetic system. Often, when students pointed out the role of the dual-adrenoceptor concept in certain pharmacological processes, Ahlquist corrected them with his famous sentence 'It depends.'[66]

64 Interview with Richard E. White PhD, Augusta, September 12, 2003, p. 5; *Speech of Ahlquist (draft) on occasion of the receipt of the Memorial Award in Therapeutics in 1973*, p. 6. Even in 1967, Ahlquist pointed out that 'Although the alpha and beta receptors have achieved international usage it should be stressed that they have only interim value until the exact nature of the responsive mechanism for adrenergic agonists is discovered'. R,P. Ahlquist, 'Development of the Concept of *Alpha* and *Beta* Adrenotropic Receptors', *Annals of the New York Academy of Sciences*, 139 (1967): 549–552, esp. 552. See also: R.P. Ahlquist, 'Agents which block adrenergic β-Receptors', *Ann. Rev. Pharmacol.*, 8 (1968): 259–272.

65 Interview with Richard E. White PhD, Augusta, September 12, 2003, pp. 1–3; Interview with Jerry Buccafusco, September 11, 2003; Interview with Armand M. Karow, PhD, September 9, 2003, Augusta, p. 4, 6. Ahlquist's teaching style was influenced by William F. Hamilton, who – like Ahlquist – preferred to teach in small student groups. See Baker et al., 'William F. Hamilton', p. 64; Dow, 'William F. Hamilton', p. 95.

66 Ahlquist, 'Historical Perspective. Classification of Adrenoceptors', pp. 104, 105 (first quotation on p. 105); *Interview with Armand M. Karow*, PhD, September 9, 2003, Augusta, p. 3; N.K. Wenger and L.M. Greenbaum, 'From Adrenoceptor Mechanisms to Clinical Therapeutics: Raymond Ahlquist, PhD, 1914–1983', *Journal of the American College of Cardiology*, 3 (1984): 419–421; *Speech of Ahlquist (draft) on occasion of the receipt of the Memorial Award in Therapeutics in 1973*, p. 6 (second quotation); Interview with Richard

Finally, it can be said that Ahlquist's difficult character was the principal factor hindering the dissemination of his concept. In addition, he was a private man and a loner, who believed in his work, but – in the words of a former student – did not 'promote himself enough'.[67] If Ahlquist succeeded in the end, it was partly thanks to William F. Hamilton, whose outgoing personality mobilized contacts among scientists. Therefore, to a certain extent, Ahlquist's work was a social construction, mainly created by the Augusta research environment.[68]

The Breakthrough: Sir James Black and the practical application of Ahlquist's concept

Although Ahlquist worked on the sympathetic system with the practical application of his results in mind, and although he was interested in clinical work and kept in contact with colleagues working in medical practice, for a long time no one saw how to apply his approach to medical practice.[69] Again inspired by Hamilton, Ahlquist participated in a postdoctoral training course for cardiovascular specialists, which between 1954 and 1963 was attended by 32 trainees from all over the world. Although there was a collaboration with the American Heart Association, dissemination of

---

E. White PhD, Augusta, September 12, 2003, p. 2, 7, 10; personal communication with Lowell Greenbaum, September 7, 2003, Augusta.

67 Interview with Richard E. White PhD, Augusta, September 12, 2003, p. 7 (quotation); Interview with Lois T. Ellison, September 9, 2003, Transcript part 2, p. 3; Interview with Jerry Buccafusco, September 11, 2003. Concerning teaching, see also the comments of his two collaborators Gerald O. Carrier and Marshal Shlafer, in 'Raymond P. Ahlquist', pp. 42–44.

68 Baker et al., 'William F. Hamilton', p. 64.

69 Interview with Armand M. Karow, PhD, September 9, 2003, Augusta, pp. 4–5; Wenger and Greenbaum, 'From Adrenoceptor Mechanisms to Clinical Therapeutics', p. 419; A.C. Witham, 'Ahlquist, Basic Research, and the Practice of Medicine. Editorial', *Journal of the Medical Association of Georgia*, 66 (1977): 637–638.

the dual-adrenoceptor concept remained weak.[70] The 'breakthrough' for Ahlquist's theory and the receptor concept in general came with the work of Sir James Black.

Black was one of the first researchers to consider the knowledge about α- and β-receptors and the corresponding blocking agents from the standpoint of medical practice, and he was the first to publish results on the matter. His story has been told and published many times, so that in the following pages we can focus on the core events.[71] Importantly, Black had received his medical training from St. Andrews University in Scotland. Between 1950 and 1957, he worked on cardiovascular diseases at the Glasgow Veterinary School, then after 1957 at Imperial Chemical Industries (ICI) near Manchester. Cardiovascular diseases became a very trendy topic in medical research in the 1960s.[72] Black's intention was to find an agent to treat angina pectoris, a very serious symptom of coronary heart disease caused by an insufficient blood supply to the heart muscle. There were two possible approaches to combat this symptom: to increase the blood supply to the heart, or to reduce the cardiac oxygen consumption. Black considered the latter, and developed the hypothesis that it might be possible to block the action of adrenaline on the heart. He performed animal experiments, measuring the heart rate as indicator of a decrease of oxygen consumption, and did work on the isolated hearts of guinea pigs and cats. In 1960, thanks to a modification to the chemical structure of DCI achieved by one of Black's collaborators, John Stephenson, compound 38,174, called 'pronethalol', was developed by ICI. Black and Stephenson described their results in *The Lancet* in 1962.[73] The announcement led to clinical trials, but it was found that the substance proved to be carcinogenic in mice. Although

70 See Baker et al., 'William F. Hamilton', p. 65.

71 See above all the article by Viviane Quirke, 'Putting Theory into Practice: James Black, Receptor Theory and the Development of the Beta-Blockers at ICI, 1958–1978', *Medical History*, 50 (2006): 69–92.

72 Ibid., pp. 78–82. On the rise of interest in cardiovascular research in the 1960s, see J. Goodman, 'Pharmaceutical Industry', in Cooter and Pickstone (eds), *Medicine in the 20th Century*, pp. 141–154, esp. p. 148/149.

73 J.W. Black and J.S. Stephenson, 'Pharmacology of a new adrenergic beta-receptor-blocking Compound (Nethalide)', *The Lancet*, II (1962): 311–314.

there were problems, it was the first 'β-blocker' to be successfully applied to patients. Because of the serious side effects associated with the substance, it was withdrawn from the market in the UK in 1965. Meanwhile, Black and his team continued with their investigation, developing improved agents, including ICI 45,520 (propranolol), which was synthesized in 1962. In the same year, Robin Shanks, Ahlquist's former research assistant, joined Black's team and tested the compound on animals. Propranolol trials in man started in early 1964. In the summer of the very same year, further clinical trials were carried out, indicating that the drug was safe and effective for the treatment of angina pectoris and certain kinds of cardiac arrhythmias.[74] Propranolol, the first clinically useful 'β-blocker', was licensed in 1965 in the United Kingdom. This was made possible, among other factors, by the good collaboration that existed between basic researchers and clinicians.[75] In years to come, the β-blockers were improved. For example, 'cardioselective' ones were developed, inhibiting only the excitatory beta-receptors, or, using Lands' terminology, the 'β1-receptors', and not the inhibitory ones, or 'β2-receptors', affecting the bronchial muscle or peripheral blood vessels.[76] One important early development was the increase of the therapeutic usage of the β-blockers: Black had suggested use of the substance, above all, in cases of angina pectoris and acute myocardial infarction. In 1964, Brian Prichard from University College Hospital London found that pronethalol had an antihypertensive effect, and he described the therapeutic use of the substance against hypertension. Prichard cited Ahlquist, as well as Powell/ Slater, Moran/Perkins, and Black/Stephenson. The substance was tested on 15 patients, among them 11 with hypertension. All showed a significant fall in blood pressure, and Prichard suggested the search for a non-carcinogenic β-blocker that could be used in therapy. This aim was reached, as together

74 Shanks, 'Discovery', pp. 48–58; J.W. Black, A.F. Crowther, R.G. Shanks, L.H. Smith, and A.C. Dornhorst, 'A new adrenergic Beta-Receptor Antagonist', *The Lancet*, I (1964): 1080–1081.

75 Shanks, 'Discovery', pp. 58. On Black's research and research attitudes, see also J.W. Black, 'Drugs, Hormones and Receptors: a personal reflection', *Hunterian Society Transactions*, 47 (1988–1989): 60–69.

76 Shanks, 'Discovery', pp. 63–65.

with his collaborator Peter M.S. Gillam, Prichard also proved the hypotensive effect of propranolol in the very same year. The substance was successfully applied to 24 patients, 16 of whom suffered from hypertension.[77] Since the early 1970s, together with diuretics, β-blockers have been widely used to decrease arterial blood pressure.[78]

It is important to consider that the success of the β-blockers was also a success for receptor theory. Further research on beta-blocking agents promoted the receptor theory itself, because new hypotheses on the mechanisms of drug binding emerged.[79] And Ahlquist's fame as the most recent major theorizer on receptors was boosted: Black did not hesitate to point out that it was Ahlquist who gave him decisive ideas for his research on β-blockers. He had found a chapter by Ahlquist on adrenergic pharmacology in Drill's Pharmacology Textbook. This chapter included Ahlquist's α- and β-receptor theory. Black used this paper for teaching medical students, but it also inspired his subsequent research.[80]

That these connections between Ahlquist's and Black's work were widely recognized was reflected in those prizes which both researches won together, namely the CIBA Award for Hypertensive Research and

77 Ibid., pp. 65–66; B. Prichard, 'Hypotensive Action of Pronethalol', *British Medical Journal*, 1 (1964): 1227–1228; B. Prichard and P.M.S. Gillam, 'The Use of Propranolol in the Treatment of Hypertension', *British Medical Journal*, 2 (1964): 725–727.

78 Shanks, 'Discovery', pp. 65–66; B. Prichard, 'β-Adrenoceptor-Blocking Agents in the Management of Hypertension', *Cardiology*, 64 (1979), Suppl. 1, 44–87; idem, 'Beta Adrenergic Blocking Drugs and the Treatment of High Blood Pressure', in K.H. Rahn and A. Schrey (eds), *Betablocker. 1. Betadrenol-Symposion Frankfurt 1977* (München, Wien, Baltimore: Urban & Schwarzenberg, 1978), pp. 126–146.

79 Quirke, 'Putting Theory into Practice', esp. p. 90.

80 M.P. Stapleton, 'Sir James Black and Propranolol. The role of the basic sciences in the history of cardiovascular pharmacology', *Texas Heart Institute Journal*, 24 (1997): 336–342; Personal comment of Sir James Black, London, 11.6.2004; Interview with Sir James Black: 'Learning by Doing', p. 141; R.P. Ahlquist, 'Adrenergic Drugs', in V.A. Drill (ed.), *Pharmacology in Medicine: a collaborative textbook* (New York: MacGraw-Hill, 2nd Edition, 1958), pp. 378–407. For the β-blocker story see also Sir James Black, 'Drugs from Emasculated Hormones: The Principles of Synoptic Antagomism', Nobel Lecture, December 8, 1988, in T. Frängsmyr and J. Lindsten (eds), *Nobel Lectures, Physiology or Medicine 1981–1990* (Singapore: World Scientific Publishing Co., 1993), pp. 418–440, esp. pp. 418–425.

the Albert Lasker Award.[81] Prichard noted that, in 1965, Ahlquist was the chairman of the first symposium ever to be held on beta-blocking agents. In 1976, when β-blockers had become fashionable in the treatment of cardiovascular diseases, the German magazine *Stern* drew links between Ahlquist's and Black's work. And in 1988, when Sir James Black received the Nobel Prize, it was speculated that, had Ahlquist still been alive, he might have joined Black. Ahlquist's colleague and old supporter Neil Moran expressed similar views, when he mentioned in the very same year: 'There is no question Ahlquist's concept had great impact ... I'm sure Ahlquist would have shared in the Nobel Prize if he were alive because they [Ahlquist and Black] shared the Lasker Award.' Moran pointed out that many winners of the Lasker Award had subsequently won the Nobel Prize.[82]

Although, as Viviane Quirke has shown in her recent paper, in the context of the pharmaceutical industry these developments helped to transform receptors 'from hypothetical entities into tangible objects' which could be used for drug design,[83] for the pharmacological community at large the receptor concept remained a construction difficult to apprehend in all its aspects.[84] Thus, in 1964, the pharmacologist D.K. de Jong wrote:

81 Deriso, 'Ahlquist laid foundation for Nobel', p. 1.

82 B. Prichard, 'Obituary Note on Raymond P. Ahlquist', in material Lowell Greenbaum; *Stern Magazine*, 1976, H.4, 15. January 1976, in: Folder on Raymond P. Ahlquist, Department of Pharmacology and Toxicology, Medical College Augusta.; S.P. Craig, 'If alive today, MCG Professor might have won Nobel', *Augusta Chronicle* (Oct. 22, 1988), press cutting, in: material Lowell Greenbaum. It seems that Ahlquist had some problems with the fame Black attracted. In 1980, Ahlquist mentioned that 'He [Robin Shanks] carried the idea of alpha and beta to ICI and was involved in the early studies of beta blockers in man'. However, the facts speak in favour of Black's version, who referred to Ahlquist's 1948 paper in his 1962 *Lancet* paper (co-authored with Stephenson). Shanks did not join Black's team at ICI until late 1962. See Letter Ahlquist to Greenbaum, May 26, 1980, in: Folder on Raymond P. Ahlquist, Department of Pharmacology and Toxicology, Medical College Augusta; Black and Stephenson, 'Pharmacology of a new adrenergic beta-receptor-blocking Compound', p. 311.

83 Quirke, 'Putting Theory into Practice', p. 89.

84 Maehle, Prüll, and Halliwell, 'The Emergence of the Drug Receptor Theory', 637–641; Prüll, Maehle, and Halliwell, 'Drugs and Cells – Pioneering the Concept of Receptors', 18–30.

> To most of the modern pharmacologists the receptor is like a beautiful but remote lady. He has written her many a letter and quite often she has answered the letters. From these answers the pharmacologist has built himself an image of this fair lady. He cannot, however, truly claim ever to have seen her, although one day he may so.[85]

In 1980, the pharmacologists William C. Bowman and Michael J. Rand made critical remarks on the receptor concept:

> Although some progress has been made in the isolation of receptors & their characterization in chemical & physicochemical terms, for most part receptors must be regarded as hypothetical entities, even though the receptor concept lies at the heart of pharmacology.[86]

Black's work did not reveal all the secrets of receptors. Nevertheless, his practical application of Ahlquist's theoretical approach made the receptor idea popular in science. In the 1960s and 1970s,[87] improved knowledge and techniques in theoretical pharmacology were not sufficient to enable a breakthrough in the receptor concept. Only its practical application and evidence of its effectiveness in medical therapy created what scientific inventions or discoveries always require: trust in their existence.

It was not easy to create such a trustful atmosphere. On the one hand, Black's work was supported by the increasingly strong interests of a community of specialists in internal medicine, especially cardiologists, and his ideas fitted well with ICI's research programme in cardiovascular diseases.[88] On the other hand, there was still a lot of resistance against the receptor concept at the time of Black's achievements in cardiovascular research. In 1948, when Ahlquist wrote his famous paper, the main thrust of research in clinical pharmacology was dominated by transmitter research and this

85 D.K. de Jong, 'Introductory Remarks', in Everhardus J. Ariens (ed.), *Molecular Pharmacology: The mode of action of biologically active compounds* (New York: Academic Press 1964).

86 W.C. Bowman and M.J. Rand, *Textbook of Pharmacology* (Oxford: Blackwell Scientific Publication 1980), p. 39.15. I am grateful to Robert Francis Halliwell for giving me hints on de Jong and Bowman.

87 See a detailed description in Robinson, *Mechanisms of synaptic transmission*, pp. 158/159, 161.

88 Quirke, 'Putting Theory into Practice', pp. 73, 84, 92.

continued to be so until the late 1960s. There was only a discrete shift from transmitters to receptors between these dates, as evidenced by the discussions of the British Pharmacological Society on the subject. However, a change of atmosphere did occur after the creation of the section on Clinical Pharmacology in 1969.[89]

Conclusion

The story of Raymond P. Ahlquist and his dual-adrenoceptor concept is an important element in the history of the receptor concept. For the first time it was possible to give plausible evidence for the existence of different receptors. Ahlquist's concept was a breakthrough in modern receptor research, as it enabled investigations on many receptor subtypes. Consequently, it also was a boost to modern biomedicine, which profited from the knowledge generated.

But the story of Ahlquist is not only important in relation to receptors, or to the development of biomedicine. It also enables us to discuss the different factors that influenced, and still influence, the development of drugs and their commercialization. Firstly, Ahlquist's dual-adrenoceptor theory was a child of its time: in the post-1945 era, there was an optimistic mood in science, with a clear vision to undertake investigations on diverse subjects with the help of tools developed in wartime. The supporters of the receptor concept benefited from this situation. In this way, Ahlquist's research was encouraged by William F. Hamilton's group in Augusta. Hamilton supported Ahlquist with instruments, ideas, teaching devices and – last but not least – with scientific contacts. Therefore, Ahlquist's dual-adrenoceptor

89 C.S. Breathnach, 'A Century of Humours and Rumours', *Journal of the Irish College of Physicians and Surgeons*, 7 (1978), No.3, offprint, in: Special Collections, Greenblatt Library, MCG, Augusta; *British Pharmacological Society*, SA / BPS / A3, A4, General Minutes, in: Contemporary Medical Archive Centre (CAMC). Library for the History and Understanding of Medicine, Wellcome Trust, London.

concept was not only the product of its creator's genius. It was also a construction, whose development and success was deeply rooted in the local conditions and the working atmosphere at the Medical College Georgia in Augusta.

Secondly, there were the specific characteristics of Ahlquist's theory itself. Not every explanation was satisfying. It had some minor flaws, it was quite vague, and also very theoretical. Although Ahlquist – in the end – undertook many painstaking experiments to develop a new and influential receptor theory, he had not originally planned to perform research work on receptors. And the theory he finally presented relied on receptors he could only indirectly give evidence for. Similarly to Ehrlich and Langley before him, Ahlquist provided no drawings and no photographs.

Thirdly, there were problems within the scientific community of pharmacologists. Ahlquist had difficulties in publishing his results, and there were difficulties in the acceptance of his theory. This resulted partly from the second point discussed above. Ahlquist's criticism of other very fashionable physiological and pharmacological theories of his time also weighed against his success. No wonder that the work of Ahlquist was not at once acknowledged by pharmacologists! Also, Ahlquist was in competition with other research groups working in the field of vasoactive substances. This competition was fuelled to a certain extent by the pharmaceutical industry. And finally, the competitive research strand on transmitters, headed by its epigone Dale, remained until the late 1960s.[90]

90 In his paper of 1948, Ahlquist gave Dale credit for his contribution to receptor research in 1906, quoting H.H. Dale, 'On some physiological Actions of Ergot', *Journal of Physiology*, 34 (1906): 163–206 (Ahlquist, 'A Study of the Adrenotropic Receptors', p. 586). But Dale's paper only contains a minor footnote on 'receptors', explaining that Elliott's term 'myoneural junction' should be used instead. This offers no support for the receptor idea, since the myoneural junction stands for the transmitter-receptor complex, thus leaving the question of drug-cell binding open-ended. Surprisingly Ahlquist did not notice Dale's hostile remarks on the receptor concept and its supporters in the 1920s and 1930s. Sir James Black has pointed out that Dale had neglected the results of receptor research, and that this had hampered the consideration of Ahlquist's work, but it was far more than only neglect. See Black, 'Drugs, Hormones and Receptors', pp. 60–69.

Fourthly, there was friction in the relationship between basic and clinical research. Ahlquist was very much associated with laboratory work, and it was hard for him and his colleagues, to some extent, to abandon this path and cross over into clinical work. This is remarkable in so far as Ahlquist himself was fundamentally interested in promoting clinical therapeutics. But what he delivered was a theory that could not be transferred at once and easily into medical practice. A new approach, which would shed new light on Ahlquist's results, was needed from outside the research labs in Augusta and Atlanta. The man who did this was Sir James Black. His work opened the door to the productive exploitation of the dual-adrenoceptor concept by the pharmaceutical industry.

Fifthly, there was the influence of Ahlquist's background and personality. His frank and direct comments on scientific matters, as well as his pessimistic and sceptical attitude, hampered his own work. Whereas Paul Ehrlich did not hesitate in painting his invisible 'pluralism' of receptors, Ahlquist was self-critical and, therefore, was not the type of man to promote himself successfully. He needed the help of others, such as Neil Moran, and it was only the practical application of his dual adrenal receptor concept by Sir James Black that made the theory public and acceptable to the scientific community of pharmacologists.[91]

Ahlquist's research on receptors illustrates the fact that scientific research, but also drug development, are not solely the outcome of the painstaking acquisition of knowledge and data from nature. On the contrary, the question of which research problem should be followed up to produce drugs, which substance should be investigated and analysed, and whose research laboratory should be supported, depends on different factors. The interplay between these factors sometimes propels specific ideas or research fields, and sometimes hinders them, or – as in the case of Ahlquist – it can cause definite delay. To understand drug development and the roles of the different actors involved in their development, one has to consider the 'zeitgeist', the social standing of the respective researchers, and the basic rules of scientific work. Very often, it is not only the quality of the research, but rather political and strategic matters that decide the fate of a project and its creator.

91 Quirke, 'Putting theory into practice', p. 92.

CLAUDE RICHARD AND MARIE-THERESE LUSSIER

6 Proposition of a Model for the Discussion of Medications during Medical Encounters

'Just as cats chase mice but don't necessarily eat them, humans chase medicines but don't necessarily take them'[1]

Introduction

Taking medication or products to help cure illness or relieve disease-related suffering is as old as time. The appearance of shamans, witch doctors and healers of all kinds probably coincided with the advent of man's social awareness, as did the social issues surrounding these practices – influence, power and privilege. Today in America and indeed in most industrialized countries, one professional group holds a virtual monopoly on the prescription of medications. While such marginal practices as alternative medicine, osteopathy, naturopathy and others continue to be popular, in these modern times where the state plays an increasingly significant role in health care, health professionals – physicians, in particular – have emerged as the sole prescribers of medications. Since health care has become a public, rather than a private, matter, it is becoming increasingly less acceptable to have products or practices which have not been subjected to scientific validation to ensure their efficacy. The state has a duty to manage public assets in a prudent manner and must justify how it spends the taxpayer's money. This is gradually leading to the elimination of what cannot scientifically be proven effective.

1 D. Misselbrook, *Thinking about Patients* (Newbury: Petroc Press, 2001).

The problem now lies not in knowing whether a product is effective, but in ensuring that it is used appropriately. For both the state and the public services, this issue has become of prime importance as, now that effective medications are available, it must be recognized that often, the patient does not take them as directed. The adherence rate ranges from 30 per cent to 70 per cent, resulting in substantial costs to society in terms of, for example, poorly treated illnesses, hospitalizations due to medication misuse, and lost work time. This situation is even more unacceptable because these costs are avoidable. Research conducted during the past 25 years has demonstrated the inadequacy of discussions of medications during medical consultations.[2] Moreover, there is a discrepancy between what physicians think they are telling the patient and what they are actually saying.[3] As a result, physicians may have some difficulty closing the information gap. There is also a discrepancy – unperceived by physicians

2 B.M. Svarstad, 'Physician-patient communication and patient conformity with medical advice', in D. Mechanic (ed.), *The Growth of Bureaucratic Medicine: an inquiry into the dynamics of patient behaviour and the organization of medical care* (New York: John Wiley, 1976); J.B. Wiederholt, B.R. Clarridge, and B.M. Svarstad, 'Verbal consultation regarding prescription drugs: findings from a state-wide study,' *Medical History*, 30 (1992): 159–173; G. Makoul, P. Arnotson, and T. Schofield, 'Health promotion in primary care: physician-patient communication and decision making about prescription medications', *Social Science and Medicine*, 41 (1995): 1241–1254; L.A. Morriss, E.R. Tabak, and K. Gondek, 'Counselling patients about prescribed medication: 12-year trends', *Medical Care*, 35 (1997): 996–1007; R. Parrott, 'Exploring family practitioners' and patients' information exchange about prescribed medications: implications for practitioners' interviewing and patients' understanding', *Health Communication*, 6 (1994): 267–280; L. Scherwitz, D. Hennrikus, S. Yusim, J. Lester, and C. Vallbona, 'Physician communication to patients regarding medications', *Patient Education and Counseling*, 7 (1985): 121–136; B. Sleath, D. Roter, B. Chewning, and B.M. Svarstad, 'Asking questions about medication: analysis of physician-patient interactions and physician persceptions', *Medical Care*, 37 (1999): 1169–1173; F.A. Stevenson, C.A. Barry, N. Britten, N. Barber, and C.P. Bradley, 'Doctor–patient communication about drugs: the evidence for shared decision making', *Social Science and Medicine*, 50 (2000): 829–840.

3 H. Waitzkin, 'Doctor–patient communication, clinical implications of social scientific research', *Journal of the American Medical Establishment*, 252 (1984): 2441–1446; H. Waitzkin, 'Information giving in medical care', *Journal of Health and Social Behavior*, 26 (1985): 81–100; Makoul et al., 'Health promotion in primary care'.

– between the nature of the information they provide regarding medications and the desire for information expressed by the patient.[4] Currently, we doubt whether patients can provide informed consent regarding the taking of medication with the information routinely supplied by health care professionals.

Thus, at present, the development of the ability to influence the behaviour of individuals to take their medications as directed (adherence) would constitute a greater benefit to society than the discovery of new molecules. Basically, we are a society in transition, where an authoritarian model of health care is gradually giving way to a more participatory model.[5] With the availability of information at minimal cost, an ever-increasing number of individuals can form their own opinions and obtain information which was hitherto the exclusive domain of the health care professional. While the modern patient does not wish to replace his physician, he does want to be more involved in managing his own health. Patients are already making up their own minds in any case; patients treat 70 per cent of their health care problems themselves, with no input from a physician, and when they do consult one, an average of 50 per cent decide to ignore their physician's prescriptions.[6]

Outside the hospital setting, treatment cannot be 'forced' on a patient. The patient must thus be convinced of the validity of the treatment and

4 D.K. Ziegler, M.C. Mosier, M. Beuenaver, and K. Okuyemi, 'How much information about adverse effects of medications do patients want from physicians?', *Archives of Internal Medicine.* 161 (2001): 706–713; F.A. Stevenson, C.A. Barry, N. Britten, N. Barber, and C.P. Bradley, 'Doctor–patient communication about drugs: the evidence for shared decision making', *Social Science and Medicine*, 50 (2000): 829–840.

5 F. Lang, 'The evolving roles of patient and physician', *Archives of Family Medicine*, 9 (2000): 65–67; D. Roter, 'The medical visit context of treatment decision making and the therapeutic relationship', *Health Expectations*, 3 (2000): 17–25; G. Elwyn, A. Edwards, and P. Kinnersley, 'Shared decision-making in primary care: the neglected second half of the consultation', *British Journal of General Practice*, 49 (1999): 477–482.

6 K.L. White, T.F. Williams, and B.G. Greenberg, 'The ecology of medical care', *New England Journal of Medicine*, 265 (1961): 885–892; L.A. Green, G.E. Fryer, B.P. Yawn, D. Lanier and S.M. Dovey, 'The ecology of medical care revisited', *New England Journal of Medicine*, 344 (2001): 2020–2025.

agree to adhere to it. Given the current lack of patient enthusiasm regarding suggested treatments, the time has now come to rethink the current approach to the prescription of medications. At present, when a physician writes a prescription, he naturally expects the patient to heed his advice and comply with the treatment. This authoritarian approach appears reasonable, given that the patient himself requested the physician's opinion. The physician thus naturally expects the patient to follow his advice. Observation, however, shows that this is simply not the case which means that, as we cannot expect to easily change the patient, we must develop an alternative approach with more room for patient collaboration.

The present chapter is devoted to the presentation of a model and a way of viewing patient-provider communication regarding medications. This model, inspired by the recent work of Lussier, Richard and Bajcar is therefore intended as a step towards a communication strategy to enable patients to truly provide informed consent regarding their medication taking and to assist them in this often complex task.[7]

The Model

The dialogue about taking medications starts long before any medication is taken. From a very early age, parents teach children how to handle medications. Often, medications are described as 'poisonous' substances to be avoided, substances which are only to be used in the presence of a responsible adult. Later, this representation of danger gradually gives way to a more qualified description of medications as most children learn to associate medications with getting better. When children and young

7 M.-T. Lussier, C. Richard and J.M. Bajcar, 'Strategies to optimize discussions and decision-making about medications during clinical encounters: teaching and changing practice', Poster Presentation, *International Conference on Communication in Healthcare* (Chicago, October 2005).

adults see and speak with people taking medications, they are told stories in which the medications play various roles and are no longer described in a one-sided manner. In short, by the time they must make their own decisions about taking medications, they have already acquired a degree of general knowledge in this regard. There is thus no specific time to start creating awareness regarding medications; it is an ongoing process, which evolves throughout the development of an individual's relationship with medications.

The medical consultation is but a single element of this larger, ongoing dialogue. During a medical visit, this dialogue takes the form of a discussion oriented towards a health objective. Such objectives are determined by the reason for consultation. There is little idle conversation during a medical consultation. Conversation is always oriented towards and focuses on the problems to be solved. One part of the consultation consists of what professionals call the therapeutic education of the patient, which accelerates the patient learning process. Here, education must be understood in the strict sense of the word – for example, how to take a newly-prescribed medication. However, specific as this information is, its presentation and the way in which it is presented take place within a social framework which complements it, gives it meaning and allows the action to extend beyond the specific act into the social sphere. Thus, the act is simultaneously the concrete manifestation of abstract concepts (such as health care) and what gives meaning to these concepts.

Naturally in the medical setting (especially when consultations are more frequent), the dialogue, as well as offering multiple opportunities for beneficial repetitions of information tailored to the condition and its treatment, can evolve and broaden in scope to cover health issues beyond the specific problem for which the patient is consulting. For example, in patients with a chronic condition such as diabetes, hypertension or asthma, it is possible for the professional to gradually develop a relationship with the patient and deal with issues that go beyond the patient's specific condition. In addition, during this process, the professional himself also changes through the repeated contact with the patient and comes to know and adapt to the patient. This knowledge enables the physician to suggest the treatments best suited to the patient.

The primary objectives of the model we are proposing are to help health care providers: 1) understand how patients approach the use of medications; 2) recognize the specific information needs patients have regarding their medications; 3) better understand the various challenges involved in communicating information regarding medications; 4) identify potential situations related to the taking of medications which can cause patient reactions and to more effectively anticipate them and finally, 5) develop communication strategies adapted to patient needs and create relationships of trust with them.

Three separate organizational levels – macroscopic, mesoscopic and microscopic – are identified in the proposed model. Each component of the model contributes to a specific dynamic which, over time, structures patient-provider exchanges regarding medication. The theatre metaphor (in which the patient and the physician are the actors) helps us understand the relationship between the three levels of the proposed model. The macroscopic level would refer to the actors' life experience, knowledge about the theatre (authors, plays, theatre houses) and their previous acting experience (none too extensive). The mesoscopic level would refer to the vocabulary we use to describe the world of theatre: the content/issues/themes the plays may address, the nature of the roles actors may personify as well as the general patterns of communication that can be identified between the actors. Finally, the microscopic level focuses on the specifics of one representation of a particular play. Although this particular play can also be analysed at the two previous levels, here we pay attention to the particularities of what happens during *this* representation: special circumstances and how the actors adapted to these without jeopardizing the unfolding of the play.

The Macroscopic Level

The macroscopic level describes the larger process of building knowledge about medications over time. It is useful to picture discussions about an individual's health as a kind of special conversation that is maintained with others throughout his/her lifetime. We suggest calling this special conversation a health 'dialogue' represented in the figure by the helicoïdal movement. This dialogue takes place intermittently, over time, with family members, friends and acquaintances, the school and social environment through information in the media and, more generally, is influenced by the culture to which the individual belongs. Thus this process of knowledge building occurs over a long period of time and is not limited to professional consultations (although such consultations are of particular interest to us here).

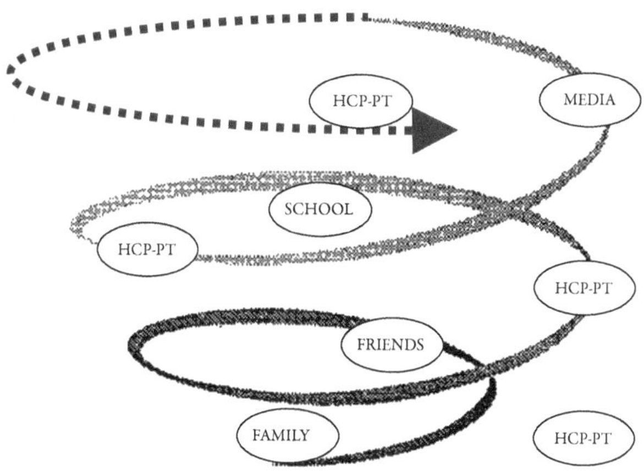

Figure 6.1: Lifelong health dialogue: learning process about medications

A series of interactions can be depicted as an ever-lengthening spiral. Throughout this series of interactions, communication strategies are used which gradually increase the patient's knowledge regarding his medications. This helicoïdal shape represents the accumulation of all of the patient's interactions regarding medications, interspersed with periodic meetings with health care professionals. Through the content of these interactions and the way in which the medication is presented, the patient will gradually acquire knowledge about medications in general and/or about a specific medication. In the event of a chronic illness, the patient may acquire more knowledge regarding the medications he must take than his physician. The patient's understanding of his medications changes over time to more closely approach or possibly surpass that of his physician. Thus, when patients interact with healthcare providers, they always come in with some prior knowledge and experience about medications enabling them to interpret the information provided. Because descriptions of medications are relatively standardized, patients can anticipate the provider's statements and sometimes even add to them as when they have had extensive experience with them as in chronic diseases.

The patient's knowledge of medications develops through an iterative process occurring over time and by the repetition of statements about medications in various contexts and circumstances. Statements regarding medications are thus affected by the patient's:

- Socio-cultural environment
- Personal environment
- Personal history and all of the prior decisions they have made regarding medications and treatments for their conditions

Thus, to summarize what is occurring at the macroscopic level, we are suggesting that we acquire knowledge based on our own use of medications and to what we hear about medications. The increasing number of references to and use of medications over time contribute towards the accumulation of general knowledge about medications. Conventional medical wisdom views adherence problems as a demonstration of patient ignorance or resistance. However, our model suggests that the patient makes active decisions, in keeping with his priorities that often differ from those of healthcare providers, needs, opinions and beliefs. Over time, as his medication-related knowledge increases, the patient becomes more capable of making informed decisions on his own, rather than having to rely on outside expertise. Rather than taking everything on faith, given time, patients can become involved in what Bajcar calls planned, rational, strategic behaviours with respect to taking medications.[8]

The Mesoscopic Level

The mesoscopic level is an abstraction level nested between the macroscopic and microscopic levels. It focuses on the more normative dimensions related to medication discussions during encounters between patients and healthcare providers. The mesoscopic level builds on the conceptual framework developed by Bajcar, and on the empirical findings of Richard and Lussier, who analysed over 500 physician-patient encounters from the standpoint of both patient-physician interaction and the content of the exchanges.[9]

At this level, the model provides us with a number of concepts to describe *general patterns* of healthcare provider and patient input regarding medications that occur during consultations. Currently, this is the most highly developed part of the model. At this level, it is possible to determine the primary medication-related topics discussed during consultations and to describe the dynamic aspects of these medication-related discussions.

8 J.M. Bajcar, 'Development of a needs-driven theory-based model for medication-taking education with plans for implementation and evaluation' (PhD thesis, Nova Southeastern University, 2003); idem, 'Task analysis of patients' medication-taking practice and the role of making sense: a grounded theory study', *Research in Social and Administrative Pharmacy*, 2 (2006): 59–82.

9 Idem, 'Development of a needs-driven theory-based model'; C. Richard and M.-T. Lussier, 'Nature and frequency of exchanges on medications during primary care encounters', *Patient Education and Counseling*, 64 (2006): 207–216.

This level of the model thus focuses on two fundamental dimensions of any communication: the 'what' and the 'how'. These are two sides of the same coin and they both contribute to building the patient's knowledge about medications.

The mesoscopic level comprises three distinct elements: 1) what the patient brings to the discussions about medications because of his personal history and use of the medication; 2) the information content of medication exchanges; and 3) the interactive elements of medication-related discussions.

Patient history and prior use of the medication

The encounter with a healthcare professional is but a discrete event in the personal life trajectory of a person. This is well documented in the descriptions of the ecology of medical care.[10] Thus what happens in the encounter must be understood within the personal trajectory of each patient.

More specifically for medications, patients' previous knowledge of, or experience with, the medication can influence the manner in which patients and healthcare providers discuss it during their consultations. To capture a part of the patient's personal history and use of medications, Richard and Lussier, as well as a few others, have distinguished medications on the basis, not only of their pharmacological class, but on the history of use of the medication by the patient.[11] They refer to these as medication statuses.

A 'newly-prescribed medication' designates the prescription of a new medication for the patient. The patient has not previously used this medication. A 're-prescribed medication' is a medication currently being taken, but which must be re-prescribed during the visit to enable the patient to

10 White et al. 'The ecology of medical care'; Green et al. 'The ecology of medical care revisited'.

11 C. Richard and M.-T. Lussier, 'MEDICODE: an instrument to describe and evaluate exchanges on medications that occur during medical encounters', *Patient Education and Counseling*, 64 (2006): 197–206. See also Stevenson et al., 'Doctor–patient communication about drugs'; Scherwitz et al., 'Physician communication to patients'.

continue taking it. The patient has personal knowledge and experience with it but its continued use requires a formal approval by the physician in the form of a written prescription. A 'currently taken medication' is currently being taken by the patient. The patient has personal knowledge and experience with it but its continued use does not require a repeat prescription by the physician during the consultation.

Our proposed model includes these three clinically relevant statuses of medications. Others are possible, such as the discussion of ruled out medications; however, these are both few in number and, from a clinical standpoint, of less importance.

Medication information content

Because of the diversity of patient populations, medical problems, medical treatments and practice contexts, the nature of the information about medications that can be discussed between the professional and the patient during their encounters can cover an extensive spectrum of topics. Richard and Lussier have drawn on the recommendations of the National Council on Patient Information and Education (2000), as well as their review of the literature regarding adherence to treatment to determine the ten topics of discussion they include in the model they propose.[12] Discussions about medications can be categorized as:

- Identification: the manner in which the medication is identified, including its brand name, scientific name, appearance or shape.
- Main effect: the desired effect when the medication is prescribed, i.e. the reason why it is prescribed. The main effect can be anticipated in the case of a new prescription or it can be observed in the case of medications that the patient is already using.

12 Richard and Lussier, in 'Nature and frequency of exchanges', have drawn on the recommendations of the National Council on Patient Information and Education (2000).

- Adverse effects: the effects which are not desired when the medication is prescribed. The adverse effects can be anticipated in the case of a new prescription or it can be observed in the case of medications that the patient is already using.
- Warnings: all of the statements designed to prevent potentially harmful situations or to identify the onset of symptoms indicating that the situation is not progressing in the direction intended.
- Dosage: instructions which must be given to ensure that the medication is taken correctly.
- When to re-consult: all of the statements related to follow-up regarding the results of taking the medication.
- Adherence: all of the statements intended to encourage the patient or to check whether he is going to use the medication prescribed or is currently taking the medication as prescribed.
- Attitudes/emotions: all of the remarks which reveal the patient's positive or negative emotions or attitudes regarding the medication discussed.

The above is not an exhaustive list of all of the possible statements regarding medications. However, in view of the general lack of medication-related content in medical interviews, we do not feel that much information is lost by confining ourselves to these categories. In fact, in the Richard and Lussier study, these categories represent a summary of the 40 categories they initially defined, the majority of which were observed in less than 10 per cent of medication discussions.[13] However, the relevance of these 10 general topics is confirmed by Bajcar's recent qualitative research, in which she interviewed patients suffering from chronic diseases about their medication taking.[14] This patient input led her to identify three tasks which an individual must perform to develop optimal medication taking: 1) Making sense of medication-taking, 2) Medication-taking acts, 3) Self-assessment of medication effects. There is a convergence between the needs patients expressed to support their medication taking behavior in Bajcar's work and

13 Richard and Lussier, 'Nature and frequency of exchanges study'.
14 Bajcar, 'Development of a needs-driven theory-based model'.

the potential topics of discussion during patient-provider encounters that Richard and Lussier have identified. Bringing together the two approaches, Richard and Lussier's more normative and Bajcar's more empirical, the 10 topics can thus be classified under three *content metacategories* derived from Bajcar's work.

- Making sense of medication-taking includes the information the patient requires to understand why he is taking medication. The discussion during the encounter of the three following topics will help the patient in this task: anticipated main and adverse effects and attitudes and emotions.
- Medication-taking acts necessary to ensure that the medication is taken appropriately require the provision of information to guide the patient in the technical aspects of properly taking his medication. Discussions about the identification, instructions (dosage), adherence to the drug regimen and warnings about its use.
- Self-assessment of medication effects refers to the information necessary for the patient to evaluate on his own the effects of taking the medication. Discussing the three following topics during the encounter will help the patient in this task: observed main and adverse effects and indications on when to re-consult the health care provider.

Richard and Lussier took into account the patient's history of usage of the medications by differentiating between the medications which are currently being taken or those which are re-prescribed or newly-prescribed at the time of the visit.[15] Here we report a summary of the recent data on the frequency of discussions of medication-related topics during primary care encounters to illustrate the situation.

The data are organized around the three content metacategories that underline the interrelationship between the macroscopic which refers to the more general life context of the patient and the mesoscopic level which focuses on the professional encounter.

15 Richard and Lussier, 'Nature and frequency of exchanges', and 'MEDICODE'.

Table 6.1 reports the frequencies of discussions of the topics that fall under the making sense of medications metacategory. Making sense of medications in Bajcar's model refers to patients' attempts to rationalize what is happening to them, to understand what is happening in their bodies and to understand their medication in the context of their illness and their daily lives.[16] The actions involved in the making sense process include relating the new information to the information they already have and then relating all of the information to their experiences with illness and medications. For patients, making sense is based on interactions with individuals who are knowledgeable about the particular illness and medication and the information derived from their experience once they start taking the medication.

Making sense	All medications	Active	Represcribed	New
	N=1492	N=643	N=243	N=158
%	%	%	%	%
Anticipated Main Effect	14.3	6.8	9.9	46.8
Possible Adverse Effect	8.2	6.8	7.0	16.5
Attitudes/ emotions	17.4	21.2	20.1	11.4

Table 6.1: Frequency of discussions between patients and physicians of the making sense metacategory of medication topics

This group of statements is found in less than 18 per cent of cases. This is of concern, insofar as a reasoned approach to medication would mean that 'reason' is what governs the patient's behaviour. It is even more remarkable that even in the new prescription category (in which the patient is

16 Bajcar, 'Task analysis of patients' medication-taking practice'.

considered to have no prior information), the frequency is relatively low regarding two of the three topics. Making sense is thus not at the core of the exchange of information, regardless of the status of the medication.

Medication-related discussions are thus not explicit in terms of the anticipated effects of the medication. There may be a structural problem related to the production of explicit information regarding these issues. Rather, it appears that the content in this regard is implicit. These are the limitations of measuring explicit discussions. However, even taking into account these reservations, we feel that there should be more explicit discussions regarding these issues.

Table 6.2 reports the frequency of the statements that support or influence patients' ability to take their medication correctly. Of the three metacategories, medication-taking acts is the one which includes the most frequently discussed topics, in particular, identification of the medication. However, the most clinically significant category – instructions – is not as frequently discussed as one might hope, particularly with respect to new medications. In this instance, the instructions should be given each time a new medication is prescribed.

Medication-taking acts	All medications	Active	Represcribed	New
	N=1492	N=643	N=243	N=158
%	%	%	%	%
Identification	73.9	82.3	90.9	53.2
Instructions or dosage	34.5	32.0	50.6	63.9
Adherence	3.8	5.3	8.2	NAP
Warnings	4.8	3.5	2.0	12.7

NAP: non applicable

Table 6.2: Frequency of discussions between patients and physicians of the medication taking acts metacategory of medication topics.

It is interesting to note that with respect to identification, the frequencies for the 'active' and 're-prescribed' statuses are higher than for the 'new prescription' status. One reason put forward to explain this is that the patient's familiarity with the medication allows the physician to refer to it by name. The topic of adherence is almost never broached in doctor–patient discussions of medications, despite the overwhelming evidence regarding the extent of problems in this regard. It is also remarkable to note the low frequency of discussions regarding 'warnings' – a topic of importance from a medico-legal standpoint.

Table 6.3 reports the frequencies of discussion of topics important to the patients' capacity for self-assessment of medication effects. Patients taking medications are aware of the effects they attribute to them and are constantly evaluating their experience with taking them. Basically, the patient must assess, on the one hand, whether the effect of the medication corresponds to the anticipated effect and, on the other, whether the adverse effects remain acceptable. The physician must discuss these issues to allow patients to monitor the course of their condition. If the changes in his symptoms are not in keeping with what is anticipated, the patient must know when it is appropriate to consult the physician.

Self-assessment of effects of medications	All medications	Active	Represcribed	New
	N=1492	N=643	N=243	N=158
%	%	%	%	%
Observed Main Effect	22.7	30.2	28.0	NAP
Observed Adverse Effect	7.6	7.3	11.9	NAP
When to reconsult MD	12.5	9.0	18.9	35.4

Table 6.3: Frequency of discussions between patients and physicians of the self assessment of the effects of medication metacategory.

Self-assessment regarding the effect of medications is more germane to the statuses of active and re-prescribed medications, particularly with respect to the perceived or observed main or adverse effects. The frequencies of discussion regarding the observed main effect are similar for all statuses, but relatively low. They are even lower with respect to the adverse effects, despite the fact that a number of earlier studies indicate that patients are most concerned about the adverse effects.[17] The topic of when to re-consult is most often discussed when a new medication is prescribed.

Interactive dimensions of medication discussions

At this level, the model we introduce an important concept that aims to capture the extent of contribution of each party to medication-related discussions during medical encounters: dialogical roles. These new measures of participation to content production are intimately related to the MEDICODE coding scheme the authors have developed and validated in which coders indicate if a medication content theme is discussed by either the physician or the patient alone or if both contribute content to the discussion.[18] Thus here, 'dialogue' regarding medications has a special meaning. Participation is envisioned as a continuum going from physician-led discussions through shared discussion through patient-led discussions.

Because increased patient participation in their care leads to better care outcomes,[19] the model proposes a series of roles (initiator, participant,

17 Makoul et al. 'Health promotion in primary care'; Stevenson et al., 'Doctor–patient communication about drugs'.

18 Richard and Lussier, 'MEDICODE' (2006).

19 D.L. Roter and J.A. Hall, *Doctors Talking to Patients. Patients Talking to Doctors* (Westport CT: Auburn House, 1993); M.A. Stewart, 'Effective physician-patient communication and health outcomes: a review', *Canadian Medical Association Journal*, 152 (1995): 1423–1433; M.A. Stewart, J.B.Brown, A. Donner, I.R. McWhinney, J.Oates, W.W. Weston, and J.Jordan, 'The impact of patient-centered care on patient outcomes', *Journal of Family Practice*, 49 (2000): 796–804; S.H. Kaplan, S. Greenfield, and J.E. Ware Jr, 'Assessing the effects of physician-patient interactions on the outcomes of chronic disease', *Medical Care*, 27 (1989): S110-S127.

information provider, listener) patients and physicians can adopt while talking about medications.[20] These roles can be grouped into two broad, opposing categories of behaviour. First, we have the roles described as dialogues, which involve the participation of two people – the initiator and the participant. Next, we have the roles which do not involve the participation of the second individual – what we call monologues – consisting of the roles of information provider and listener. This vocabulary can be particularly useful in the context of medical education to help make the professional aware of his behaviour in medication-related exchanges and how to adopt behaviors that favour patient participation. Promoting the roles of initiator and participant, rather than those of information provider or listener, will encourage the development of true dialogue regarding medications.

Dialogue roles:

INITIATOR: an individual who initiates the medication-related dialogue.

PARTICIPANT: an individual who participates in the medication-related dialogue, initiated by the other party.

Monologue roles:

INFORMATION PROVIDER: an individual who provides information with no input from the other party regarding content.

LISTENER: an individual who listens to the information provided by the other party.

Richard and Lussier analysed the roles that general physicians adopted when discussing medication-related topics during primary care encounters. Table 6.4 presents the average frequencies with which each role was adopted taking into account the status of the medication.

20 C. Richard and M.-T. Lussier, 'Measuring patient and physician participation in exchanges on medications: dialogue ratio, preponderance of initiative and dialogical roles', Poster Presentation, *Each International Conference on Communication in Healthcare* (Bruges, Belgium, September 2004).

		Physician's role			
Set of themes	Status [a]	Listener	Info. giver	Part.	Insti.
All themes combined	Active	32.1	43.2	7.8	17
	Represcribed	23.6	46.0	7.7	22.7
	New	10.5	69.7	4.1	15.8

[a] Active Discussed, n = 643; Represcribed, n = 243; New Prescriptions, n = 158.

Table 6.4: Physician dialogue role as a mean proportion (%) of discussions for all themes combined and by status

Their results demonstrate that the roles physicians adopt while discussing medications during primary care encounter vary depending on the medication content theme under discussion and the status of the medication. However, there is a predominance of the role information giver that indicates that most of the time these physicians discussed medications in a monologue fashion. They are, nevertheless, capable of adopting an instigator role that invites patient participation.

Microscopic Level: the medication-related communication process in a specific encounter

The microscopic level of the model focuses on the specificities of each of those unique moments of contact between patients and healthcare providers that periodically occur during this lifelong learning process about medications. The physician-patient relationship falls into the general category of a service-type relationship.[21] Basically, the physician is there to respond to a request for care by the patient and makes his expertise available to the patient. This includes helping the patient to seek care, while taking into account the patient as a whole person.

At the microscopic level of analysis we discuss the specific problems related to the discussion of medications by a professional and a patient, as well as the communication solutions to address these problems. Here once again we build and expand on Bajcar's model of medication-taking as we transpose it to the actual physician-patient encounter and discuss specific communication strategies that are helpful in 'the here and now' of the encounter.

Medication-related events and patient/physician responses to them

Patient meetings with health care professionals are not always the same. However, generally speaking, the purpose of a meeting with a physician is to obtain a professional service, such as a check-up, a routine monitoring of the patient's blood pressure or cholesterol levels, or to discuss an acute problem, such as the presence of pain or discomfort. In all these cases, the physician may have to report and discuss an 'event' which, from the professional's perspective, may or may not be serious (new debilitating illness, onset of new symptoms, necessity of a new treatment because the illness is not responding to the current treatment), but which will nonetheless prompt a reaction from the patient such as, for example, shock, sadness, anguish, anxiety, worry, or concern.

The physician-patient relationship is punctuated by this type of 'news-event', where the physician must report and the patient must understand and then accept. When such an event occurs during a consultation, the normal

21 Y. St-Arnaud, *L'Interaction professionnelle: efficacité et coopération* (Montréal: Presses de l'Université de Montréal, 1995); L. Giroux, 'Les modèles de la relation médecin-patient', in C. Richard and M.-T. Lussier (eds), *La Communication professionnelle en santé* (Montreal: ERPI, 2005); J.D. Robinson, 'Asymmetry in action: sequential resources in the negotiation of a prescription request', *Text*, 21 (2001): 19–54.

course of physician-patient interaction is affected because the psychological impact of the event on the patient can be substantial. The introduction of a new medication (particularly in the context of a chronic illness) or an increase in the dose of medication are 'events' likely to upset the patient.

The event may have no impact on the individual's cognitive or affective ability to continue his interaction with the physician. However, it may and then it is not unusual for the patient to temporarily be at a loss and thus unable to continue the interaction as usual (given his decreased ability to understand the information and instructions he is being given).

A dialogical approach

When medication-related events occur during a consultation, the patient may present different types of responses. However, when the physician respects the general rules of effective medical communication as described by Kurtz, Silverman, and Draper,[22] and favours roles that support dialogue he can attenuate the patient's reactions.

Patient responses that favour comprehension and retention of medication-related information

When the physician strives to understand the event from the patient's perspective, he adopts the role of initiator (rather than information provider) and the patient can be a participant (rather than simply a listener) and share his thoughts and concerns about the medication. The information discussed is relevant to the patient's situation and he is then able to reconcile his understanding of his problems with the information he is receiving regarding the action of the medication.

22 S. Kurtz, J. Silverman and J. Draper, *Teaching and Learning Communication Skills in Medicine* (Abingdon: Radcliffe Medical Press, 2nd Edition, 2005).

Generally, it is recommended that the patient's beliefs regarding his medication and, in particular, his understanding, acceptance, motivation to take the medication and the degree of control he wishes to have in this regard be explored. When the situation permits, the physician must also promote a shared decision-making process. In addition, he must ensure that all of the essential information has been communicated in a manner conducive to patient understanding and retention. Where the self-administration of medication is involved, the best treatment is the one with which the patient will comply. It is especially important to ensure patient cooperation and, if possible, to establish a partnership with the patient. To achieve this, the patient must be convinced of the usefulness of taking his medication. Thus, the best strategy consists of ensuring that the patient explicitely agrees with the treatment. The emphasis here is on communication, interaction and the relationship established with the patient. In addition to discussing adherence and taking into consideration the patient's attitudes/emotions regarding medications, the recommendations from the National Council on Patient Information and Education are useful in guiding how the information is presented.[23] To ensure that the 10 major topics when prescribing medications are covered (particularly important when prescribing a new medication), the physician must create a moment where he interrupts the flow of conversation to create an opportunity to mention and explain this new medication.

Some physicians fear losing control of the interview and thus leave little room for patient input. This loses sight of the fact that patients will rarely confront their physicians; they'll simply let them talk. However, once they get home, they will not follow the treatment. Everyone loses in this situation – society (which pays for the service), both the physician and the patient (who waste their time) and, finally, the patient (who will delay treatment).

23 National Council on Patient Information and Education (2000).

Patient responses that hinder comprehension and retention of medication-related information and doctor responses

The patient is in a context where he is ill and often vulnerable. His reaction to a piece of news may be intense and affect his understanding of the physician's suggestions. The presence of anxiety, fear, strong emotions can hinder the process of learning and retaining medication-related information. Many events are potential sources of patient responses that will momentarily limit the patient's capacity to effectively interact with the healthcare provider. Here are some examples related to the illness: presentation of a new diagnosis of a chronic illness, discussion of inadequate control of the illness; and others related to the medication: presentation of a major change (type of Rx, dose, i.e. an increase) in the medication, discussion of risks or serious side effects associated with the medication. In such circumstances, the physician will be on the look-out for signs from his patient indicating a state of shock such as the absence of reaction when a reaction would be expected, loss of eye contact, silence, the appearance of tears. Less intense emotional responses can be signalled by verbal expressions of surprise or hesitations, the direct expression of concern or doubts about the medication (its necessity, efficacy, side-effects or long term effects) as well anticipation of problems. Concerns or hesitations may also be expressed non verbally for example by physical body withdrawal, facial expressions of concern or disappointment.

Fear, shock or ambivalence are states which, at worst, can completely prevent understanding or, at best, cause confusion.[24] The patient's emotional state can completely overwhelm him. When this is the case, the patient is not capable of receiving new information about the medication and its dosing. In this event, the patient's subsequent course may be compromised, particularly with respect to adherence with treatment. The first thing the physician must do is to deal with this emotional state to make

24 R.W. Miller and S. Rollnick, *Motivational Interviewing: preparing people to change addictive behavior* (New York: The Guildford Press, 1991); Bajcar, 'Task analysis of patients' medication-taking practice'.

the patient capable of receiving new information about the medication and its dosing. Smith has suggested a simple, effective way of managing the patient's emotions during the interview.[25] He suggests dealing with the perceived emotion directly by *N*aming it, by specifically indicating to the patient (legitimizing the reaction) that his reaction is understandable (*U*nderstand), by showing him respect for his perceived ability to manage this problem (*R*espect) and by assuring him of support (*S*upport). Smith suggests the following mnemotechnic method to identify the components of managing emotions during the interview – NURS. Only when the professional clearly recognises the affective nature of the problem can he resume the usual approach of exchanging information.

Some events trigger difficulties that are not of an emotional nature but more of a conceptual or cognitive nature. Such problems may occur at any time and are sometimes less obvious than the patient's emotional reactions. The patient does not understand the physician's suggestions – either because he does not understand the vocabulary used or because the concepts are foreign to him. He does not have the knowledge required to understand what the physician is explaining to him or else is unable to make the connection between what the medication is supposed to do and what he understands about his illness or symptoms. For example, when the information is presented in a jargon or a manner which is too complex. The patient is not a specialist; the information provided must be simplified and given in layman's terms. The patient may not be able to read, may have cognitive limitations or there may simply be too many things to remember. When the problems are conceptual (i.e. when things are not logical or intelligible for the patient), everything must be explained taking into account and integrating into the explanations provided what the patient already knows. If the patient's erroneous beliefs are the source of the problems, they must be corrected, while ensuring that the new explanations intended to replace the erroneous beliefs fulfil the same functions as the old ones.

25 R. Smith, *The Patient's Story: integrated patient–doctor interviewing* (Boston, MA: Little Brown & Company, 1996).

Kurtz et al. (2005) delineate the communication skills that are useful when interacting with a patient to minimize misunderstandings and to optimize patient comprehension:

1. Provide information in easy-to-manage increments
2. Assess the patient's knowledge prior to providing the information and adapt the information according to his response
3. Ask the patient what other information would be useful
4. Provide the information at the appropriate time
5. Organize the information to be provided regarding the medication
6. Use specific categories (topics) and announce them
7. Use repetition and summary
8. Avoid medical jargon
9. Use visuals or written material to supplement your explanations
10. Check the patient understanding[26]

Another cause of this type of patient response may be that the information provided by the physician contradicts the information the patient already has acquired from his experience, his beliefs regarding the medication or his knowledge regarding his illness. This underlines the interplay of the macroscopic and microscopic elements of the model. When the information he holds comes from a consultation with another provider, the situation is especially difficult for the patient, as he must choose between two professional opinions. In this event, the patient often abstains from making a choice and will either consult someone else or wait. Whatever the case may be, we are heading towards a problem with patient adherence. On a more subtle level, the patient may resist the physician's opinion because he does not understand the physician's explanations or rationale in prescribing the treatment. Thus, the goal must be to find the common ground between the patient's perspectives and those of the health care provider.

Although presented as separate responses, there may be overlap between emotional and cognitive responses as when the lack of common ground

26 Kurtz et al., *Teaching and Learning Communication Skills.*

between the patient and the health care provider causes the patient to feel confused, uncomfortable or even anxious about his medication.

Strategies to Manage Knowledge about Medications: an iterative approach

It is obvious that the physician's task is both vast and complex with respect to how he can contribute to the patient's learning about medications and encourage him to take his medication appropriately. The physician must accept that this process is iterative in nature – i.e. it must be developed over a number of visits, as the patient becomes more familiar with the medication over time. The physician must have short-, middle- and long-term objectives, prioritizing the information, sequentializing the tasks and using specific communication strategies adapted to the patient's particular problems. The success of this venture requires adjustments on the physician's part – in his perception of his role in his patient's medication-taking behaviour, as well as in his management of the consultation and his communication skills.

When a new medication is first introduced, and when the medication is known not to produce any irreversible effects, the physician can confine himself to introducing the subject in a relatively simple manner (taking into account, of course, how urgently the treatment needs to be started). He can limit the information to what is required to take the medication correctly and suggest a follow-up visit in the near future. During that visit, he asks whether the patient had any problems taking the medication and, if so, to explain what they were. The physician should be as clear as possible and, if possible, use the same terms during repeated explanations. Sometimes, simply using an equivalent, but different, term can be perceived as an inconsistency on the physician's part. The physician also checks whether the information given previously has been retained and understood. Only then can he add new information, repeating the procedure until he and the patient have covered the essentials to reach a true informed consent regarding the matter or the patient indicates that he does not wish to have any additional information.

The physician must help create a network of convergent medication-related information focusing on supporting medication-taking (adherence) by the patient. The information must include as many convergent references as possible (such as the illness itself and its course, the patient's life style) and be presented in several ways and by several professionals. The repetition of the information in various, complementary forms and coming from several sources makes it difficult to forget.

Conclusion

The exchange of information during medical consultations remains the primary way of informing the patient, to enable him to give his informed consent and to teach him how to take his medication correctly. We believe that the health care provider can build on the knowledge already acquired by the patient, as well as on his experience, to increase prescription-related dialogue. Doing so will encourage patient participation and participation is known to be associated with better health outcomes.

This model proposes a global approach to medication-related discussions, allowing us to choose the appropriate level of abstraction. For more in-depth comprehension of the phenomenon and to put it into context, the use of more macroscopic concepts is desirable. These concepts encompass a philosophy, a perspective regarding medication, which can guide the actual interventions with a patient. For example, if a patient's history with various medications is especially dramatic and has resulted in mistrust or cynicism, it is desirable to discuss these issues, as they may represent an obstacle to patient adherence. At the intermediate level of abstraction, we provide a number of concepts which lend themselves well to the study of prescribing behaviours. These concepts ask what must be said and done by the physician. At this level as well, we can study the styles of relationships with the various roles the physician can play when discussing medications. At the more microscopic level, we suggest intervention approaches and tools. The higher-level concepts can be used retroactively at lower levels to guide the practitioner in his process of change. However, the conducting of the interview must involve much more specific strategies to guide the practitioner. The macroscopic and intermediate aspects must be included in these specific strategies, which are both the example and its embodiment during the interview. Within these concrete strategies, we will find the expression of the social and cultural aspects, the perceptions and the history of the patient regarding medications. Through these strategies, we will also establish the review of the topics to be covered and the roles to be adopted and will allow ourselves more or less medication-related dialogue during the interview. Thus, at this level, we can take into account whether or not the situation is problematic, if the patient shows signs of emotional or cognitive problems regarding the taking of medication.

It is our opinion that this set of interrelated and mutually supportive concepts will help the practitioner to understand how patients approach medication and how to recognise their specific informational needs. By explaining certain communication mechanisms, we feel that the patient's behaviour becomes more understandable and predictable. This makes it easier to develop interventions adapted to the particular needs of each patient. This model thus suggests an approach to informed consent to be provided by the patient. It develops over time and has its place in both the patient's relationship of trust with the professional and his ability to understand and assimilate new information. This approach is in contrast to a solely rational approach, where an informed decision requires analysis by the patient in order to choose the best medication. The concept of trust must be reintroduced and promoted. It should be remembered that the patient often receives information under difficult circumstances, when he cannot make sense of the information provided.

Acknowledgement

The authors extend their thanks to M. Luc Carrier, head of the Professional Educational Department of AstraZeneca Canada Inc. for the translation from French to English of this chapter.

PART 3

Developing, Selling and Representing Drugs

TONY CORLEY

7 Twentieth-Century American Contributions to the Growth of the Beecham Enterprise[1]

Introduction

In a dynamic world, the entrepreneur's key role is that of an innovator, forever combining tangible and intangible resources in new and more lucrative ways to improve the firm's performance.[2] The present article discusses how by means of constant innovation, the Beecham enterprise in Britain raised itself to leadership status, first of all in patent medicines and then in pharmaceuticals, and how far it benefited over the years from advanced knowledge obtained in the United States. Beecham's earlier achievements – and American connections – in pill-making are briefly recorded below. However, as this article is in the main concerned with its pharmaceutical innovations in the twentieth century, the innovatory path it followed is first of all placed in an analytical context.

According to Michael Burstall and John Dunning, 'to a quite unusual extent the [world] pharmaceutical industry organizes its activities along transnational lines'.[3] Since many drug companies rely for their progress on a two-way traffic with other countries, the industry's structure and overseas operations can usefully be analysed through the theory of the multinational

1 My thanks are due to Viviane Quirke and Judy Slinn for some helpful comments.
2 J.A. Schumpeter, *The Theory of Economic Development* (Cambridge MA: Harvard University Press, 1934), pp. 65–66.
3 M. Burstall and J. Dunning, 'International Investment in Innovation', in N. Wells (ed.), *Pharmaceuticals among the Sunrise Industries* (New York: St. Martin's Press, 1985), p. 185.

enterprise (MNE) or transnational corporation. Firms involve themselves internationally through trade across borders, licensing, or foreign direct investment (FDI); the last-named entails the creation of assets abroad, which remain under their control.

Dunning, in the eclectic paradigm which underlies his MNE theory, sets out the motives for venturing abroad in terms of ownership, location and internalization advantages. On the ownership side, entrepreneurs will take the initiative to deploy organizational, marketing and/or technological skills as well as finance. As to location, the home country may have useful economic assets, for instance an appropriate labour force, or intangibles, whether know-how or an amenable government or legal system. The internalization advantages involve controlling foreign activities rather than leaving to outside agents the distribution of exports or to licensees the marketing of their products. The more costly and elaborate the goods concerned, the greater the incentive to internalize through FDI.[4]

Further, Dunning distinguishes between the roles of countries, industries and firms. The industry characteristic is a vital one here, as pharmaceutical enterprises are overwhelmingly composed of the intangibles mentioned above. That distinctive feature has been crudely, but arrestingly, illustrated for the UK's Glaxo Wellcome, which in 1998 had a market capitalization – the value which investors placed on its operations – of £60 billion; yet its tangible assets came to less than £5 billion.[5] Likewise, a country such as the United States, which has had the educational, cultural and market size advantages to create and apply commercially new technological knowledge, will have the power to dominate a sector such as the pharmaceutical industry.[6] That domination is greatly strengthened when the economy concerned is also the world's largest market for drugs.

4 J.H. Dunning, *Multinational Enterprises and the Global Economy* (Wokingham: Addison Wesley, 1993), pp. 81–84.

5 Parliamentary Papers (PP), House of Commons Science and Technology Committee, Session 1997–1998, 3rd Report, 'Glaxo Wellcome and SmithKline Beecham: the Merger Proposals' (HC 627, 3 June 1998), p. 7.

6 J.H. Dunning, *American Investment in British Manufacturing Industry* (London: Routledge, Revised Edition, 1998), pp. 191–192, 255.

The UK's disadvantage has been in providing the inappropriate kind of education, designed for future rulers of the British Empire or for low-grade clerical jobs from its grammar and technical schools. Neil McKendrick, in three hard-hitting essays composed in the later 1970s, demonstrated how the older British universities and the UK literary world alike have not only ignored the wealth-creating value of business but have also been moved to express violent antipathy to business and commerce in general.[7] Economically, Britain's population was far smaller than that of the United States, depriving all too many of its enterprises of large-scale economies at home.

This article begins its account of the Beecham company's growth by applying certain aspects of the foregoing analytical framework to a real-life case. It stresses the ownership advantages of utilizing innovation and effective marketing to overtake rivals. By linking the country and industry characteristics, it suggests that the built-in disadvantages related above could well hinder attempts to create a strong British-owned pharmaceutical industry. As the United States possessed so many of the essential advantages, that hindrance might be offset by means of assistance from across the Atlantic in order to overcome national deficiencies at home. The chronological account below traces Beecham's experiences in this field, principally during the twentieth century.

7 N. McKendrick, 'In Search of a Secular Ideal', in C. Trebilcock (ed.), *The Vickers Brothers: armament and enterprise, 1854–1914* (London: Europa Publications, 1977), pp. 9–34; idem, 'Literary Luddism and the Businessman', in P.N. Davies (ed.), *Sir Alfred Jones: shipping entrepreneur par excellence* (London: Europa Publications, 1978), pp. 9–56; idem, 'The Enemies of Technology and the Self-made Man', in R. Church, *Herbert Austin: the British motor car industry to 1941* (London: Europa Publications, 1979), pp. 9–50.

Beecham: a family firm, 1859–1924

The Beecham enterprise is unique in British corporate history as the only major UK healthcare company to transform itself from a patent medicine manufacturer into a pharmaceutical giant. Its founder, Thomas Beecham (1820–1907)[8] made his proprietary pills from 1859 onwards in the Lancashire town of St. Helens. During the twentieth century, no longer a family business but managerially run, it diversified into a broad range of proprietary medicines, for example Beecham's powders, and consumer products including health drinks such as Lucozade and Ribena. Only during the 1950s did it begin the manufacture of pharmaceuticals. In 1989, as a medium-sized company in global terms, it merged with an American corporation to become SmithKline Beecham. This account concludes with its subsequent merger with Glaxo Wellcome in 2000 to become part of GlaxoSmithKline.

From the outset, Beecham built up a comparative advantage over British rivals through its policy of innovation, by mechanizing and enforcing rigorous quality control. It thereby outstripped competitors whose pills were poorly mixed and of irregular sizes. Its marketing strategy reinforced that advantage, under the direction of the firm's Anglo-American general manager, Charles Rowed (1855–1933), noted for combining 'Lancashire grit with more than Yankee cuteness' or flair.[9] A constant outpouring of light-hearted advertisements made the Beecham name celebrated throughout Britain; hence by the 1880s it had become the country's leading patent medicine manufacturer.

From that decade onwards, the Beechams saw off most attempts by American patent medicine adventurers to export to and then manufac-

8 T.A.B. Corley, 'Thomas Beecham (1820–1907)', in H.C.G. Matthew and B. Harrison (eds), *Oxford Dictionary of National Biography* (Oxford: Oxford University Press, 2004), Vol. 4, pp. 798–799; A. Francis, *A Guinea A Box: a Biography* [of Thomas Beecham] (London: Hale, 1968).

9 *Chemist and Druggist*, 44 (12 May 1894), p. 659.

ture their nostrums in Britain.[10] Having previously made entirely different products such as typewriters or iron safes, the latter expected to succeed in creating a substantial market there by hard selling and attractions such as chariots used in US medicine shows. The high initial sales soon plummeted, and their companies usually collapsed. Thus low-tech American products such as proprietary medicines and biscuits made little progress against innovative British rivals. In theoretical terms, country-specific advantages were offset by the lack of industry-specific advantages, in the patent medicine case a sophisticated understanding of consumers' tastes in the recipient or host country.

Meanwhile, Beecham's domestic advantage was extended overseas by the founder's son, (Sir) Joseph Beecham (1848–1916).[11] Location-specific factors governed the pattern of his firm's involvement abroad. As Beecham's pills were scarcely demanded outside English-speaking countries, most exports went to the United States and territories in the British empire. He entered the American market after an international convention secured the protection of trade marks in 1887. Three years later, he began manufacture in New York, after the US administration imposed a tariff equivalent to half the pills' price.

Through regular transatlantic visits, Joseph Beecham actively exploited for his own firm's purposes the locational advantages of the United States, in particular by adopting certain American innovations for use in the UK. He introduced electricity into his town of St. Helens as well as into the modern factory he built there during the 1880s. Up-to-date machinery, for sifting ingredients, came from the United States, as did the pill boxes, manufactured in New England of white birch wood and superior to British makes for having tight-fitting lids. In 1910 he installed in his New York factory the latest American labour-saving plant. He did not trouble to

10 T.A.B. Corley, 'Interactions between the British and American patent medicine industries 1708–1914', *Business and Economic History*, 2nd series, 16 (1987): 111–129.

11 T.A.B. Corley, 'Sir Joseph Beecham (1848–1916)', in H.C.G. Matthew and B. Harrison (eds), *Oxford Dictionary of National Biography* (Oxford: Oxford University Press, 2004) Vol. 4, pp. 797–798; idem, 'The Beecham Group in the World's Pharmaceutical Industry 1914–70', *Zeitschrift für Unternehmensgeschichte*, 39 (1994): 18–30.

replace the older plant in St. Helens as wages were low enough to make a change uneconomic.

However, Joseph failed to introduce American cost-accounting methods, such as allocating costs and earnings to departments or functions. Preferring to run his firm on a cash-only basis, he was never able to differentiate between home and overseas profits. After 1900 he ceased to be an innovator at home, spending much of his energy and fortune in supporting the musical activities of his son, the conductor Sir Thomas Beecham (1879–1961). Having inadvisedly committed himself to purchasing for £2 million the Covent Garden market in London, he died in 1916 leaving behind a heavily indebted estate. That had to be administered by the UK Court of Chancery, which did not release it until 1924.

Beecham's Early Years as a Managerial Company, 1924–1951

In 1924 the firm, no longer under family control, was acquired by an investment company, the Midland Industrial and General Trust. By then fully owning the Covent Garden market, it took the name of Beecham Estates and Pills Ltd., with £2.8 million capital. The leading director was Philip Hill (1873–1944),[12] a London property magnate who became its third innovator after Beecham father and son. He introduced its first new product, Beecham's powders, and set up a laboratory.

Hill also acquired the Veno Drug Company Ltd., which made a cough remedy by means of American-style technology. Its founder, the Scottish-born Sir William Veno (1866–1933), had begun his business career in Pittsburgh, Pennsylvania, by establishing a company to manufacture his 'lightning cough cure'. In 1888 he returned to Britain and built a technically

12 T.A.B. Corley, 'P.E. Hill (1873–1944)', in H.C.G. Matthew and B. Harrison (eds), *Oxford Dictionary of National Biography* (Oxford: Oxford University Press, 2004), Vol. 27, pp. 169–171.

advanced plant in Manchester, hailed as the world's sole entirely self-sufficient factory. It had not only its own printing department but also a subsidiary nearby to manufacture the bottles. Hill tried but failed to instil into Beecham managers an American entrepreneurial spirit by moving senior staff to Manchester and attempting to integrate the two companies' operations. Antagonisms and obstruction on both sides caused the experiment to fail after seven frustrating years; yet when an extension to the St. Helens factory was built in 1935, it almost precisely followed the Veno design.[13]

In 1928, Hill separated the profitable patent medicine operations from the low-earning property business by hiving off the latter into the Covent Garden Properties Company. He then established the £2.25 million Beechams Pills Ltd., planned to be the nucleus of Britain's largest medicinal-cum-consumer product company, which he would create through a programme of acquisitions. His most spectacular purchases, in 1938–1939, were of Macleans (toothpaste), Eno's (fruit salts), Brylcreem (a hair product) and Lucozade (a glucose drink). Macleans' founder was Sir Alexander Maclean (1872–1948) from New Zealand, who had emigrated to the United States, becoming a salesman for the Spirella Corset Company at Niagara, New York. Having established a British branch of Spirella, he departed to establish his own toothpaste business in Brentford, near London. His lengthy sales experience allowed him to build Macleans into a progressive marketing company. Where he obtained knowledge about manufacturing is not known, but in 1935 his factory was celebrated for the scientific basis of its production, utilizing American continuous flow methods.[14]

Hill's technical consultant was the eminent biochemist, Sir Charles Dodds (1899–1973), who had learnt about the latest North American medical techniques during an extended tour.[15] On Dodds' advice he authorized

13 *Chemist and Druggist*, 96 (24 June 1922): 190–193; ibid., 119 (8 July 1933): 38–39; ibid., 126 (30 May 1936): 629–632.

14 'Modern Methods in Pharmaceutical Manufacture', *Industrial Chemist, Pharmaceutical and Cosmetic Supplement* (October 1935), pp. 118–123, ibid. (November 1935), pp. 133–139; *Industrial Chemist* (June 1937), pp. 234–237.

15 F. Dickens, 'Edward Charles Dodds 1899–1973', *Biographical Memoirs of Fellows of the Royal Society* (1975) Vol. 21, pp. 237, 253–4.

the setting up of a central laboratory in 1942 for basic research and development (R&D). However, after Hill's death two years later his successors until 1951, in charge of the now renamed Beecham Group, judged that the coming of the National Health Service (NHS), established in 1948, would spell the end of patent medicines. Instead of financing the group's research facilities adequately, they put funds into acquiring a number of wholesale groceries, and thereby sacrificed the opportunities for providing advanced medicines required by the NHS. Hence British subsidiaries of American pharmaceutical firms, led by Parke Davis and Merck Sharp & Dohme, were by the 1950s supplying half the NHS's prescription drugs, while continental European producers contributed a further one-eighth.[16] Beecham's financial performance deteriorated to such an extent that in 1951 the institutional investors, through one of its directors, Kenneth (later Lord) Keith (1916–2004), promoted a boardroom coup. Leslie Lazell (1903–1982),[17] the fourth innovating entrepreneur, was appointed chief executive.

The Lazell Era, 1951–1968

Lazell, a trained accountant, had formerly been a manager with the research- and marketing-based Macleans. Having come to know executives in the UK branches of American firms, such as those noted above, he recognized that Beecham and other British companies were vulnerable to take-over or relative decline unless they matched those rivals' superior efficiency through innovative R&D and structural reforms. On becoming chief executive in 1951, his first priority was to create a thriving sales market in the United

16 Dunning, *American Investment*, p. 40.

17 H.G. Lazell, *From Pills to Penicillin: the Beecham story* (London: Heinemann, 1975); T.A.B. Corley, 'H.G.L. Lazell (1903–1982)', in H.C.G. Matthew and B. Harrison (eds), *Oxford Dictionary of National Biography* (Oxford: Oxford University Press, 2004), Vol. 32, pp. 940–941.

States, being convinced that Beecham would never succeed elsewhere in the world unless its products made good there.

As part of his campaign to market Brylcreem and Maclean's toothpaste throughout America, in 1954 he built an extension to the Eno's factory in Bloomfield, New Jersey, in order to manufacture those products locally.[18] That branch was renamed Beecham Products Inc. in 1960. By intensive advertising, he made some inroads into that highly competitive market with Brylcreem and Maclean's toothpaste, but lacked a sufficiently strong management on the spot to achieve a break-through.

To reinvigorate the group at home, Lazell adopted a number of American business practices. In marketing, his plan was to co-ordinate all sides of the company into a system for discovering and then gratifying consumers' wants. He made top management directly responsible for implementing marketing policy along those lines, using the consumer surveys of the US market research company, A.C. Nielsen Co. He also consolidated into functional divisions the group's 100 or so subsidiaries, which Hill had wastefully allowed to remain as separate entities. That restructuring followed closely the ideas being developed by Alfred Chandler of Harvard Business School, who had tracked the progress towards divisionalization of the United States' largest corporations. Although Chandler's *Strategy and Structure* was not published until 1962,[19] Lazell may have learnt about his preliminary researches from the American management consultants, Booz-Allen & Hamilton, whose expertise he sought on organizational questions.

Lazell made the most far-reaching contribution to Beecham's future progress by authorizing R&D work on penicillins in 1954. Instructed by (Sir) Ernst Chain (1906–1979), a pioneer in antibiotics, three years later his scientists discovered the hitherto elusive penicillin nucleus. They developed the world's first semi-synthetic penicillins by subjecting the nucleus

18 J.H. Dunning, 'Revisiting UK FDI in US manufacturing and extractive industries in 1960', in G. Jones and L. Gálvez-Muñoz (eds), *Foreign Multinationals in the United States: management and performance* (London: Routledge, 2002), esp. p. 56.

19 A.D. Chandler, *Strategy and Structure: chapters in the history of the American industrial enterprise* (Cambridge MA: MIT Press, 1962).

to a chemical acylation process. Broxil, launched in 1959, was followed by four further and improved brands by 1970. Although that discovery was a purely British innovation, Lazell had to make use of American technology, provided by Bristol-Myers of Syracuse, New York, in order to move forward from pilot plant to full-scale production. Bristol-Myers also planned Beecham's first antibiotic factory, at Worthing, Sussex, and trained its leading scientific and managerial staff.[20]

To set up a marketing organization for penicillins, Lazell appointed (Sir) Graham Wilkins (1924–2003),[21] a scientist by training and lately a Beecham manager in North America. Until he could establish the group's own sales subsidiaries around the globe, Wilkins concluded a series of individual agreements, licensing local overseas companies, such as Bristol-Myers in the United States, to manufacture and distribute those drugs. Then in 1966–1977 he oversaw the establishment of a $4.5 (£1.6) million antibiotic plant at Piscataway, New Jersey, to supply the whole of the Western Hemisphere. The prospect of greatly improved marketing opportunities there for penicillins helped to alleviate Lazell's disappointment that US sales of Brylcreem and other consumer products accounted for less than 15 per cent of Beecham's turnover. His aspiration to see the group's presence in America a hundred times larger that that in Britain remained an unfulfilled dream. Yet the mainly intangible contributions he had received from the United States had helped to make Beecham into a noteworthy pharmaceutical company, offering products that were extensively demanded throughout the rest of the world.

20 T.A.B. Corley, 'Beecham and the Development of Semi-Synthetic Penicillins 1951–1970', Centre for Industrial Performance, University of Reading, Discussion Paper 2003–01 (2003), pp. 1–26; for the reactions of a disappointed American scientist, see J.C. Sheehan, *The Enchanted Ring: the untold story of penicillin* (Cambridge MA: MIT Press, 1982), pp. 126–197.

21 T.A.B. Corley, 'Sir G.J. Wilkins (1924–2003)', in L. Goldman (ed.), *Oxford Dictionary of National Biography 2000–2005* (Oxford: Oxford University Press, 2009), pp. ; G.J. Wilkins, 'A Record of Innovation and Exports', in G. Teeling-Smith (ed.), *Innovation and the Balance of Payments: the experience in the pharmaceutical industry* (London: Office of Health Economics, 1967), pp. 14–21.

Beecham after Lazell, 1968–1990

Sir Ronald Edwards (1910–1976),[22] Lazell's successor from 1968–1975, was recruited from outside the group, having been chairman of Britain's Electricity Council. Unlike Lazell, he had no business contacts with the United States, and indeed took a personal interest in markets further afield, for example by striving to build up sales in Japan, the second largest consumer of drugs after the US. Moreover, he actively campaigned for British-owned companies, then supplying between them only 7 per cent of the world's pharmaceuticals, to merge themselves into larger units so as to compete on a more equal footing against global giants.

In 1971, he therefore launched a £290 million hostile bid for Glaxo, the leading British drug company and even more research-orientated than Beecham was.[23] Glaxo then had little direct involvement in the United States, where until 1978 it confined itself to licensing and co-partnership agreements. Hence the US did not become an issue in the intensive war of words between the two companies that followed. The Glaxo board opposed the bid with persistence and vigour, following which the UK Monopolies Commission definitively rejected it. Edwards thereafter abandoned his efforts to consolidate the industry around a nucleus of British companies.

How far Beecham penetrated the US market over the next two decades cannot be precisely measured, as its published annual sales and profit data were for the Western Hemisphere as a whole, Australia and New Zealand being included in that area from 1964 to 1973. From then until 1988 the data covered North and South America together. Those reveal that from

22 M. Ackrill, 'Sir Ronald Edwards (1910–1976)', in H.C.G. Matthew and B. Harrison (eds), *Oxford Dictionary of National Biography* (Oxford: Oxford University Press, 2004), Vol. 17, pp. 962–963.

23 PP 1971–1972 XXII: *Monopolies Commission: Beecham Group Ltd. and Glaxo Group Ltd. Report on Proposed Mergers.* HC 341, July 1972; E. Jones, 'The Beecham Bid', *The Business of Medicine: the extraordinary history of Glaxo* (London: Profile Books, 2001), pp. 173–194.

1971/1972 onwards total sales in the Western Hemisphere were overtaken by those in continental Europe, already earning a higher profit margin, no doubt owing to the rapid growth in trade among the European Economic Community's member states since its formation in 1958.

Beecham did, however, improve its prospects in the US on both the pharmaceutical and the medicinal sides by its purchase in 1971 of S.E. Massengill Co. of Bristol, Tennessee, a maker of both prescription drugs and over-the-counter (OTC) or proprietary medicines. The group thereby acquired American managerial and marketing skills, and was encouraged to double the capacity of the Piscataway factory and to set up a second antibiotic factory on the Massengill site.

Edwards was succeeded from 1975 to 1984 by Sir Graham Wilkins, who had set up the global marketing organization for Beecham's penicillins, and by Sir Ronald Halstead (b. 1927) from 1984 to 1985. Halstead had, like Wilkins, held a senior post with the group in North America, and had also taken a marketing course in his spare time at New York University. He shared the tough American business philosophy of tolerating only success rather than the 'soft English [*laissez-faire*] approach', especially to marketing. Later, in charge of the Consumer Products division in Britain, he had introduced popular novelties including mixer drinks, such as tonic water and fruit juices, to use with alcoholic beverages.[24]

During Wilkins' and Halstead's periods as chairmen, Beecham made only seven acquisitions in the United States of medicinal companies as against 16 of those making consumer products; the latter increasingly fell into non-core categories such as cosmetics and home improvement (do-it-yourself) requisites. Halstead justified such disproportionate numbers of consumer product purchases as being equivalent to R&D in pharmaceuticals, on the grounds that they secured new techniques and other useful information for the group.

24 G. Foster, 'Beecham Takes its Medicine', *Management Today* (October 1985), pp. 51–59; J. Saunders, *Nightmare: Ernest Saunders and the Guinness affair* (London: Arrow Books, 1990), pp. 21–25.

To rebut charges that Beecham was dispersing its corporate strength too thinly over so many varied activities, Halstead pointed to the United States, where investors readily supported such profitable diversified-drugs corporations as Bristol-Myers and American Home Products. In fact, within Beecham that diversity was harmfully widening the gulf between the pharmaceutical and consumer product sides of the business, so that mutual incomprehension and hostility were rife even at board level. By 1985 such discord undoubtedly contributed to the group's almost stagnant performance. Earnings from pharmaceuticals were declining, while consumer products scarcely generated enough profits to pay current R&D costs. Kenneth (now Lord) Keith, the vice-chairman – who had been responsible for the earlier coup in 1951 – therefore organized a boardroom putsch which unseated Halstead. Keith openly declared that a younger and more dynamic management was needed, and that such dynamism must come from the top; Halstead was widely held to have spent too much time on outside commitments.

Unable to find a replacement chairman in Britain, Keith looked to the United States: a move he welcomed, as he judged leading UK executives to be 'sleepier, less adventurous, less mobile and less ambitious' than their American counterparts.[25] In 1986 he appointed Robert Bauman (b. 1931),[26] a US citizen who turned out to be Beecham's fifth innovative leader. Bauman had taken an MBA at Harvard, and soon revealed ambitious plans to shake up the Beecham Group by employing Harvard Business School methods. That school's curriculum involved analysing realistic case-studies, as a means of developing leadership and teamwork skills. Its students were trained in decision-making and as team leaders in how to win round those with conflicting views. Having taken over at Beecham, which he found to be both hierarchical and in need of reform, Bauman pursued

25 *Financial Times* (12 November 1985); *Investors Chronicle* (22 January 1988), p. 18.

26 D.G. Hyde and P. Haspeslagh, 'The Making of the *Simply Better* Healthcare Company: SmithKline Beecham (A) and (B)', Fontainebleau, INSEAD, 1994; R.P. Bauman, P. Jackson and J.T. Lawrence, *From Promise to Performance: a journey of transformation at SmithKline Beecham* (Boston MA: Harvard Business School Press, 1997); for Harvard Business School case method see www.hbs.edu/case/case-print.html.

the goal of creating a single corporate ethos, so as to heal the crippling division between pharmaceuticals and consumer products, and thereafter to motivate staff at all levels. He encouraged, bullied and cajoled managers into striving for progressively higher performance targets. Maintaining that innovation in every department of the group was a continuous process, he gave employees no time to relax after each phase of his recovery programme before imposing further demands on them.

To assist him in his objectives, Bauman engaged the US firm of management consultants, Booz-Allen & Hamilton (which had advised Lazell in the 1960s) to devise a plan for building up Beecham, a medium-sized company, into an industry leader. The consultants reported that the group could never reach a dominant position by organic growth alone, and recommended that it should seek a merger of equals, by selecting an American corporation as a future partner.

Beecham's choice was SmithKline Beckman of Philadelphia, then sixth in the global pharmaceutical sales league to Beecham's 23rd. SmithKline had profited massively from the anti-ulcer Tagamet, the world's top-selling drug until overtaken in 1983 by the rival Zantac, which Glaxo, still Beecham's leading UK pharmaceutical competitor, aggressively co-marketed in the United States with Hoffmann-la Roche of Basle, Switzerland. Instead of reinvesting the Tagamet earnings in further R&D at its research centre in Upper Merion, Pennsylvania, SmithKline had acquired a non-core scientific instrument and diagnostic company, Beckman Instruments. Then the regulators raised problems over its second most popular drug, Dyazide, a diuretic for treating hypertension; all the corporation's mistakes and setbacks caused sales to plummet. When its previously higher market capitalization declined to the level of Beecham's, it agreed to a merger.

In 1989, SmithKline Beecham (SB) was registered under British ownership, being then the world's second largest drugs company after Merck Sharp & Dohme. SmithKline's former chairman, Henry Wendt (b. 1931), became non-executive chairman and Bauman chief executive. A logical decision was to site the operational headquarters in the United States, where 40 per cent of SB's combined £3.7 billion sales took place, earning nearly 41 per cent of trading profits there, compared with 15 per cent and 10 per cent respectively in Britain. Beecham's 750 representatives would be added to SmithKline's sales force of 1,150, a total exceeding that of any American competitor. The range of their respective products and R&D activities was broadly compatible, with few overlaps.

As all too many mergers, especially between pharmaceutical companies, had not in the past yielded higher long-term value for investors, Bauman resolved to pursue what the *Economist* graphically termed a 'wrenching top-to-toe reorganization designed to eliminate, once and for all, any vestiges of the two separate companies'.[27] That integration process, overseen by the prestigious New York management consultants, McKinsey & Co., entailed reducing the 63,000 work-force by a tenth. The direct cost of effecting the merger was £77 million, with an extra £500 million provision for restructuring expenses. SB's borrowings therefore rose to £1.75 billion, which took three years to be paid off.

SmithKline Beecham in the 1990s

The resignation of a disaffected head of the pharmaceutical division prompted Bauman in 1990 to recruit from outside two senior executives, the sixth and seventh innovators in Beecham's and SB's history. Both were of European stock but had given long service to American pharmaceutical corporations. Jan Leschly (b. 1950) – the sixth innovative entrepreneur – , a Dane and former president of Squibb Corporation of Princeton, New Jersey, became head of pharmaceuticals, and proved to be a rigorous taskmaster in the Bauman mould. Jean-Pierre (J.P.) Garnier (b. 1947), French-born but a naturalized US citizen and until lately a senior executive in Schering-Plough of Madison, New Jersey, was appointed head of North American operations.[28] In 1993 they astutely closed the gulf between

27 *The Economist* (4 November 1989), p. 122.

28 For Leschly see *Sunday Times* (25 January 1998); A. Davidson, 'J.P. Garnier', *Management Today* (March 2004), pp. 38–43.

pharmaceuticals and consumer products by restructuring the latter division under the name of Healthcare Products. In addition to OTC medicines, those would comprise health drinks such as Lucozade, Ribena and Horlicks, as well as oral products to include Macleans' toothpaste and mouthwashes. Articles designed for hair and body adornment, for example Brylcreem – Hill's inspired purchase of 1939 – shampoos and bath additives, had to be sold off. When Bauman retired in 1994, Leschly took over the chief executive's post, Garnier becoming his deputy as chief operating officer. Wendt, who retired at the same time, was succeeded as chairman by the British Sir Peter Walters (b. 1931), former chairman of British Petroleum (BP) and the Midland Bank, familiar with American business methods from two periods in BP's New York office.

However costly in financial resources and human effort, SB's successive periods of American-led reorganization since Bauman's arrival gave it a corporate strength which Beecham had conspicuously lacked before 1985. Leschly and Garnier were therefore well placed to meet the three major challenges of the 1990s: the genetic revolution, the managed care revolution, and the rival company Glaxo's hostile take-over of Wellcome, which created the world's second pharmaceutical leader, still behind Merck Sharp & Dohme.

The genetic revolution had as its core the sequencing of the human genome, expected to uncover the genetic role in people's susceptibility to disease, and hence to permit the development of more specific and efficacious medicinal remedies.[29] In 1993, SB therefore bought a 7 per cent share in Human Genome Sciences Inc. of Rockville, Maryland, to gain access to its database, the most extensive in the world. Having been the first British pharmaceutical company to involve itself directly in genomics, SB later licensed data to competitors including Glaxo.

The managed care revolution occurred because national providers of prescription drugs, such as governments in Britain and Japan and insurance

29 PP, House of Commons Science and Technology Committee, Session 1999–2000, 4th Report, 'Glaxo Wellcome and SmithKline Beecham', HC 207-II (2000), Vol. II, pp. 9–10.

companies in the United States, were striving to keep down those drug costs.[30] In America, pharmaceutical benefit management (PBM) companies therefore operated as wholesalers able to secure the most economical deals from producers through high-volume purchasing. Leschly acquired a leading PBM company, Diversified Pharmaceutical Services Inc. (DPS) of Minnesota. He intended to market SB products through that channel, and also to obtain valuable marketing data about the pattern of demand for drugs by its 6 million members. Then the Food and Drug Administration outlawed exclusive deals of that kind, a move which sharply reduced the profitability of DPS. That company was finally sold off in 1999.

Meanwhile Glaxo, having in 1971 defeated Edwards' bid, had restructured itself along multidivisional lines and disposed of all its non-pharmaceutical products. In 1978 it moved directly into the United States by acquiring Meyer Laboratories, of Fort Lauderdale, Florida, and four years later started up extensive R&D operations at Research Triangle Park, near Raleigh in North Carolina. Lately, as stated above, it had done well out of the substantial profits earned by its anti-ulcer blockbuster drug, Zantac. However, Sir Paul Girolami (b. 1926), its chairman from 1985 to 1994, had rejected all plans to grow by means of outside acquisitions. After his retirement, Sir Richard Sykes (b. 1942) became chief executive and further increased Glaxo's presence in the US, most notably by additions to marketing capacity there. He had worked in Squibb Corporation from 1977 to 1986, and subsequently as chairman appointed an American, Robert Ingram (b. 1942), as chief executive. In 1995 Sykes reversed Girolami's anti-acquisition policy by making a hostile bid for Wellcome, a British company standing 18th in the world sales league, and thus some way behind Glaxo's number two global position, Merck Sharp & Dohme remaining the leader. As the patent on Zantac was due to expire in 1997, the acquisition of Wellcome would raise expenditure on R&D, and Sykes's hope of evolving replacement drugs, by a third, to more than £1 billion annually.[31]

30 Hyde and Haspeslagh, 'Making of *Simply Better* Healthcare' (B), pp. 5–8.

31 PP, House of Commons Science and Technology Committee, Session 1994–1995, 2nd Report, 'The Glaxo Bid for Wellcome plc' (1995), pp. v–ix; 'Sir Richard Sykes', *Management Today* (December 1995), pp. 56–60.

Towards GlaxoSmithKline, 1995–2000

The formation of Glaxo Wellcome, which precipitated subsequent mergers of pharmaceutical giants around the world, brought home to SB its vulnerability to take-over, as it steadily declined to 11th in the global ranking of pharmaceutical sales. In 1996, therefore, Leschly secretly instructed its financial advisers, Morgan Stanley, to search for a fresh corporate partner.[32] As Beecham had found in 1989, SB on its own stood no chance of becoming a dominant player in the industry.

Glaxo Wellcome was the most obvious choice as SB's candidate for a merger. Leschly therefore approached Sykes, whose unacceptable condition was that SB must be prepared to reduce its size to one-third of a potential combine. Because that would involve its selling off non-pharmaceutical assets, including many healthcare products, Leschly declined and approached another name on Morgan Stanley's list. That was the diversified-drugs corporation, American Home Products (AHP) of Philadelphia, at the moment ranked seventh globally. Their merger would create an undoubted world leader with a market capitalization of £75 billion, a record figure in any industry.

However, AHP was widely regarded as bureaucratic and financially conservative, whereas the SB managers were said to be an 'egotistical bunch' who wanted 'to conquer the world'.[33] Even so, Leschly and Garnier pursued that merger, perhaps because both had residences in Philadelphia where they spent most of their working lives. British investors and financial commentators, on the other hand, were afraid that SB would pass out of UK control as an amalgamation with AHP would inevitably transfer the combine's ownership to the United States.

That fear was echoed by the UK prime minister, Tony Blair, according to industry reports which Glaxo Wellcome never denied. To avoid the loss to the economy of important corporate assets in what was acknowledged

32 *Times* (2 February 1998).
33 *Times* (21 January 1998).

to be one of Britain's most successful industries – the other being oil –Blair apparently contacted Sykes and requested him to find some means of keeping SB in British hands.[34] At any rate, Sykes at once interrupted the negotiations with a telephone call to Leschly, asking 'Jan, is it too late [to discuss a merger]?'[35] Leschly's response was to break off talks with AHP. A high-level team, comprising Sykes, two other Glaxo Wellcome executives and Leschly and Garnier for SB, began virtually non-stop talks which within a few days hammered out an agreement to merge, based on a 59.5 to 40.5 per cent equity split in Glaxo Wellcome's favour. Sykes was to be executive chairman and Leschly chief executive. Then three weeks later Sykes abruptly called off the deal, claiming that it would not work.

The totally unexpected collapse of the merger agreement, which precipitated a fall of £13.5 billion in the combined value of the two companies' shares on the London Stock Exchange, was commonly attributed to personal rivalries, namely the clash of Sykes's and Leschly's egos. More recent evidence suggests a more fundamental reason: a struggle over the direction which the pair should take, whether to pursue British- or American-type strategies.[36]

Sykes's basically British approach was to spend as much as possible, in his case 15 per cent of turnover, on R&D: a key element in his programme to meet the substantial costs of genetic research and of the new computer-assisted methods of drug design; combinatorial chemistry involved testing thousands of molecules for biological activity. In May 1999 – when he and Leschly were widely believed to be too far apart to resolve their differences – he expressed his hope of 'two big successful organizations coming together, not to protect future earnings growth but actually to increase critical mass to really outperform the industry [especially in the

34 *Sunday Times* (1 March 1998).

35 PP, House of Commons Science and Technology Committee, Session 1997–1998 (see fn.5), p. 23.

36 S. Randles, 'Complex Systems Applied? The Merger that Made GlaxoSmithKline', *Technology Analysis and Strategic Management*, 14 (2002): 346–349; B. Bátiz-Lazo, 'GSK – a merger too far?', *ICFAI Journal of Mergers and Acquisitions*, Vol. 1, No. 4, pp. 75–89.

United States] … The more effort, the more money, and the more power you can put [in]to research, the stronger the [merged] company is going to be'.[37] Thus a policy of maximizing research activity was expected to take precedence over financial and marketing considerations.

Leschly, on the other hand, pursued a more money-saving policy, in the main following the American-based Pfizer's recent practices. That included concentrating production into five therapeutic categories, such as the central nervous system, the gastro-intestinal area and anti-infectives: ones he deemed most appropriate for tackling the American market, in which SB made 51 per cent of its sales, compared with Glaxo Wellcome's 42 per cent. SB, by then operationally an American corporation, sought to economize on R&D expenditure by the in-sourcing of compounds from small biotech companies for further development and ultimate marketing, and co-partnerships with larger rivals to purchase from them marketable drugs in therapeutic areas it did not comprehensively cover. Leschly also planned to increase productivity within existing R&D outlays by dividing research scientists into small teams, so as to encourage among them greater flexibility and personal initiative.

Contrary to popular belief, after the breakdown of 1998 Sykes and Leschly not only remained on good terms but also collaborated in harmonizing the two companies' working practices. For example, as their production methods differed widely, Sykes adopted SB's manufacturing supply chain system, of preparing therapeutically active compounds in the UK and then shipping them to factories in selected parts of the world for making up into finished drugs. Leschly in turn followed Glaxo's earlier example of concentrating on pharmaceuticals; while not sacrificing health drinks such as Lucozade and Horlicks, in 1999 he sold off the non-core businesses, Diversified Pharmaceutical Services and SB's clinical laboratories. The two men's decision to accept each other's best practices, according to later evidence by Garnier, 'ended up [in] transforming our companies'.[38]

As the year 1999 wore on, it became clear that Leschly's American-based strategy was yielding more tangible results than the British route

37 *Pharmaceutical Executive* (May 1999), p. 37.

38 PP, House of Commons Science and Technology Committee, Session 1999–2000 (see fn. 27), Vol. II, p. 5.

pursued by Sykes. In the four years since 1995, SB's turnover had increased by 20 per cent and its trading profit by almost 40 per cent, while Glaxo Wellcome achieved only an 11 per cent increase in turnover, trading profit improving by less than 2 per cent. Sykes's efforts to launch replacements to Zantac – off-patent since 1997 – suffered a setback because the US Food and Drug Administration blocked on safety grounds three drugs, a gastro-intestinal one, an antibiotic and one to combat diabetes, while his anti-influenza drug Relenza lost sales when it turned out to be less effective than predicted. Sykes allowed the number of his employees to increase by no less than 14 per cent in 1998–1999, thereby increasing overheads. SB by contrast reduced its overall workforce by nearly 9 per cent, the losses being mostly in manufacturing and administration, whereas employment in R&D and in sales and marketing was permitted to rise.

Thus two crucial developments had taken place: the two companies' organizations were moving closer to each other, and SB was catching up in relative terms. Towards the end of 1999, the two leaders seem to have reached a self-denying compact in order to facilitate an eventual merger. Rumours were already in circulation that Sykes would be accepting an outside post that precluded him from serving as a full-time executive chairman. Then in December SB announced that Leschly, due to retire at 60 by the end of 2000, would be stepping down in April. Garnier, then just 52, would succeed him as SB's chief executive. After that announcement, Sykes lost no time in inviting Leschly, Garnier and their team to a secret rendez-vous in France for merger discussions. There the two sides reached an agreement in mid-January 2000 to form GlaxoSmithKline (GSK), under Garnier as chief executive.[39]

Sykes, disclosing that he would shortly become rector of Imperial College in London, was to be non-executive chairman. Robert Ingram, whom he had earlier earmarked to be GSK's chief executive, was no more than Garnier's deputy as chief operating officer. Likewise, Tachi Yamada of SB, a American citizen resident there, became head of R&D, while Sykes's previous nominee for that post, James Niedel, was appointed chief science and technology officer. Thus although Glaxo Wellcome held 58.75 per cent of the shareholding stake, to SB's 41.25 per cent – a marginal adjustment

39 *Sunday Telegraph* (5 December 1999); *Financial Times* (17 January 2000).

of 0.75 per cent over the 1998 figure – SB people held the key executive positions in the new company's top management.

Before the merger could be completed, the two companies had to satisfy the demands of the Federal Trade Commission (FTC), the American competition regulator. The FTC insisted on certain drugs, namely those to treat herpes and nausea suffered by cancer patients, being sold off to rival firms. It then claimed that the companies' smoking cessation drugs would create a monopoly in that therapeutic area, but then dropped further insistence on divestment after an appeal. In December 2000, GSK at last came into being. Its head office was in Britain, to enjoy the advantage of easy access to the London stock market and other financial services there. However, on Garnier's insistence, the operational headquarters would be at SB's former installation in Philadelphia. Further evidence of American influence in the new GSK was that about half of the combined £2.4 billion annual R&D expenditure would be spent in the United States. Thus it appeared as if a policy of progressively adopting American business practices had delivered to SB a decisive say in the administration of GSK, in preference to Glaxo Wellcome's essentially British strategy of concentrating on the 'critical mass' of R&D.

Conclusion

The theoretical framework, at the beginning of the present article, found that entrepreneurs, in their roles as innovators, would strive to succeed by exploiting the competitive advantages open to them. Those advantages could operate at the country or industry level, or indeed spring from the entrepreneur's own innate or trained skills. Thus the country which, thanks to educational, cultural and entrepreneurial-intensive factors, holds many of the advantages required for business success, is the United States, enjoying a very extensive home market for goods and services.

As to the industry-specific considerations, American enterprise can be expected to enjoy a crucial position in pharmaceuticals. The country's educational and other advantages listed above are likely to act as a spur to inventiveness there, in the present case the development of new types of medicinal drugs and then the vigorous marketing of them. In 2000, no fewer than five of the world's 10 largest pharmaceutical corporations were American, or six if GlaxoSmithKline's operations, run from that country, are included.[40]

Yet even an innovative firm based in a country not so well endowed with competitive advantages can make good in a technologically-based industry such as pharmaceuticals. Firm-specific factors are by definition bestowed randomly, since one firm may pursue innovations and the next be content to be merely a follower of others. In the economically less advantaged UK, Beecham by innovatory strategies raised itself from obscurity to industry leadership, at first in patent medicines, and between the 1950s and 2000 scaling the high entry barriers into pharmaceuticals.

Even so, in that last half-century, Beecham could not have achieved its noteworthy status without reliance on American know-how. Britain did have a strong tradition of scientific discovery on both the organic chemistry and biomedical sides, to such an extent that most overseas pharmaceutical companies set up research facilities in the country so as to benefit from that expertise. Yet when Beecham's scientists succeeded in discovering the penicillin nucleus, it had to look for assistance from the American-based Bristol-Myers to manufacture its first brand of semi-synthetic penicillins commercially and build an entirely new antibiotic factory in Britain for that purpose. Beecham's need for up-to-date technical knowledge spurred it on to form a research establishment in the United States, which would secure rapid information about the latest American discoveries.[41]

After 1989, Beecham by merger with SmithKline transformed itself by gaining access to the latter's tangible and intangible resources, extensive enough for the resulting combine, SmithKline Beecham, to base its operational headquarters in the United States. It then recruited Leschly and Garnier, both highly experienced in pharmaceuticals there. To balance the foregoing data about global pharmaceutical corporations, in 1988 no fewer than six of the top 10 drugs dispensed in the United States were the products of British-owned firms, Beecham's broad-spectrum antibiotic

40 *Sunday Times* (16 January 2000).

41 Beecham Group, *Annual Report for 1966/7* (1967), p. 9.

Amoxil heading the list.[42] In 1995, ten of the world's 35 major prescription drugs had been developed in UK laboratories, often by the overseas branches referred to above.[43]

Oddly enough, although from the 1960s onwards Beecham was intensively exporting to and investing in continental European countries – assisted by the American consultants, Booz-Allen & Hamilton – and in 1969 erected a factory for antibiotics in Belgium, it never seems to have sought technical information from the Swiss, German and other pharmaceutical companies on the continent that were patently as innovative as itself.

To sum up, certain overriding questions remain. Why did British pharmaceutical companies in general succeed in holding their own against American giants which at the country, industry and firm levels held so many key competitive advantages? How far was their success due to persistent searching for techniques and ideas from the United States? A broader investigation is badly needed, to study the transatlantic relationships of Glaxo, Wellcome, ICI (now Astra-Zeneca) and smaller UK drug companies. The present study of Beecham's experiences may throw up some provisional findings of assistance to further research.

42 J. Taggart, *The World Pharmaceutical Industry* (London: Routledge, 1993), p. 155.

43 T.A.B. Corley, 'The British Pharmaceutical Industry since 1851', in L. Richmond, J. Stevenson and A. Turton (eds), *The Pharmaceutical Industry: a guide to historical records* (Aldershot: Ashgate, 2003), p. 28.

SERGIO SISMONDO

8 Linking Research and Marketing, a Pharmaceutical Innovation

> In 1991, William C. Steere, Jr., the new CEO of Pfizer, started a re-organization of the company. 'He told me he had three priorities,' recalls [Steve] Conafray [then head of political affairs at Pfizer] ... 'The first one was get marketing and research closer together. The second one was get marketing and research closer together. And then he said the third one was get marketing and research closer together.' Everyone got the message.[1]

Introduction: a phenomenon

Let us begin by looking at a recent drug, rofecoxib, vigorously marketed as Vioxx by Merck. Rofecoxib has attracted considerable media attention because in 2004 Merck voluntarily withdrew the drug, noting new evidence that its users faced increased chances of cardiac arrest. That action prompted many lawsuits, some closer looks at the research on rofecoxib, and a claim that Merck may have inappropriately withheld data. This paper ignores these issues. Instead, it draws attention to a way in which rofecoxib is a typical high-volume drug of our era, or at least was until it was withdrawn from the market. In particular, rofecoxib is not most importantly characterized by shortage of research on its defects, but by a super-abundance of research, mainly clinical research, on its benefits.

1 G. Critser, *Generation Rx: how prescription drugs are altering American lives, minds and bodies* (New York: Houghton Mifflin, 2005), p. 91.

Rofecoxib is a COX-2 inhibitor. The drug was approved by the US Food and Drug Administration (FDA) in 1999, for relief of osteoarthritis, acute pain, menstrual pain; approval by other national regulatory bodies followed. It was the second COX-2 inhibitor approved by the FDA, after Pfizer's (then Pharmacia's, before subsequent mergers) celecoxib (Celebrex) in 1998. Efforts to create large markets, and competition between the two – and a further Pfizer COX-2 inhibitor, valdecoxib (Bextra) – led to substantial advertising and marketing campaigns. Both rofecoxib and celecoxib were 'blockbuster' drugs by 2001, boasting $2.6 billion and $3.1 billion in sales respectively that year. Blockbuster drugs are usually defined as ones with sales of more than US $1 billion per year, and in recent years the pharmaceutical industry has been dominated by attempts to create blockbusters; they will likely remain important despite pressures to create more 'niche' drugs.[2]

A Medline search of core clinical journals (performed in March, 2006) for the keyword 'rofecoxib' pulls up 161 articles published between 1999 (the year the drug was first approved) and 2003, and a search for the keyword 'celecoxib' pulls up 196 articles in that same period; core clinical journals are approximately 130 journals chosen for the Abridged Index Medicus, on the basis of their importance and because they publish articles on clinical research of wide interest, either generally or within major specialties.[3] Though many of the articles turned up by that search are other types of systematic analyses, such as reviews or reports of meta-analyses, the majority of them report on clinical trials involving these drugs. And almost all of the articles are centrally *about* these drugs, perhaps among others, in some important sense.

Here we have an interesting phenomenon: blockbuster drugs are blockbuster research targets! The phenomenon is not limited to COX-2 inhibitors, but can be as easily seen among other classes of drugs. For example, a keyword search on the cholesterol-lowering agent atorvastatin (Lipitor) brought up 512 articles in core clinical journals in a similar period, and the

2 See for example G. Carpenter, 'The Notable Marks', *Pharmaceutical Marketing* (6 September 2006) http://www.pmlive.com/

3 Abridged Index Medicus (AIM) Journal Titles. http://www.nlm.nih.gov/bsd/aim.html (accessed 5 October 2006).

proton-pump inhibitor for acid-related gastrointestinal problems omeprazole (Losec, Prilosec, Nexium) had 1,155 articles in core clinical journals in a similar period. New drugs that are 'orphans,' for which there are few potential patients, are associated with small or only modest amounts of published research.

Thus there is a fairly good correlation between the number of clinical research articles published in medical journals and the market status of a drug. The rest of this paper discusses some of the reasons for that correlation.

Large Potential Markets Beget Research

This phenomenon has existed for at least 30 years, though it may be not much older than that. Taking the period 1975 to 1979 as a comparison, there were approximately 100 articles in core clinical journals on each of a number of well-prescribed drugs, such as ibuprofen (then Brufen or Motrin), represented by 83 articles, though there are a few very prominent drugs featured in considerably more articles: diazepam (Valium) sees 400 articles in core clinical journals, and for methylprednisolone (Medrol) there are 313.

There are obvious structural reasons why there should be some correlation between numbers of articles published and market status. Clinical researchers want to do research, and it is much easier to do a study of a drug for which there are many potential consumers than for which there are few. Thus it is, perhaps, unsurprising that there would be few studies of genuine 'orphan' drugs: such studies are problematic to run, as it is difficult to recruit patients. But while bottlenecks because of small numbers of patients might explain cases in which there are few studies, they do little to explain much of the rest of the variation.

Between 1999 and 2003 there were only three articles in core clinical journals on the new drug cevimeline, used to treat Sjögren's syndrome, an auto-immune disease with an incidence rate of between one-third and

1 per cent of the population of Western countries (an NIH estimate).[4] Compare these numbers with the 1,155 articles on omeprazole, indicated for gastroesophageal reflux disease (GERD), affecting an estimated 3 per cent of Western populations – though many more people are affected by conditions that are potential precursors to GERD, such as frequent heartburn (again, an NIH estimate).[5] The major difference between these two pairs of drugs/illnesses is not the difference between the number of potential patients with Sjögren's syndrome and the number of potential patients with GERD, but that omeprazole has become very widely prescribed, especially in the United States, for heartburn and other problems involving excess stomach acid. This wide usage is a *result* of marketing by AstraZeneca (and earlier marketing by Glaxo of Zantac for GERD); AstraZeneca's direct-to-consumer advertising for the 'purple pill' has taken on an iconic status, commonly cited by both marketers and critics as among the most aggressive and successful campaigns for any drug.[6] Meanwhile, clinical research on omeprazole is presumably not driven by an intrinsic medical interest in heartburn. We gain a coherent picture if we see both the number of prescriptions and the number of articles as driven by efforts to fill the largest potential market. This phenomenon is particularly well studied in the area of psychopharmacology, where work by such researchers as David Healy and Charles Medawar and Anita Hardon has established that the growth of interest in depression and other recently important psychiatric conditions has been fuelled by the promotional work of pharmaceutical companies.[7] But the phenomenon is widespread: returning to the case of rofecoxib, large numbers of articles were being published at the same time that Merck was heavily advertising Vioxx, creating the potential for it to be

4 CureResearch.com. 2006. 'Conditions by People Affected'. http://www.cureresearch.com/lists/preval.htm (accessed 1 April 2006).

5 Idem.

6 See C.W. Schmidt, 'Matters of Money', *Modern Drug Discovery*, 7 (March, 2003). http://pubs.acs.org/journals/mdd/index.html (accessed 5 October 2006).

7 D. Healy, *Let them Eat Prozac* (Toronto: James Lorimer, 2003); idem, 'Shaping the Intimate: influences on the experience of everyday nerves', *Social Studies of Science*, 34 (2004): 219–246. C. Medawar and A. Hardon, *Medicines out of Control? Antidepressants and the conspiracy of goodwill* (Amsterdam: Aksant Academic Publishers, 2004).

seen as a general painkiller, rather than simply a drug for managing osteoarthritis. At least some of the time, then, important drugs and conditions become important because commercial interests support them.

PhRMA, the pharmaceutical industry lobbying agency in the United States, claims that its members spent US$39 billion (thousand million) on research and development in 2005,[8] the vast majority of which was spent on clinical trials. The entire budget of the National Institutes of Health in that year was US$28 billion, some of which was spent on research unrelated to pharmaceuticals, and much of which was spent on research other than clinical trials. The pharmaceutical industry therefore provides a very high percentage of the support for clinical trials in the United States, and probably the majority of that support. Consequently, the industry has the means to have considerable influence on which drugs and conditions are worth studying.

Articles are Marketing Tools

According to a study commissioned by the *New England Journal of Medicine*, physicians view reprints from NEJM as qualitatively superior to those from similar journals. The survey, conducted by the Matalia Group, questioned 3,400 physicians and residents in cardiology, family practice, hematology/oncology, infectious diseases, internal medicine and oncology. The study concluded the following:

- Next to peer-reviewed journals, reprints are the most reliable information source for physicians.
- NEJM reprints influence physicians' behaviour.

8 PhRMA. 2006. 'R&D Investments by America's Pharmaceutical Research Companies Near Record $40 Billion in 2005'. http://phrma.org/ (accessed 5 May 2006).

- In most specialties surveyed, physicians recall receiving more reprints from NEJM than any other journal.
- NEJM reprints are considered significantly more relevant and valuable to physicians' practices than reprints from other journals.[9]

As the above notice in the Massachusetts Medical Society's 'Vital Signs' indicates, reprints of medical articles are valued for their ability to influence physicians' behaviour. The notice is effectively an advertisement for a product sold by Massachusetts Medical Society, reprints from its publication, the *New England Journal of Medicine*. Pharmaceutical companies and their agents are the largest buyers of reprints, buying them in numbers ranging from the large to the very large. Richard Smith, a former editor at the *British Medical Journal*, reports that pharmaceutical companies can spend more than a million dollars on reprints of a single article.[10] Even if that figure represents an exceptional case – it suggests a print run in the hundreds of thousands – it remains that NEJM can quote standard prices on orders of 10,000 reprints of an eight-page article in black and white (US$15,974, quote emailed to author). Most other medical journals also sell reprints, and their publishers advertise them, and associated services, to pharmaceutical companies. Blackwell, for example, claims to be able to deliver reprints within 10 working days of an order, and to be able to add company logos, advertisements, or product information.[11]

Reprints are deployed by drug representatives in their visits to physicians, or, as a last resort, are mailed to physicians as part of marketing campaigns. They provide the scientific evidence to back up drug representatives' and advertisers' claims. When holding a reprint of a scientific article, drug

9 Massachusetts Medical Society, 'Reprints Study Shows Value of New England Journal of Medicine Content'. http://www.massmed.org/Content/NavigationMenu3/BackIssues/2003/May2003/InsideMMS/Reprints_Study_Show.htm (accessed 1 April 2006).

10 R. Smith, 'Medical Journals Are an Extension of the Marketing arm of Pharmaceutical Companies', *PLoS Medicine*, 2 (2005): e138.

11 Blackwell 2006 Advertisement for Blackwell Reprints: http://www.blackwellpublishing.com/MedStudent/healthcare/reprints.asp (accessed 1 April 2006).

representatives stand simultaneously as promoters of products and as communicators of scientific information. Reprints thus facilitate companies' claims to be part of an 'ethical pharmaceutical industry' distinguished from the patent medicine companies of an earlier age (and their descendants, still ably competing today).

There are 90,000 drug representatives (officially 'pharmaceutical sales representatives') in the United States,[12] directing their attention to some subset of the approximately 700,000 active physicians in the US[13]. Earlier labels, 'detail men' or 'detailers,' suggested the scientific role drug representatives were supposed to play: as communicators of detailed information, as translators of medical research into terms ordinary physicians can understand. They do much more than that, of course: they are best known for dispensing free samples, for taking physicians and their staff to lunch, and for leaving well-chosen reminders – from pens with logos to fine wines – of them and their products. Because doctors generally do not want to see themselves as influenced by mere gifts or advertising pitches, the scientific information is particularly important. Thus already in the 1950s handbooks for detailers established rules of behaviour that placed them as scientific colleagues of doctors.[14]

As has been noted many times, the randomized, controlled trial (RCT) is relatively new to medicine. Although one can find a number of fore-runners, credit for the first RCT in medicine is often given to Austin Bradford Hill, for his 1946 trial of the effect of streptomycin on tuberculosis, and for his advocacy of RCTs in medicine.[15] The rise of the RCT over the following few decades, to become the 'gold standard' of clinical research by the 1990s, followed such events as the birth defects caused by thalidomide,[16]

12 C. Elliot, 'The Drug Pushers,' *The Atlantic Monthly* (April 2006): 82–93.

13 US Census Bureau. 2006. http://www.census.gov/ (accessed 2 May 2006).

14 J. Greene, 'Attention to "Details": etiquette and the pharmaceutical salesman in postwar America,' *Social Studies of Science*, 3 (2004): 271–292.

15 S. Timmermans and M. Berg, *The Gold Standard: the challenge of evidence-based medicine and standardization in healthcare* (Philadelphia: Temple University Press, 2003); H.Marks, *The Progress of Experiment: science and therapeutic reform in the United States, 1900–1990* (Cambridge: Cambridge University Press, 1997).

16 Timmermans and Berg, *The Gold Standard.*

and extensive advocacy by statisticians and statistically-minded medical researchers.[17] Since the 1970s, physicians have been repeatedly told that they should pay attention to only one kind of reliable information, reports of randomized, controlled trials.[18] Other forms of information, to the extent that they are reliable, are built on top of those RCTs: meta-analyses and review articles are analyses of RCTs.

The intersection of research and marketing, then, is dominated by the intersection of randomized controlled trials and marketing. If pharmaceutical companies are to approach physicians to convince them of the merits of their products, they are almost forced to do so using the results of RCTs. Physicians – not to mention most other people who buy or control the buying of drugs – want their drugs to be part of a rational world centred on health. Any visible aspect of drug research, development, or promotion that is not part of a logic of health is immediately suspect. Thus pharmaceutical companies have a strong incentive to ensure that there is useful clinical research on products they wish to sell widely.

It is not merely that pharmaceutical companies support research for independent reasons, and reap the benefits of having articles for their representatives to deploy and opinion leaders who support their interests. The companies recognize the value of research, and support it for that value. In the early 1990s, Pfizer wanted to sell its selective serotonin reuptake inhibitor (SSRI) sertraline (Zoloft) in a market dominated by fluoxetine (Prozac) and a few other early SSRIs. To catch up with the competition, Pfizer applied its CRAM ('Central Research Assists Marketing') program to the problem. The company commissioned research to show that sertraline had a cleaner breakdown route, and was associated with higher quality of life, than its competitors.[19] The data could then be used at seminars, sales events, and continuing medical education (CME) to mark distinctions between sertraline and other SSRIs. Meanwhile, Pfizer created programs to educate physicians about depression, which could be expected both to increase the number of diagnoses of depression and the number of

17 Marks, *The Progress of Experiment.*

18 H. Sacks, T.C. Chalmers and H.Smith, Jr,, 'Randomized versus Historical Controls for Clinical Trials'. *The American Journal of Medicine*, 72 (1982): 233–240.

19 Healy, *Let them Eat Prozac*; Critser, *Generation Rx.*

prescriptions for that product: unbranded disease awareness campaigns have substantial effects on sales of particular brands, perhaps because of consilience between unbranded and branded information.[20]

Opinion Leaders and the Use of Research

Medical journal articles themselves, even without being reprinted, even without being transported to doctors' offices, must have some marketing value. This value of research in isolation is limited, given the few physicians who have the time or inclination to read widely in medical journals. But presumably a significant bulk of publication will have some effect, if it conveys a sense that a drug is attracting attention and being taken seriously by the research community. And if articles are read and assimilated by 'opinion leaders' – who may serve as consultants on research, as members of speakers bureaus, or as educators – and then translated into forms by which their conclusions reach more prescribers, then their value as marketing tools increases. Of course if articles are read and translated by journalists, their value is even higher.

Influencing opinion leaders is a key part of many promotional campaigns. An advertisement for a report (*Pharmaceutical Thought Leaders*, priced at US $6,995) by the research firm Cutting Edge Information, says:

> The most lucrative thought leader relationships are steeped in cutting edge scientific research that helps the company promote its treatments while allowing the physician to advance her interest – and influence – in a given area.[21]

20 G.W. 't Jong, B.H.C., Stricker, and M.C.J.M. Sturkenboom, 'Marketing in the Lay Media and Prescriptions of Terbinafine in Primary Care: Dutch cohort study', *British Medical Journal*, 328 (2004): 931.

21 Cutting Edge Information, Advertisement for *Pharmaceutical Thought Leaders: Brand Strategies and Product Positioning*, 2004, http://www.pharmathoughtleaders.com/index.htm#body (accessed 1 April, 2006).

Opinion leaders are courted starting as early as Phase I research does, and by one estimate pharmaceutical firms spend an average of US $40 million per brand supporting opinion leaders.[22] Expenses come in such varied forms that firms may not know how much they are paying even individual opinion leaders: an article in *Pharmaceutical Executive* worries that not knowing how much is being paid to individual opinion leaders could result in public relations disasters.[23] The problem is a difficult one because 'the number of physicians available to play these roles cannot keep pace with demand,' and as a result individuals play many roles, from clinical trial investigator to expert speaker to advisory board member.[24] The language of supply and demand suggests opinion leaders' status as resources.

Opinion leaders are not given only monetary reward. They are given status, or support for status, by being given forums. In some of their roles opinion leaders are given research support, allowing them to conduct research that leads to published articles. They may also be given access to valuable data, collected in studies in which they may or may not have participated.

Most importantly, though, opinion leaders are given direct access to publication in the form of ghost-written articles. The honorary authors who append their names to the tops of the articles simultaneous sponsor the research and pad their publication records; they stand as guarantors of quality, even though they may have had little control over the claims that are made under their names.

Drug companies pay for research by contract research organizations, analyse the research in-house, and hire medical education and communication companies (MECCs) to plan publications. Those MECCs hire ghost-writers to put together manuscripts, ask academic researchers who have had little or no prior connection to the research to serve as authors on those

22 Idem.

23 K. Chaudhry and A. Love, 'Key Opinion Leaders Interactions with Pharma', *Pharmaceutical Executive*, 1 October (2005), http://www.pharmaexec.com/pharmaexec/articleDetail.jsp?id=197784&pageID=1 (accessed 1 April, 2006).

24 Idem.

manuscripts, and then shepherd the manuscripts through to publication.[25] This whole process is known as 'publication planning', though we might better see it as the 'ghost management' of research and publication.[26]

Let us return to rofecoxib briefly. As a result of court cases, Merck's ghost management of the medical literature on the drug became available for scrutiny. Documents reveal that Merck was involved in the production of 96 articles on the drug, most of them published in core clinical journals.[27] Most of the published articles acknowledged Merck's sponsorship, but none revealed the process by which Merck and its agents were intimately involved in their production. The very high number (96) of articles accords well with the number of articles on sertraline (Zoloft) that were part of a Pfizer publication plan that similarly became available as the result of litigation. In the late 1990s, 85 manuscripts on sertraline were coordinated for Pfizer by the MECC Current Medical Directions; those constituted approximately 40 per cent of the medical literature on sertraline, and tended to be published in more prominent journals with more prominent authors than apparently independent research on the drug.[28] Thus, blockbuster drugs are blockbuster research targets because their makers contribute considerable amounts of that research.

25 S. Sismondo, 'Ghosts in the Machine: publication planning in the medical sciences', *Social Studies of Science*, 39 (2009): 171–198; A. Barnett, 'Revealed: how drug firms "hoodwink" medical journals', *The Observer* (Sunday 7 December 2003). http://observer.guardian.co.uk/uk_news/story/0,6903,1101680,00.html (accessed 16 August, 2004); Healy, *Let them Eat Prozac*.

26 S. Sismondo, 'Ghost Management: how much of the medical literature is shaped behind the scenes by the pharmaceutical industry?', *PLoS Medicine*, 4, 9 (2007): e286; A. Matheson, 'Corporate Science and the Husbandry of Scientific and Medical Knowledge by the Pharmaceutical Industry', *Biosocieties*, 3 (2008): 355–82.

27 J.S. Ross, K.P. Hill, D.S. Egilman, and H.M. Krumholz, 'Guest Authorship and Ghostwriting in Publications Related to Rofecoxib: a case study of industry documents from rofecoxib litigation', *Journal of the American Medical Association*, 299 (2008): 1800–12.

28 D. Healy and D.Cattell, 'Interface between authorship, industry and science in the domain of therapeutics'. *British Journal of Psychiatry*, 183 (2003): 22–27.

Pharmaceutical companies and their agents support ghost management because of the value of independent research – and in the support of ghost management we can again see that the companies *recognize* the marketing value of research.

Active Researchers are Intermediaries

Pharmaceutical companies operate in many arenas at once: medical, research, educational, financial, regulatory, and legal. To a greater or lesser extent, the expertise of academic medical researchers is relevant to all of these arenas. Thus the companies cultivate researchers whose views support their interests, and cultivate views that support for their interests among researchers.

It is well established that there are associations between research funding and research results, and evidence that those associations can be extended to views as well. In a review of studies on the relations of support and results, Joel Lexchin and co-workers,[29] show that industry-funded clinical studies report positive results more heavily than non-industry-funded studies, with an odds ratio of 4.05. The ghost management of research helps to explain that correlation, though undoubtedly sponsorship by itself has some effects on results.

An observation from the regulatory debates around COX-2 inhibitors suggests that funding influences researchers' views. After Merck's voluntary withdrawal of rofecoxib, and evidence showing that a Pfizer COX-2 inhibitor, valdecoxib (trade name Bextra) also increased heart risks, an FDA advisory committee made recommendations about the three COX-2 inhibitors that had been approved. On that committee 10 members had

29 J. Lexchin , L.A. Bero, B. Djulbegovic, and O. Clark, 'Pharmaceutical Industry Sponsorship and Research Outcome and Quality: a systematic review', *British Medical Journal*, 326 (2003): 1167–70.

declared conflicts of interest because of financial associations with Merck and Pfizer. Those 10 members voted 9 to 1 in favour of keeping valdecoxib on the market, and 9 to 1 in favour of bringing rofecoxib back on the market. The other members voted 12 to 8 against and 14 to 8 against on the two issues.[30] Thus there was in this case a very high correlation between association with the industry and votes in its favour, suggesting that these experts with industry associations were, probably without recognizing it, serving as pharmaceutical companies' representatives in FDA meetings.[31]

Jennifer Fishman describes how researchers on female sexual dysfunction acted as mediators between pharmaceutical companies, the FDA, physicians, and potential consumers.[32] For example, in 2001, researchers organized a consensus conference on 'Androgen Deficiency in Women', designed to establish the definition of and diagnostic criteria for this developing disorder. The conference was supported by grants from several pharmaceutical companies in the process of developing testosterone products for women, and was important to the prospects for success of these products, because the FDA only approves drugs that treat established medical disorders. The conference's consensus document, then, was a key step in establishing the regulatory legitimacy of female sexual dysfunction in the form of 'female androgen insufficiency syndrome'.[33] In addition to looking at documents, the FDA turns to researchers like the conference organizers and participants in order to judge the documents: they have the relevant expertise to contribute to the agency's decisions. It is telling that the FDA was represented at the conference, giving a presentation on preferred endpoints of trials.

30 G. Harris and A.Berenson, '10 Voters on Panel Backing Pain Pills Had Industry Ties', *New York Times* (electronic edition), 25 February 2005.

31 See also Lurie et al. for a consistent but less dramatic analysis of FDA advisors more generally. P. Lurie, C. Almeida, N. Stine, A.R. Stine, and S.M. Wolfe, 'Financial Conflict of Interest Disclosure and Voting Patterns at Food and Drug Administration Drug Advisory Committee Meetings', *Journal of the American Medical Association*, 295 (2006): 1921–1928.

32 J. Fishman, 'Manufacturing Desire: the commodification of female sexual dysfunction', *Social Studies of Science*, 34 (2004): 187–218.

33 Ibid, 193.

Published research, continuing medical education, and opinion leaders are all useful for expanding markets, either via 'disease mongering' or the promotion of 'off-label uses'.[34] Disease mongering, as in the case of female sexual dysfunction,[35] requires not only academic and regulatory recognition of medical conditions, but also physician and patient recognition of those conditions. Academics thus also serve as mediators between pharmaceutical companies and practicing physicians. The conference Fishman uses as an example included a CME component for local physicians. While they earned CME credits, they learned about products they could prescribe off-label for women's complaints about low sexual desire. This is not unusual in CMEs: in 2001 the pharmaceutical industry claimed to provide funding for half of all CME courses in the US.[36]

Doctors may prescribe a drug for whatever condition they think it warranted, even while regulatory agencies only approve drugs for particular uses. Prescriptions for unapproved uses, which are common, are off-label, and may form an important part of the market for a drug. Because off-label promotion by drug companies is illegal, research, education, and word of mouth – through published articles, CME, and opinion leaders – are vectors by which information about alternative uses of drugs can be disseminated. In a suit challenging Parke-Davis's marketing of Neurontin, an executive was found to have said to marketing managers:

> Dinner programs, CME programs, consultantships, all work great, but don't forget the one-on-one. That's where we need to be, holding their hand and whispering in their ear, Neurontin for pain, Neurontin for monotherapy, Neurontin for bipolar, Neurontin for everything.[37]

Academics who jump early onto developing disorders, then, receive research support, attention from their colleagues and the media, and have signifi-

34 R. Moynihan and D. Henry, 'The Fight against Disease Mongering: generating knowledge for action', *PLoS Medicine*, 3 (2006): e191.

35 See also L. Tiefer, 'Female Sexual Dysfunction: a case study of disease mongering and activist resistance, *PLoS Medicine*, 3 (2006): e178.

36 A.F. Holmer, 'Industry Strongly Supports Continuing Education', *Journal of the American Medical Association*, 285 (18 April 2001): 2012.

37 Quoted in Critser, *Generation Rx*, pp. 105–106.

cant influence on the shape of regulatory and medical decisions – on the shape of medical disorders.

Pharmaceutical companies also support research by non-academic physicians, who serve as intermediaries in so doing. What is called 'post-marketing research' is done after approval by regulatory agencies. While the term has wide use, it is commonly applied to studies designed to familiarize physicians with products, to encourage prescriptions, or to allow drug representatives more access to prescribers. 'Seeding trials' pay physicians to prescribe specific drugs as part of trials, but are clearly aimed at increasing prescriptions. A US Federal investigation revealed that in some of Schering-Plough's post-marketing trials of its hepatitis C treatment Intron A, liver specialists were paid between US$1,000 and $1,500 per patient enrolled in the trials.[38] There were, doctors claimed, few efforts to ensure that doctors would collect and forward data to the company. Such a trial can be a successful seeding trial because the Intron A was not provided for free, as it would be in a Phase III trial, but was paid for by patients and their insurers, at an annual cost of more than $20,000 (when packaged with Schering-Plough's ribavarin). The physicians who participate may gain status, as well as money, by becoming researchers and not merely prescribers. They may give, and not only receive, continuing medical education.

For the purposes of this chapter, what is relevant about post-marketing research is that it is research, and at least sometimes is published. It thus provides a route via which promotion efforts turn into academic articles, contributing to the large numbers of articles on commercially important drugs.

38 G. Harris, 'Medical Marketing — Treatment by Incentive; as doctor writes prescription, drug company writes a check', *New York Times* (27 June 2004).

Conclusion

Clinical research on pharmaceuticals, then, is not merely research. It serves a number of functions at once, only some of which are importantly related to increasing the knowledge held by the medical research community. The pharmaceutical industry needs to be able to display scientific research when it markets its products, in order to be part of scientific medicine. It needs to support researchers, who can then act as intermediaries between it and various key groups and agencies. It uses studies as tools for encouraging prescriptions of its products, and as tools for gaining influence with researchers and physicians. And it has the resources to encourage the development of research areas and literatures. The phenomenon with which we started, then, large numbers of journal articles about particular commercially important products, is exactly what we should expect to see given this constellation of pressures.[39]

39 I would like to thank Viviane Quirke and Judy Slinn for very helpful comments and crticisms of an earlier draft. The writing of this paper was supported by a grant from the Social Sciences and Humanities Research Council of Canada.

MANON NIQUETTE[1] AND WILLIAM J. BUXTON

9 Relieving Twentieth-Century Excesses: the socialization of antacid and laxative uses through advertising[2]

Introduction

The history of pharmaceuticals and that of advertising are closely linked to each other. The development of advertising is one of the most important

1 Manon Niquette wishes to thank Claude Giroux for sharing his thoughts on the history of pharmaceutical advertising. See C. Giroux, 'La contribution synergique des pratiques de publicité pharmaceutique à la transformation du rôle contemporain du médicament', *Revue Internationale sur le Médicament*, 1 (2007) online: http://chaine.uqam.ca/revue_RIM/RIM1/PDF/RIM1-3-Giroux33–77.pdf, consulted on 6 May 2009. The authors also wish to express their thanks to Rick Shriver of the vintage print advertising archive at the Associate program of Applied Science in Electronic Media, Ohio University – Zanesville, and to Marge Post of Bayer HealthCare LLC for their assistance in the reproduction of the illustrations included in this text.

2 This paper has been produced as part of an international research program for the study of the 'pharmaceutical life cycle' ('chaîne du médicament' in French), directed by Catherine Garnier, and funded by the Social Sciences and Humanities Research Council of Canada. The aim of this program is to study and explain how the 'pharmaceutical life cycle' works, from the moment of the conception and the production of the pharmaceuticals to their consumption. For an overview of the research activities of the members of the research program, see J.J. Lévy and C. Garnier (eds), *La Chaîne des médicaments: perspectives pluridisciplinaires* (Québec: Les Presses de l'Université du Québec, 2007). See also http://www.geirsomedicaments.uqam.ca/index.en.php, consulted on 6 May 2009. For more details on the concept of 'pharmaceutical life cycle', see C. Garnier, 'La chaîne du médicament: Lieu de rencontre des systèmes de représentations sociales', *Journal international sur les représentations sociales*, 1 (2003) online: geirso.uqam.ca/jirso/Vol1_Sept03/Garnier_c.pdf, consulted on 6 May 2009.

aspects of the growth of the pharmaceutical industry in the twentieth century. Conversely, the business of patent medicines in the nineteenth century, along with that of cosmetics and body-care products, can be said to have given birth to modern advertising.[3] The commercial success of some over-the-counter pharmaceutical products – such as Bromo-Seltzer, Eno's Fruit Salt, Phillips' Milk of Magnesia, Listerine disinfectant, and Lambert's cough syrup – deeply marked Western popular culture in the last century. The promotional discourses that contributed to these great pharmaceutical brands did more than merely prescribe uses; they also enhanced the relationship that users had with others through medication. It would be tempting to examine these discourses in terms of self-care, self-help, or body discipline, as it is often done within the fields of cultural studies, sociology of the body, or health communication. However, in doing so, we would be ignoring the fact that the twentieth-century quest for wellbeing is fundamentally one of social inclusion: at work, in love, with friends, and so on. Instant relief is a matter of social responsibility. Therefore, the making of social representations of pharmaceutical uses in modern advertising has to be thought of not only as a mere operation of top-down manipulation – planned in a pure objective fashion in order to condition individual behaviour – but also as an inter-subjective process of meaning production.[4]

3 See C. Goodrum and H. Dalrymple, *Advertising in America: the first 200 years* (New York: Harry N. Abrams, Inc., 1990), pp. 17, 24–29; L. Côté and J.-G. Daigle, *Publicité de masse et masse publicitaire: le marché québécois des années 1920 aux années 1960* (Ottawa: Presses de l'Université d'Ottawa, 1999), pp. 119–123. See also D. Goulet, *Le Commerce des maladies: la publicité des remèdes au début du siècle* (Québec: Institut québécois de recherche sur la culture, 1987). Juliann Sivulka explains that the refusal of newspapers to give free reins to the advertising of patent medicines forced their producers to find innovative ways to promote their remedies. J. Sivulka, *Soap, Sex and Cigarettes: a cultural history of American advertising* (Belmont, CA: Wadsworth Publishing Company, 1998), p. 38.

4 As Richard Ohman put it, advertising 'carried on a conversation among the like-minded'. R. Ohman. *Selling Culture: magazines, markets, and class at the turn of the century* (London/New York: Verso, 1998).

This paper is not primarily a history paper. While it is based on historical material, it mainly draws on insights from the history of advertising, the sociology of media uses, and communication studies. Our contribution is above all a discussion of how the socialization of the uses of popular remedies is reflected in the representations of pharmaceutical uses through advertising. Particular attention is given to the emergence of the normative idea, in Western commercial discourse, that health could be a matter of choice and lifestyle. While the products discussed in this paper are mostly American in origin, the significance of their representations transcends the borders of the United States. The theatre of consumption became a global experience through the 'Great Exhibitions', which had started at the Crystal Palace, in Hyde Park, London, in 1851, followed by others opened in major cities in Europe and North America throughout the nineteenth and twentieth centuries.[5] British patent-medicine makers were pioneers not only in advertising, but also in advertising in foreign markets, especially in the emerging colonial markets where new remedies were in high demand.[6] In mid-Victorian England, patent medicines were already leading advertised products.[7] James Crossley Eno, a former official pharmacist at the Newcastle Royal Infirmary, and the manufacturer of Eno's fruit salts, relied heavily on advertising for the development of his business. Eno's salts were among the first commercial products to have full-page advertisements in the press.[8]

5 As Rosalind Williams has noted, the sensual pleasures of stylized layouts triumphed over the interest in production at the Paris Exhibition of 1900. R. Williams, *Dream Worlds* (Berkeley: University of California Press, 1982), pp. 59–60.

6 For a discussion of how English medicine packaging had an influence on American drug advertising, see J.H. Young, *The Medical Messiahs: a social history of health quackery in twentieth-century America* (Princeton: Princeton University Press, 1967).

7 See Sivulka, *Soap, Sex and Cigarettes*, p. 38.

8 See Royal Pharmaceutical Society of Great Britain, 'James Crossley Eno (1827–1915)', online exhibition *Pharmacy – the mother of invention?*, produced by the museum of the Royal Pharmaceutical Society of Great Britain for National Science Week (2003). http://www.rpsgb.org.uk/informationresources/museum/exhibitions/themotherofinvention/eno.html, consulted on 6 May 2009.

We have chosen to focus on antacids and related laxatives. Their increasing popularity – subsequent to their appearance towards the end of the nineteenth century – is closely related to the expansion of advertising. The commercials of the 1930s and the 1940s are of particular interest as they correspond to the beginning of modern professional advertising. These two decades were also the golden years of antacids, laxatives, and other medications related to digestive problems. In our view, the notion of an 'upset stomach', used widely in the late 1940s to talk about gastric pain, was the perfect embodiment of how the feeling of insecurity left by war, the accelerating rhythm of modern life, and the overeating of greasy comfort meals and modern processed food, were affecting people. As will be explained in the first part of the paper, these pain relievers are closely linked to twentieth-century excesses, that is to the need to recover from happy get-togethers in order to fulfil one's social duties, or conversely, to relieve one of social obligations in order to be ready for partying. Although this time period was not yet one of abundance, advertising prepared the ground for the creation of a discourse of affluence, freedom of choice, and reward. As Avner Offer has put it, self-control takes time, and the flow of novelty that characterizes affluent Western societies undermines the capacity for personal and social commitment to the pursuit of wellbeing.[9] The pharmaceutical promise of instant relief somehow contradicts the sustained effort needed for healthy living, that is for resisting overeating and fighting a sedentary lifestyle.

Most of the material discussed in this paper is taken from the public histories of pharmaceutical companies, antiquarians' websites, university special collections, and private advertising archives. There is a wide variety of material in the history of advertising, and in the history of pharmacy, based on oral narratives and published reminiscences.[10] Since the quality of the material available may be variable, we used it with caution as to the reliability of the information provided. As William O'Barr remarked, because

9 See A. Offer, *The Challenge of Affluence: self-control and well-being in the United States and Britain since 1950* (Oxford: Oxford University Press, 2006), pp. 1–4.

10 See S. Anderson, '"I remember it well": oral history of pharmacy', *Social History of Medicine*, 2 (1997): 331–343.

the nature of commercials is ephemeral, the making of social history using this kind of material is like archaeological work: discovery is often a matter of opportunity.[11] This is even truer for cultural analysis. The most meaningful documents in Western commercial culture can be found in a wide-range of sites, including print and electronic media. This is particularly the case for radio or television advertisements. Our aim is not to present an exhaustive account of the media content to which a limited audience was exposed, but to provide a frame of interpretation for particular tropes of representation, in conjunction with the socialization of pharmaceuticals for excess-related conditions. The paper is divided into two parts. The first part provides the reader with some contextual elements that allow one to understand the role played by over-the-counter antacids and laxatives in socializing bodily experiences at the inception of a *new society of excessive consumption* (rather than just 'mass consumption'). The second part suggests a framework for understanding how social relations are embedded, in different ways, in the social representations of these popular pharmaceuticals.

Socializing the Regulation of Bodily Experiences in a New Society of Excessive Consumption

Advertisements for remedies that purport to cure problems related to social habits such as overeating, over-drinking, over-working, over-celebrating, and over-indulging, span the entire twentieth century. As will be further explained in this paper, popular medicines such as Bromo-Seltzer were promoted well into the 1930s to help one return to normal after consuming too much alcohol. Its association with food excesses came only later, in parallel with the growth of the food industry. For instance, in 1934, an advertisement for Phillips' Milk of Magnesia proclaimed: 'If you eat

11 W.M. O'Barr, 'A brief history of advertising in America', *Advertising and Society Review*, Vol. 6, supplement unit 2, (2005) http://muse.jhu.edu/journals/asr/v006/6.3unit02.html, consulted on 6 May 2009.

starches, meats, sweets – read this: They're All Necessary Foods – But All Acid-Forming. Hence Most of Us Have "Acid Stomach" at Times. Easy Now to Relieve'.[12] The Norwich Pharmaceutical Company, the original producers of Pepto-Bismol, sponsored the drama *The Fat Man*, which was aired on ABC between 1946 and 1951.[13] The program's main character, Brad Runyon, played by J. Scott Smart, was a hard-nosed, overweight, private detective who, unlike contemporary heroes, could overeat, drink to excess, and be obese.

According to Pollay, the most profound effect of modern advertising is that it induces people to keep working in order to buy.[14] However, the basic message of pharmaceutical advertisements during the period spanning the Depression and the Second World War was formulated the other way around: one had to consume in order to be more productive, or more precisely, women had to consume in order to keep men more productive. Buying painkillers would help women to keep men on track. In a 1939 audio advertisement (known as an 'articulated message') created for Emerson's Bromo-Seltzer, the sound of a talking steam engine went 'Wheeeooo, wheeeooo, whe, wheeeooo, Bromo-Seltzer, Bromo-Seltzer, Bromo-Seltzer, Bromo-Seltzer, Bromo-Seltzer'.[15] The workforce had to hop on the fast train to relief, and efficiency was the key word. The caption of a 1934 advertisement for Bromo-Seltzer states 'You can't afford a dull head in busi-

12 Distress related to overindulging was not confined to adults. A 1947 Pepto-Bismol advertisement shows the picture of a young boy looking at an enlarged reflection of himself in a deforming mirror, with an amusement park in the background. The heading reads: 'Boy, I'm in a bad way', and the text states in plain words: 'You don't need a "funny mirror" to tell you when you've had too much feasting. You have inside information …'

13 See C.D. Laughlin, *J. Scott Smart, a.k.a. the Fat Man: a biography* (York, ME: Three Faces East Press, 1995), and idem, 'The Fat Man', *The Old Time Radio (OTR) WWW pages*, http://www.old-time.com/sights/fatman.html, consulted on 6 May 2009.

14 See R.W. Pollay, 'The distorted mirror: reflections on the unintended consequences of advertising', *Journal of Marketing*, 50 (1986): 25.

15 According to Steve F. Howard, this advertisement was created by Harry Edward Stewart and Gretchen Ida (Sissell) Ross while they were working at Scrit Writer Co. See http://www.yogiyorgesson.com/bio.html, consulted on 6 May 2009. See also D. Goodwin, 'Chatty Choo-Choo Praises Blue Bottle Pain Reliever', http://www.old-time.com/commercials/1940's/Chatty%20Choo-Choo.htm, consulted on 6 May 2009.

ness, today', along with an illustration showing a secretary who brings her boss a glass, presumably filled with the mixture. During the same period, the caption of an Alka-Seltzer advertisement (shown in Figure 9.1) poses the question 'Do You Want to Lose Your Job?', along with the illustration of a man who asks his wife 'Tell the boss I won't be there today', and a second illustration showing the boss, who hangs up the phone and complains: 'That's the fourth day he's missed this month'.[16]

A standard character in the 1940s' Pepto-Bismol campaigns was a man unable to work after having partied the day before. In order to emphasize the contrast between the work status of the main protagonist and his health condition, all the men in these advertisements wore classic dark suits with a tie, an embodiment of modern American male style, associated with occupations linked to formality, class, and distinction. Hence, a 1940 Pepto-Bismol advertisement shows the picture of a gentleman in a dark suit, holding his belly, with the caption 'He was the life of the party'. Exposing the reveller in action was even punchier. A 1941 advertisement for the pink medicine shows the picture of a mature man wearing a party hat and blowing a whistle, along with the contrasting image of the same man, holding his stomach the day after. The text explains: 'Elmer was a scream last night' ... 'Too much to eat, too much to drink ... and last night's party is no party today!'[17]

In any case, Pollay's point of view is still relevant from a consumerist perspective; buying is a form of labour, which, in a large part, has been traditionally undertaken by women.[18] People have to buy pharmaceuticals in order to recover from having consumed too much, this is in order to be in shape for continuing to consume even more! One could read in a wartime

16 Alka Seltzer (missing work), in *The Story of Money and Medium Exchanges*, 1939, vintage print advertising archive, Associate program of Applied Science in Electronic Media, Ohio University – Zanesville, http://www.zanesville.ohiou.edu/emedia/Advertising%20archive/index.htm, consulted on 6 May 2009.

17 Pink elephants were the ultimate symbol for over-drinking. An advertisement of the same period shows a man lying down, holding his belly, but surrounded this time by small drawings of elephants. The title reads: 'Pink elephants last night?'.

18 See D.D. Hill, *Advertising to the American Woman 1900–1999* (Columbus: Ohio State University Press, 2002).

advertisement for Alka Seltzer (in Figure 9.2): 'When you go Shopping Remember Alka-Seltzer for Shoppers Fatigue' (sic).[19] The text explained 'When two or three hours of shopping have tired you out and given you sore and aching muscles – stop at a convenient soda fountain. Relax and refresh yourself with a glass of peppy, sparking ALKA-SELTZER. In just a few minutes, you begin to feel better. More than this, fatigue acid is neutralized and you are ready to go on with renewed energy'.

What's more, a 1949 advertisement for Phillips' Milk of Magnesia, which shows two pictures of a young woman – one at night and the second the morning after – reads: 'Don't let "too much Christmas" rob you of sleep! Formula for morning brightness in your happy New Year'. The text states: 'Lots of gay parties, rich foods and sparkling liquids help the Yuletide season'. The point of the message is easy to understand: the end-of-year festivities must continue.

This type of discourse tells us much about the difficulty people had in coping with the mass production of goods, and, more specifically, with the direct corporeal experience of what we may call *a new society of excessive consumption*. To some extent at least, advertisements for antacids and laxatives shaped the frame of reference that was needed for the formation of the concept of 'healthy choices', and therefore the production of lifestyle drugs. As Anthony Giddens has observed, in late modernity the body came to be 'a phenomenon of choices and options'.[20] Every activity related to bodily self-management requires that individuals be capable of sustaining routinized control of their bodies for day-to-day social interaction.[21]

In Foulcauldian terms, one could talk of antacids and laxatives as a means of regulating and disciplining bodies for the governing of modern

19 Alka Seltzer (shoppers fatigue), in *The Story of Money and Medium Exchanges* (1939), vintage print advertising archive, Associate program of Applied Science in Electronic Media, Ohio University – Zanesville, http://www.zanesville.ohiou.edu/emedia/Advertising%20archive/index.htm, consulted on 6 May 2009.

20 See A. Giddens, *Modernity and Self-Identity: self and society in the late modern age* (Stanford: Stanford University Press, 1991), p. 8.

21 Ibid., p. 56.

Figure 9.1: Alka-Seltzer advertisement entitled 'Do you want to lose your job?'

Source : *The Story of Money and Medium Exchanges* (Miles Laboratories, 1939). Reproduced with the permission of Bayer HealthCare LLL, and with the courtesy of the vintage print advertising archive of the Associate program of Applied Science in Electronic Media, Ohio University-Zanesville.

society.[22] Concepts such as 'nervous stomach' or 'irregularity' could be considered as facets of a discursively constituted formation,[23] giving direction to the meanings surrounding the physical and psychological discomforts of modern life. From this perspective, one may even wonder to what extent pharmaceutical advertisements, as discourses on the afflictions of advanced modernity, prepared the ground for the popular understanding of the concept of stress and its wide penetration within popular culture. Pharmaceutical advertisements provide good material for a poststructuralist analysis[24] of the ways in which the emerging experience of the self has been bound up with the medicalization of modern life. The scope of our reflection is in some way related to this kind of analysis, but our focus is not on how pharmaceutical discourses have participated in the construction of medical conditions *per se*, but rather on the ways in which the

22 Michel Foucault has analysed the development of sophisticated ways of disciplining bodies as the construction of a mode of individuality that enables people to perform their duties within the new economic and political forms of modern organizations. See M. Foucault, *Discipline and Punish: the birth of the prison* (New York: Random House, 1975). In the third volume of *The History of Sexuality*, Foucault explains that self-care cannot be reduced to the egoistic pursuit of strictly individual gain. He rather sees it as a search for life according to an established order that assures the durability of society. M. Foucault, *The History of Sexuality: the care of the self* (New York: Random House, 1986).

23 The concept of 'discursive formation' has been developed in Foucault's *The Archæology of Knowledge*. It refers to a system of discursive relations through which statements are organized in respect to one another. These relations may be ones of similarity as well as of difference, and the considered elements are not restricted to linguistic units, but can be extended to various forms including practices, institutions, and architectures. M. Foucault, *The Archæology of Knowledge* (New York: Pantheon Books, 1972).

24 The term 'poststructuralism' refers to a movement of thought that responds to the scientism underlying the systematicity of structuralist methods. Although poststructuralism shares many affinities with structuralism – such as a critical view of the humanist subject and a strong belief in the forces of hidden structures – its antifoundational epistemology and its emphasis on interpretation and diachronic analyses contrasts with the realism and the prominence of synchronic analysis characteristic of structuralism. See M. Peters, *Poststructuralism, Politics and Education* (Westport, Connecticut and London: Bergin & Garvey, 1996); see also M. Poster, *Critical Theory and Poststructuralism: in search of a context* (Ithaca: Cornell University Press, 1989).

Figure 9.2: Alka-Seltzer advertisement entitled 'When you go shopping'.

Sources : *The Story of Money and Medium Exchanges* (Miles Laboratories, 1939). Reproduced with the permission of Bayer HealthCare LLL, and with the courtesy of the vintage print advertising archive of the Associate program of Applied Science in Electronic Media, Ohio University-Zanesville.

pharmaceutical product is represented in relation to the transformation of social life. In more concrete terms, we examine the construction of social representations of pharmaceuticals and how it is rooted in the everyday relationships people have with medication, and with others through medication. This perspective differs from the poststructuralist perspective in two respects:

- The use of medication is not only dependent on prescriptive or promotional discourse; Discourses gain influence when they have a timely relevance for action. Pharmaceuticals are above all physical objects capable of transforming people in a changing world, and as such, they have become an essential requirement for work, intimacy, family and social life.
- The use of non-prescription drugs is not only what one may call in Foucauldian terms a 'technology of the self';[25] it is a social practice, and in this sense it has to be thought of in terms of interaction with others.

It is in the commercial interest of the pharmaceutical industry to frame health-care as 'self-care'. The Nonprescription Drug Manufacturers Association of Canada refers to self-care as having been driven historically 'by necessity, given the cost and scarcity of medical services'.[26] This association between the pharmaceutical view on self-care and the introduction of health in the realm of commodities is consonant with Jackson Lears's perspective as expressed in his cultural history of advertising in America. Lears attributes the popularity of televised medicine shows and the com-

25 See M. Foucault, 'Technologies of the self', in L.H. Martin, P.H. Hutton and H. Gutman (eds), *Technologies of the Self: a seminar with Michel Foucault* (Amherst: University of Massachusetts Press, 1988), pp. 16–49. Transcript of a faculty seminar presented at the University of Vermont in 1982.

26 Nonprescription Drug Manufacturers Association of Canada, 'The Evolution of Canada Self-Care Health Products Industry' (2005), http://www.ndmac.ca/index.cfm?fuseaction=main.DspSubPage&PageID=11&SubPageID=45&fkMainPage=0, consulted on 6 May 2009.

mercial success of patent medicines at the end of the twentieth century to a modern version of the desire for a magical metamorphosis of the self. Furthermore, he sees in this quest for individual transformation a key element in the development of a hedonistic culture of consumption.[27]

Lears's study illuminates the link between the social values of the patent-medicine era and the idea of intimate wellbeing that lies behind the management ethos promoted by the contemporary pharmaceutical industry. But it overlooks people's struggle with sickness and their actual need for pain relief. One should keep in mind that living with a painful body is not only an individual experience; it also embodies the sharing of a common experience of suffering. Pain is above all, as suggested by Jean Jackson, 'an emotion already permeated with cultural meanings',[28] and the study of advertisements for drugs such as analgesics, antacids, or laxatives, allows one to understand the ways in which social life is embodied in the experience of the sufferer.

Most popular over-the-counter drugs are not intended to cure diseases, but to relieve the collective experience of illness. Kleinman makes a fundamental difference between illness and disease: the term 'illness' refers to the experience of disease by sick persons and by members of their families or wider networks.[29] The experience of illness not only includes the feeling of suffering and the problems that symptoms create in everyday life, but also the social and cognitive activities involved in the understanding and sense-making of physical disability, in addition to the close monitoring of bodily processes such as excretory functions, sleeping habits, body temperature, emotional moods, and sexual performance.[30] The distinction between illness

27 See J. Lears, *Fables of Abundance: a cultural history of advertising in America*, New York: Basic Books, 1994), pp. 43–46.

28 J. Jackson, 'Chronic Pain and the Tension between the Body as Subject and Object', in T. Csordas (ed.), *Embodiment and Experience: the existential ground for culture and self* (Cambridge: Cambridge University Press, 1994), p. 211.

29 A. Kleinman, *The Illness Narratives: suffering, healing, and the human condition* (New York: Basic Books, 1988), pp. 3–4.

30 J. Weinman, M. Heijmans, and M. Joao Figueiras, 'Carer perceptions of chronic illness', in L.D. Cameron and H. Leventhal (eds), *The Self-Regulation of Health and Illness Behavior* (New York: Routledge, 2003) pp. 207–219.

and disease can also be extended to the difference between relief and healing. The etymology of the word 'relief' derives from the French word *relief*, which literally means 'raised', related to the verb *relever*. *Relever* also refers to the idea of 'helping people back on their feet'. Therefore, the experience of relief also implies standing up, becoming active and preparing to perform in the social arena. In this sense, looking for relief is a modern obsession. Moreover, it has to be a quick relief or, as in the name of the promotional character who appeared in hundreds of radio commercials in the 1950s, a speedy relief (provided by 'Speedy Alka-Seltzer'). Nowadays, anything that may cause exclusion, marginalization, social isolation, or the loss of work-family roles, calls for quick-relief medication, such as for pain, stress, anxiety, sadness, attention deficit, or menopausal symptoms.

Over-the-counter medicines are, by definition, non-prescribed. They are no less subject to the recommendation of an intermediary who, consequently, may lead its user to experience a need for belonging, a pressure to comply with some normative expectations of feeling good. Wellness belongs to the same category as fitness. In the same sort of dualism as that involving illness and disease, or relief and healing, Zygmunt Bauman's distinction between fitness and health may as well be extended to the difference between wellness and health.[31] According to Bauman, fitness is to the society of consumers what health is to the society of producers. The art of selling goods is not focused on fulfilling desires, but on preventing purchasers from achieving complete satisfaction. As is the case with any other commodified desiderata, the ideal of fitness has no limit; one can never reach it and enjoy what one has. In our view, the same may apply to the notions of relief or wellbeing. Nobody can ever be completely or definitively relieved from the sorrows of the modern way of life. If there is any distinction to be made between fitness and wellness, it is that if fitness provides more desirable bodies, wellness certainly provides more desirable companionship. It both cases, the commodification of self-transformation is at stake, but the pursuit of wellness calls for a greater sense of social responsibility. Thus, the constitution of media representations for the sell-

31 Z. Bauman, *Liquid Life* (Cambridge: Polity Press, 2005), pp. 93–95.

ing of pain relievers has to be thought of as an inter-subjective process of meaning production, rather than a mere attempt to persuade individuals through the production of forceful discourses.

The Social Representations of Pharmaceuticals: an inter-subjective process of meaning production

The framework developed by Bernard Blandin[32] for the study of the social construction of others through the use of objects may be very helpful for examining how social relations were embedded in the constitution of the representations of pharmaceuticals at the inception of what we called in the previous section of this paper 'a society of excessive consumption'. According to Blandin, the various aspects of a subject's relation with an object affect its uses, and each of these aspects implies an interaction between three terms: the user, the object, and the other. Blandin's work is rooted in the sociology of technology and the so-called 'French sociology of uses'.[33] Researchers in this subdiscipline are concerned mainly with the uses of new technologies, but not so much with how the uses of these technologies are depicted in the media. Working within a constructivist framework, Blandin incorporates the idea that the uses of an object are mediated by social representations, but he does not address the question of the ways in which the meanings conveyed by these social representations may be rooted in the relationship one has with this object or with others through this object. This is the perspective adopted in our paper. In our view, the practices related to the use of medication depend on an inter-subjective process of formation and mediatization of the meanings given to pharmaceuticals, within a particular social context.

32 B. Blandin, *La Construction du social par les objets* (Paris: Presses Universitaires de France, 2002).

33 In French: '*sociologie des usages*'.

In what follows, we will focus on four aspects of the relationship a user may develop with pharmaceuticals, as revealed by the content of advertisements: instrumental, affective, cognitive, and symbolic. For each of these four aspects, we will provide examples that show the ways in which social relations are embedded in the constitution of social representations of the uses of remedies such as antacids and laxatives.

The instrumental aspects

The instrumental aspects of the relationship users have with pharmaceuticals is not a necessary condition. Medicines are only useful if there is a need for them. Development in the pharmaceutical industry is often based on what Cohen, March and Olsen have called the 'garbage-can model of organizational choice': organizations create solutions for which they have to find a problem.[34] What's more, the nature of the 'problem' depends on a particular context of social relations: the kind of relief offered depends not only on market-oriented relations, but also in great part on cultural changes in the shared meanings of illness, and on the position of the organization relative to competitors.

In the first part of the twentieth century, the boundaries between today's well-known antacid mixtures and other popular pharmaceuticals, such as analgesics and laxatives, were rather blurry. Over-the-counter antacids, such as Bromo-Seltzer, Alka-Seltzer, Pepto-Bismol, and Rolaids, are a blend of three basic salts – magnesium, calcium, and aluminium – combined with hydroxide or bicarbonate ions. Antacids are known to relieve heartburn and other mild symptoms related to gastro-oesophageal reflux disease. Their side effects may vary. The use of magnesium salt can lead to diarrhoea, and aluminium salts can cause constipation. Before well-known antacids began to proclaim that they could soothe nervous stomachs, these pharmaceuticals were sold for many other kinds of discomforts, including

34 M.D. Cohen, J.G. March, and J.P. Olsen, 'A garbage can model of organizational choice', *Administrative Science Quarterly*, 17 (1972): 1–25.

diarrhoea or constipation (said today to be side effects of the mixtures), headaches, nausea, indigestion, insomnia, aching muscles, 'jumpy nerves', cold and flu symptoms, 'discomfort' from smoking too much, and even feelings of being depressed.[35]

James Crossley Eno sold his Fruit Salts preparation to sea captains and sailors as a safeguard 'against sea-sickness, fever, and change of climate' for their journeys abroad.[36] Bromo-Seltzer, a preparation combining acetaminophen, sodium bicarbonate, and citric acid, is known today for the relief of pain occurring together with heartburn, sour stomach, or acid indigestion. The original remedy was patented by Isaac Edward Emerson, who graduated as a chemist from the University of North Carolina in 1879.[37] It contained potassium bromide, established in the nineteenth century as a treatment for nervous conditions.[38] In 1895, Bromo-Seltzer was promoted to cure the 'headaches of socialites', a euphemism for 'hangovers'.[39] Depicted on a trade card as the 'brain cooler' that one needs 'after a night with the boys', Bromo-Seltzer was said to be better than the big competitor of its time, 'a bag of ice on the head', without any mention of digestive problems. In the early 1930s, following the Great Depression, headaches were still the first cause of pain against which Bromo-Seltzer was aimed. Facing strong

35 It is interesting to note that, today, GlaxoSmithKline promotes TUMS tablets as a good supplement of calcium for the prevention of osteoporosis after the menopause.

36 See Royal Pharmaceutical Society of Great Britain, 'James Crossley Eno (*1827–1915*)', online exhibition *Pharmacy – the mother of invention?*, produced by the museum of the Royal Pharmaceutical Society of Great Britain for National Science Week (2003). http://www.rpsgb.org.uk/informationresources/museum/exhibitions/themotherofinvention/eno.html, consulted on 6 May 2009.

37 Isaac Edward Emerson's papers and business papers related to the Emerson Drug Company are located at the Manuscripts Department of the library of the University of North Carolina at White Chapel, Southern collection, # 4744.

38 See the paper written by Joe Schwarcz, Director of McGill University's Office for Science and Society, 'Take a Bromide', *Canadian Chemical News* (September 2003), p. 13, http://www.accn.ca/index.cfm/ci_id/2970.htm, consulted on 6 May 2009.

39 Nonprescription Drug Manufacturers Association of Canada (1996),' Celebrating 100 Years of Responsible Self-Medication', in NDMAC Self-Medication Digest, vol. 2, no 2, spring 1996, online: http://www.ndmac.ca/index.cfm?fuseaction=main.DspSubPage&PageID=10&SubPageID=795&fkMainPage=10), consulted on 6 May 2009.

competition from similar products such as Alka-Seltzer, Emerson Drug, the maker of Bromo-Seltzer, broadened its scope by enumerating the multiple kinds of headaches its product could remedy in its advertising campaigns. Bromo-Seltzer's main promotional statement was that it fights headaches in three ways: it relieves headache pain, neutralizes acidity in the stomach, and calms jumpy nerves. This was probably in response to its competitor of the time, Alka-Seltzer, which only claimed to give relief in two ways. While the 1930s advertisements mention in passing that Bromo-Seltzer does not have the adverse effect of upsetting the stomach – in contrast to Alka-Seltzer which supposedly could irritate the stomach because it contained Aspirin[40] – the campaigns of the late 1940s described the soothing of an upset stomach as the main benefit the product could bring.

Pepto-Bismol is another interesting case. It was originally developed in the beginning of the twentieth century in New York to cure cholera infantum. Made from pepsin, bismuth salicylate, zinc salts, salol, and oil of wintergreen, it was called 'Bismosal: Mixture Cholera Infantum' and was made pink by adding some colouring to it.[41] With the development of new sanitary advances, infant diarrhoea became a much less important preoccupation. In 1919, Norwich Pharmaceutical Company gave the name of Pepto-Bismol to its product in order to open new markets, and promote its use among adults. In the 1940s, the product was claimed to work 'where Soda and Alkalizers fail', before it was positioned more radically as being 'not antacid, not laxative'. Its distinctive benefit was the soothing effect its coating provided both to the stomach and to the intestine. Pepto-Bismol's positioning statement was 'Not in the stomach alone, but also in the many feet of intestines BELOW the stomach, where trouble often occurs'. Today, the current producer of Pepto-Bismol, Proctor & Gamble,[42] claims that

40 The same argument is still used today. See http://www.numarklabs.com/index.php?src=gendocs&link=BromoSeltzer&category=Over-the-Counter%20Products, consulted on 6 May 2009.

41 See the history section of the Procter & Gamble official web site for Pepto-Bismol, http://www.pepto-bismol.com/history.shtml, consulted on 6 May 2009.

42 Procter & Gamble took over Pepto-Bismol as part of its 1982 acquisition of Norwich Eaton Pharmaceuticals.

'Pepto-Bismol is the only medicine that relieves five stomach problems' (nausea, heartburn, indigestion, diarrhoea, upset stomach).

The case of Phillips' Milk of Magnesia is also worthy of note. As with Eno, Milk of Magnesia was promoted both as a laxative and as an antacid, but in contrast with Eno, it was advertised as being a laxative in the first instance, and having an antacid effect as a subsidiary advantage. The most amusing twist in the story is that, in the 1930s, Milk of Magnesia was also sold in the form of a cream that could be applied to the face. Two products were actually created: a cleansing cream, and a so-called 'texture cream'. The claim was that these creams could neutralize excess acid on the face, just as Milk of Magnesia did in the stomach.

The cognitive aspects

The gestures involved in the use of medication, whatever the form of the drugs – tablet, effervescent drink, milky mixture, or cream – are highly socialized. Often taken for granted, the 'routinization' of these gestures actually requires the formation of a whole set of mental utilization schemes. These schemes are the products of social learning.[43] The cognitive relation users have to pharmaceuticals is not only determined by the material characteristics of the medication (such as its shape, size, colour and texture). This relation is constructed by an interactive process that leads both subject and object to co-produce each other, through the mediation of social behaviour.[44] It takes place during childhood, as well as at the adult stage,

43 Based on the works of Jean Piaget on the genesis of instrumental knowledge, Pierre Rabardel has developed the notion of 'social schemes of utilization'. See P. Rabardel, *Les Hommes et les technologies : approches cognitives des instruments contemporains* (Paris: Armand Colin, 1995), p. 115.

44 Conversely, objects are mediators in the constitution of social relations, through the attempts people make to produce meanings around their uses. See S. Tisseron, *Comment l'esprit vient aux objets* (Paris: Aubier, 1999), pp. 181–182. On the dialectic of objects as being both producers and products of social relations, see A. Semprini, *L'Objet comme procès et comme action: de la nature et de l'usage des objets dans la vie quotidienne* (Paris: L'Harmattan, 1995).

through a process of assimilation and accommodation. People learn to use medication in their everyday lives, as part of their routine, each time manipulating objects which help them to develop appropriate schemes of utilization. This means, for instance, that being at ease with sugar-coated pills or patches presupposes some prior learning with the consumption of products which require the same kind of gestures: one may think of Smarties-like candies for sugar-coated pills, and plasters for the patches.

The approach used by the Emerson Drug Company to encourage consumers to drink its effervescent mixture proved to be quite ingenious. Bromo-Seltzer could be served through a soda fountain just like soda pop. Seltzer water was known to be the favourite beverage of many Eastern European immigrants, who found drinking soda pop to be a good way of demonstrating their Americanization.[45] Furthermore, Emerson Drug developed a fun, fruit-flavoured drink that could be consumed in the same way as Bromo-Seltzer. The product was called 'Fizzies', and was made in seven different flavours: grape, cherry, orange, punch, berry, lemon-lime, and root beer. Fizzies were presented as a 'new sparkling drink sensation in handy tablet form'. While it cannot be proved that Emerson Drug created Fizzies as a way of preparing young consumers to make use of Bromo-Seltzer, the company certainly provided a fruitful and bubbling context for the effervescent fame of its sparkling medication.

The affective aspects

The social aspects of the uses of medication depicted in pharmaceutical advertisements enable one to understand how the instrumental, the cognitive, and the affective relations that people develop with these objects are intertwined. The formation of cognitive mental schemes for the use of pharmaceuticals is concomitant with the social production of

45 M. Feldberg. 'Beyond Seltzer Water: the Koshering of Coca-Cola', *Jewish Life: Chapters in American Jewish History*, Jewish Federations of North America, http://www.jewishfederations.org/page.aspx?id=80335, consulted on 4 January 2010.

unhealthy habits. At the beginning of the century, both Emerson's Bromo-Seltzer and Miles Laboratories' Alka-Seltzer were promoted through popular-song booklets. Thanks to pain relievers, people could gather in good company. Moreover, they did not have to fear being 'under the influence'. The heavy promotion of antacids with slogans such as 'If you keep late hours for Society's sake, Bromo-Seltzer will cure that headache' reveals the extent to which drinking and having to deal with mornings-after were rooted in social conventions; it shows how pharmaceuticals may often both be the source and product of social relations.

This being said, our study also questions how pharmaceuticals were represented as a tool for the transformation of social relations. The construction of social representations of pharmaceuticals depends on the affective aspects of the relationship people have with medication, and with others through medication. The variety of advertisements showing people in need of relief in order to be able to comply with others' expectations is endless: at work, in love, with friends. Pharmaceutical advertisements do not talk in the first place about medication, but about sociability.[46] They addresses problems such as loving a woman, becoming more efficient at work, being in a good mood with friends and colleagues, the ending of feuds between brothers or sisters, dealing with family life, and living up to men's expectations of how a wife should behave. This is still true today. On this point, the advertisements of the 1930s and the early 1940s are particularly interesting, as they take the form of what Roland Marchand calls *social tableaux*.

> Within this category fall all advertisements in which persons are depicted in such a way as to suggest their relationships to each other or to a larger social structure.

46 R.J. Welch Cline and H.N. Young note that visual cues in most direct-to-consumer prescription drug advertisements offer identity or relational rewards. See R.J. Welch Cline and H.N. Young, 'Marketing drugs, marketing health care relationships: a content analysis of visual cues in direct-to-consumer prescription drug advertising', *Health Communication*, 16 (2004): 131–157.

> The depiction of a single person may qualify if that person is placed in a setting suggestive of social relationships with others.[47]

As many of these *social tableaux* correspond to a period when intimacy within couples became valued (at least for women), numerous advertisements for quick-relief medication resemble Viagra commercials. Curiously, antacids and laxatives were positioned as guarantors of hetero-normacy. In a 1936 advertisement for Bromo-Seltzer, a man explains to his wife: 'Of course I love you-but I'm crazy with this head'. A 1951 advertisement for the same product shows a romantic picture of a couple celebrating the birthday of the woman, and the caption: 'Don't ever let headache interfere' (sic). A 1948 advertisement for Phillips' Milk of Magnesia, entitled 'It's tough being a woman', and presented in the form of a cartoon, depicts an overwhelmed housewife who declines her husband's invitation to dance to their favourite song, because she 'feels just awful' (Figure 9.3). The man shows a bottle of Phillips' remedy to his wife and says 'This is what your need, baby! It'll make you feel fine in no time'. The day after the woman stands in an elegant dress, ready to have a good time with her man. It is worth mentioning that Phillips's Milk of Magnesia creams were advertised as 'Things of Tomorrow's Loveliness'.

Clearly, this kind of advertisement not only provides information about past social realities; it also reveals how pharmaceutical products became objects of mediation, commodities whose value is not embodied in their production and exchange, but in their consumption as a tool for relationship management.

The symbolic aspects

The creation of cultural representations plays an important role in the consumption and social acceptability of pharmaceuticals. As with any object,

47 R. Marchand, *Advertising the American Dream: making way for modernity, 1920–1940* (Berkeley: University of California Press, 1985), p. 165.

pharmaceuticals may become signs,[48] but unlike most consumer goods, their ultimate finality is not symbolic. The effects of pharmaceuticals are much more meaningful than the product *per se*. This is why pharmaceutical advertisements tend to place greater emphasis on the social and affective consequences of feeling better than on the physiological effects themselves. In any case, over-the-counter drugs and other pharmaceuticals that are facing tough competition from generic drugs or similar products need to be perceived as offering greater benefits. In this context, increasing the symbolic value of the product appears to be the preferred strategy.

Norwich Pharmaceutical, the maker of Pepto-Bismol, deployed a very distinctive strategy to promote the use by adults of its children's mixture, and to differentiate it from effervescent antacids. The cotton-candy look of the preparation and its sweet flavour certainly had an enormous influence on how the relationship between the medication and its users would be defined. Pink is a tricky colour; it can easily become nauseating rather than being soothing. But what is often generically called today the Pepto-Bismol pink was just right. Pepto-Bismol became a medication for the inner child within each adult. Rather than representing people interacting with one another, the advertisement had a one-to-one relationship with its viewers. The pink liquid appeared as a *simulacrum* of parental care: its protective coating supposedly calmed down the stomach and intestinal walls; the promise was to 'feel good again' (and not to have instant relief).

According to Blandin, the symbolic aspect and the instrumental aspect of the relationship one has with an object are not necessarily interrelated. Even though automobiles are among the consumption goods that are most invested with meaning, people may make use of a car without any concern for its symbolic value. At first sight, the possible dissociation between instrumental and representational uses seems to apply to pharmaceuticals too. The whole point of the placebo-controlled clinical trial is to make sure

48 About the communicative function of objects, see M. Csikszentmihalyi and E. Rochberg-Halton, *The Meaning of Things* (Cambridge: Cambridge University Press, 1981), pp. 20–54; W. Leiss, S. Kline, S. Jhally and J. Botterill, *Social Communication in Advertising: consumption in the mediated marketplace* (New York and London: Routledge, 2005), pp. 225–259.

Figure 9.3: Phillips advertisement entitled 'It's tough being a woman!'

Source: *Time* (10 May 1948), reproduced with the permission of Bayer HealthCare LLL.

that the beneficial effect of the drug is free from any symbolic meaning. However, since the drug transforms the user, and, according to Goffman,[49] it affects the performance the subject plays in the drama of social life, there are good reasons to believe that social relations are central to the connection between the instrumental uses of medication and its symbolic representations. This is especially true for a psychoactive drug or any other medication that is sold with the promise of keeping its user active. The meanings attributed to this consumption by others differ according to the kind of substance that is taken into the body, whatever the quality of the performance achieved: being able to work or to love someone with the help of an antacid does not have the same negative connotations as having to rely on antidepressants. The interplay between the uses of pharmaceutical drugs and what these uses represent for a person who feels inadequate cannot be fully understood without considering how people react to the knowledge that this person's performance depends on a chemical boost.

Conclusion

The basic premise of this paper is that the consumption of pharmaceutical drugs has to be considered not solely in terms of self-care, but also in terms of interaction with others. This is particularly true for the use of remedies that relieve the discomforts of modern life. The study of how social relations were embedded in the commercial representations of these remedies at the inception of a society of *excessive consumption* has led us to draw four conclusions:

- The use of medication is often determined by cultural changes in the socialization of illness and the position of the commercial organization relative to competitors.

49 E. Goffman, *The Presentation of Self in Every Day Life* (Doubleday: Garden City, New York, 1959).

- The routinization of the gestures involved in the use of non-prescription drugs is a result of the socialization process that occurs when people manipulate objects that help them to develop appropriate utilization schemes.
- The themes that appear in pharmaceutical advertisements are typical of the relational requirements characterizing the transformation of intimacy in modern life: pharmaceuticals are represented as tools for relationship-management.
- The beneficial effects of quick-relief medication cannot be completely separated from the symbolic meanings of this kind of product, because individuals engage in impression-management through the consumption of popular remedies.

Antacid and laxative advertisements provide a wonderful source for studying the socialization of bodily experiences and their regulation in a world in which a growing middle class learned new ways of socializing through eating, drinking, smoking, and other addictive habits. The material is rich in meanings, images, and humour. The baby-boomers, who have been exposed to these advertisements at the peak of the anal phase of their psychosexual development,[50] enjoy the memories of these funny jingles and hilarious images, as much as they love remembering how they resisted their parents' efforts to force vile-tasting liquids down their throats. These advertisements certainly represent the first cultural heritage of the globalized collective experience of Western consumption. It is not without reason that a large number of private collectors and historical societies invest considerable time and money in their preservation and circulation. Their legacy, if there is any, is to remind us that despite the rise of individualism and the intensifying cult of self-development, the experience of relieving twentieth-century excesses will always be a profoundly communal one.

50 Sigmund Freud believed that during the age of toilet-training, the child enters a phase in which concerns are focused on defecation and control. This stage is characterized by a conflict between erotic pleasure derived from the expulsion or the retention of the fæces, and societal pressure for bowel control. The child learns how to respond positively to authority, or on the contrary, how to show opposition to it. See S. Freud, 'Three essays on the theory of sexuality', in P. Gay (ed.), *The Freud Reader* (London: W.W. Norton Company, 1989), pp. 239–296.

PART 4

Drug Regulation and its Limits in the United States

JOHN P. SWANN

10 Reducing with Dinitrophenol: self-medication and the challenge of regulating a dangerous pharmaceutical before the US Food, Drug, and Cosmetic Act[1]

Scientific interest in dinitrophenol began with studies of yellow coal tar dyes, especially in France and the US in the late-nineteenth and early-twentieth centuries. Later researchers examined its pharmacology in the context of the French munitions industry, where it served as a stabilizer for picric acid. Explosives workers suffered unusual effects from dinitrophenol, characterized in part by elevated temperature and metabolism. While it is unclear if dinitrophenol achieved widespread use as a reducing agent in France, both this product and an analogue, dinitrocresol, had some popularity there beginning in the 1930s as insecticides and herbicides. John Parascandola and others have discussed this early history of dinitrophenol.[2] However, my particular interests in the history of this drug – in the elaboration of its weight reduction properties and the rise and fall

1 Presented by John P. Swann, FDA History Office, Rockville, Maryland, USA, at the Perspectives on Pharmaceuticals Conference, Oxford University, Oxford, UK, 14–16 July 2005.

2 J. Parascandola, 'Dinitrophenol and Bioenergetics: an historical perspective', *Molecular and Cellular Biochemistry*, 5 (1974): 69–72, and idem, 'Dinitrophenol and Problems of Drug Control Prior to 1938', in *Drug Regulation in the United States: historical perspectives*, colloquium to mark the 75th anniversary of the passage of the Food and Drug Act, University of Illinois Medical Center, Chicago, Illinois, 12 May 1981, transcript, Files, FDA History Office, pp. 28–31; E. Thorpe, *A Dictionary of Applied Chemistry* (London: Longmans Green, 1928), vol. 3, pp. 24–25, 28–32; and 'Dinitrophenol Not Acceptable for N.N. R.', *Journal of the American Medical Association*, 105 (1935): 31 (this includes citations to the studies discussed here).

of its use by the American public, can be traced to the work of clinician Windsor C. Cutting and pharmacologist Maurice L. Tainter at Stanford to investigate further the nitrophenols for their possible therapeutic application beginning in 1931.[3]

In May 1933 they reported that a single injection of dinitrophenol at a dose of 10 mg. per kilogram of body weight increased metabolism by 50 per cent, from which a return to normal metabolic levels could take up to 48 hours. Eight clinical subjects that were given oral dosages less than half that amount all lost weight, up to twenty pounds in ten weeks. While none experienced any harmful effects, the authors realized that this was still a short-term exposure. They cautioned readers that, '... for the present, dinitrophenol be used only as an experimental therapeutic procedure in carefully selected patients under close observation by the physician'.[4] This would, indeed, become a mantra in many professional articles published on the clinical aspects of this drug.

In November the Stanford group presented an update on 113 cases of obesity that received dinitrophenol for one to four months, about half of whom had previously had little success with dietary regimens or desiccated thyroid, the latter a reduction treatment in use for decades. All were adults, and all but fifteen were women. The dosage, which was increased slowly from 100 mg. the first week, was based on the amount required to yield a weekly loss of two to three pounds, a dose that averaged 300 mg. per day. A quarter of the patients perspired profusely, but not so much as to withdraw. A few lost their ability to detect sweet and salty tastes – perhaps an added benefit considering the indication – but all recovered within three weeks after the drug was withdrawn. The most notable side effect was found in 7 per cent of the group, who experienced a painful urticarial rash, which

3 M.L. Tainter, W.C. Cutting, and A.B. Stockton, 'Use of Dinitrophenol in Nutritional Disorders: a critical survey of clinical results', *American Journal of Public Health*, 24 (1934): 1045, and W.C. Cutting and M.L. Tainter, 'Actions of Dinitrophenol', *Proceedings of the Society for Experimental Biology and Medicine*, 29 (1932): 1268–1269.

4 W.C. Cutting, H.G. Mehrtens, and M.L. Tainter, 'Actions and Uses of Dinitrophenol: promising metabolic applications', *J. Am. Med. Assoc.*, 101 (1933): 193–195 (quote from p. 195).

diminished within five days of removal of the drug; half of those patients resumed therapy without recurrence of those symptoms.

Their conclusions are worth quoting, for certainly many others did, many whom the authors did not have in mind when they published their findings. Tainter and his colleagues said that, '... dinitrophenol can be used therapeutically to reduce the weight of obese patients regardless of the etiology of the obesity ... the loss of weight takes place predominantly from the hips and abdomen ... the drug can generally be taken without discomfort to cause a loss of from 2 to 3 pounds weekly over extended periods'. Those that so frequently invoked these results, unfortunately, paid no attention to the first sentence in the following paragraph, which cautioned that, 'aside from the skin reactions, the chief danger is in the indiscriminate or careless overdosing that may result from its sale to the public'.[5]

Concern with this drug among the clinical medical community began to rise in 1934, as some fatalities and toxic reactions began to be associated with dinitrophenol. Agranulocytopenia was a fairly recently identified pathological syndrome at this time, and few cases had been reported after its delineation in 1922. This symptom complex was characterized by depletion of granular leukocytes from the blood, intensely sore throat, inflamed mucous membranes, and possibly sepsis and then death. In the late 1920s and early 1930s, however, a substantial number of cases developed that became linked to a febrifuge called Pyramidon, or amidopyrine.[6] So, clinicians in the mid-1930s were, if nothing else, more than prepared to appreciate the gravity of any product that manifested this sort of effect. At least five cases of dinitrophenol-linked agranulocytopenia were published in 1934. While the first two involved patients self-medicating for weight reduction – and thus the dosage and early warning signs were not

5 M.L. Tainter, A.B. Stockton, and W.C. Cutting, 'Use of Dinitrophenol in Obesity and Related Conditions', *J. Am. Med. Assoc.*, 101 (1933): 1472–1475 (both quotes are from p. 1474).

6 M. Wintrobe, *Clinical Hematology* (Philadelphia: Lea & Febiger, 6th Edition, 1967), pp. 1215–1217.

carefully monitored – the others were under the care of physicians, and one of these patients died.[7]

The same year, Tainter and his colleagues acknowledged these reports but reported no blood dyscrasias of any kind in their patients or in laboratory studies.[8] Late in the following year they reiterated this finding, concluding that 'from these observations of a rather large series of patients dinitrophenol did not affect the blood cells, contrary to the assertions of others'. But given the reactions reported by several, they suggested dinitrophenol serve as a 'court of last appeal' after other more traditional means of reduction have failed.[9]

Others shared their belief that dinitrophenol still had a role in weight reduction under a physician's care. One practitioner, reviewing his 90 patients on the drug, felt dinitrophenol was a 'convenient means of losing weight' when used in small doses, and with the understanding that it simply was not suitable for some patients, perhaps nearly half. Some would develop skin rashes, others would have an idiosyncratic reaction, a few simply would not be able to lose weight on this drug, and others would be scared off when friends and family frightened them with dire predictions of taking dynamite for reduction.[10] Another clinician believed that untoward effects could be controlled by using doses about half the recommended

7 A.M. Hoffman, E.M. Hunt, and N.G. Hickey, 'Neutropenia Following Amidopyrine: Preliminary Report', *J. Am. Med. Assoc.*, 102 (1934): 1214 (despite the title, reference was made to a patient on dinitrophenol); S.S. Bohn, 'Agranulocytic Angina Following Ingestion of Dinitrophenol', ibid., 103 (1934): 249–251; W. Dameshek and S.L. Gargill, 'Studies in Agranulocytosis. IV. Report of Two Cases of Agranulocytosis Following the Use of Dinitrophenol', *New England Journal of Medicine*, 211 (1934): 440–443 (these authors refer (p. 442) to a personal communication from R.R. Kracke, who had done considerable work on agranulocytopenia and amidopyrine, that he had received reports of an additional 16 cases related to dinitrophenol, but this was never published to my knowledge. See Wintrobe, *Clinical Hematology*, pp. 1215–1216); and S. Silver, 'A New Danger in Dinitrophenol Therapy', *J. Am. Med. Assoc.*, 103 (1934): 1058.

8 Tainter et al., 'Use of Dinitrophenol in Nutritional Disorders', 1050–1051.

9 M.L. Tainter, A.B. Stockton, and W.C. Cutting, 'Dinitrophenol in the Treatment of Obesity', *J. Am. Med. Assoc.*, 105 (1935): 334 (first quote) and 336 (second quote).

10 J. Markowitz, 'The Use of 2:4 Dinitrophenol in the Treatment of Obesity', *Bulletin of the Academy of Medicine, Toronto*, 8 (1934–1935): 75–76 (quote).

starting dose, and to be especially discriminating when choosing the source of the drug.[11] A St. Louis physician reminded his readers that, as with morphine, strychnine, and other powerful drugs, some people simply were too sensitive to dinitrophenol, and failure to control the dose of any such drug could lead to disastrous circumstances, too. So, 'under properly controlled conditions there should be no fear in instituting a course of treatment with this powerful addition to our armamentarium'.[12] At the end of 1934, the several reports of blood disorders apparently notwithstanding, even the editors of the *Journal of the American Medical Association* acknowledged that dinitrophenol had a place, albeit a place of last resort, in cases where a physician deemed reduction of obesity important.[13]

Many other toxic reactions associated with dinitrophenol began to trickle in, such as reversible peripheral neuritis, ear and hearing complications, asthma, and in animal models, teratogenicity.[14] However, it was difficult for medical professionals to continue advocating the use of this promising metabolic accelerator once reports of a more widespread serious adverse event emerged in 1935. In July of that year, three San Francisco physicians revealed that six women who had been taking dinitrophenol – some under medical supervision, some not – suffered rapidly developing and unstoppable cataracts, fully formed to the point of blindness.[15]

11 Wallace D. Hunt, 'Dinitrophenol: clinical experiences', *Northwest Medicine*, 33 (1934): 211–212.

12 O.S. Jones, 'What May We Expect of Dinitrophenol?', *Journal of the Missouri State Medical Association*, 32 (1935): 196 (quote).

13 'Dinitrophenol in Obesity', *J. Am. Med. Assoc.*, 103 (1934): 1950–1951. See also 'The Treatment of Obesity with Dinitrophenol', editorial, *New Orleans Medical and Surgical Journal*, 86 (1933–1934): 808, which opined that, 'given by a physician there is no reason why dinitrophenol should not be employed to combat obesity'.

14 See H. Dintenfass, 'An Ear Complication from Dinitrophenol Medication', *J. Am. Med. Assoc.*, 102 (1934): 838; J.E. Nadler, 'Peripheral Neuritis Caused by Prolonged Use of Dinitrophenol', *J. Am. Med., Assoc.*, 105 (1935): 12–13; M.H. Noun, 'Asthma from Dinitrophenol Sensitization', *Journal of the Iowa State Medical Society*, 25 (1935): 610–611; and E.M. Josephson, G. Cameron, and M. Freiberger, 'Genetic Dangers of Dinitrophenol Reduction Therapy', *Eugenical News*, 21 (1936): 34–35.

15 W.W. Boardman, 'Rapidly Developing Cataract After Dinitrophenol', *J. Am. Med. Assoc.*, 105 (1935): 108, and W.D. Horner, R.B. Jones, and W.W. Boardman, 'Cataracts Following the Use of Dinitrophenol', ibid., 108–110.

Within months cases began to spring up across the country, in Massachusetts, Illinois, Michigan, and other states.[16] A survey of nearly three dozen dinitrophenol-associated cases in the San Francisco area alone demonstrated that cataracts formed regardless of the duration of treatment. Sometimes the occlusions developed during therapy, and other times as long as one year after the therapy was discontinued.[17]

Could physicians justify treating obesity with a drug that manifested such a litany of serious and debilitating side effects? Was obesity at this time perceived to be a serious enough problem for physicians to expose patients to these sorts of risks? Though many obese and especially morbidly obese women and men might have bristled at the thought, the editors of *The Lancet* captured what many practitioners believed when they posed this question in the summer of 1934: 'Is it justifiable in dealing with a condition which almost always responds readily to a little wholesome discipline, to employ instead a drug of unquestioned potency but uncertain toxicity?'[18]

A plethora of therapies (most of no value) plus a constant reminder in advertising, fashion, and the cinema of desirable thin body images conveyed a widespread sense that obesity was something to be either avoided or eliminated in 1930s America. It was a trend long in the works.[19] The marketplace was rife with treatments, from specialized diets, reduction creams, spas, devices, and of course pharmaceuticals, a particularly popular unscientific approach to weight loss according to Arthur Cramp of the American Medical Association, who catalogued America's obsession with

16 B.F. Glowacki, 'Acute Cataracts: presumably due to Dinitrophenol therapy', *Journal of the Michigan State Medical Society*, 34 (1935): 535; D.G. Cogan and F.C. Cogan, 'Cataracts and Dinitrophenol', *New England J. Med.*, 213 (1935): 854–856; and N.K. Lazar, 'Cataract Following the Use of Dinitrophenol', *J. Am. Med. Assoc.*, 105 (1935): 794.

17 F.H. Rodin, 'Cataracts Following the Use of Dinitrophenol: a summary of thirty-two Cases', *California and Western Medicine*, 44 (1936): 276–279.

18 Quoted in G. Edsall, 'Biological Actions of Dinitrophenol and Related Compounds: a review', *New England J. Med.*, 211 (1934): 388.

19 See R.P. Seid, *Never Too Thin: why women are at war with their bodies* (New York: Prentice Hall, 1989).

Figure 10.1: Cover illustration from 'Obesity Cures'.
Source: American Medical Association's Nostrums and Quackery
(FDA History Office)

health quackery of all kinds in several massive volumes.[20] Desiccated thyroid, for example, was popular, and somewhat like the case with dinitrophenol, thyroid had been appropriated from physician's reports from the 1890s forward for widespread distribution directly to the public.[21]

With the revelation of the growing number of drug-induced cataracts, one researcher clarified the issue rather tersely. W.W. Boardman, who first reported the link between cataracts and the drug, concluded that the drug should definitely be contraindicated.[22] Condemnation of the drug did not begin with news of cataracts, but it cascaded soon afterward with calls that, despite the good intentions of physicians, the drug simply had to be discontinued because it was so unsafe and yet the indication was, many felt, comparatively benign.[23] The AMA Council on Pharmacy and Chemistry, invoking dinitrophenol's trail of toxicity, the scattered record that questioned its effectiveness, the widening global efforts to restrict and warn against its use, and the Council's own ongoing concern with the proliferation of the drug for self-medication, voted against including the drug in *New and Nonofficial Remedies*.[24]

But whether or not physicians were going to give up on dinitrophenol was a moot point, as far as the public health impact of the drug was concerned. The market was becoming flooded with direct-to-consumer preparations of the drug; within weeks of Cutting and Tainter's announcement of their initial clinical results, over a half-dozen were available, and almost twenty by the Fall of 1934. What follows are two case studies to

20 A.J. Cramp, *Nostrums and Quackery and Pseudo-Medicine*, (Chicago: American Medical Association, 1936), vol. 3, p. 174.

21 J.P. Swann, 'Marmola and the Battle Over Authority in Obesity Treatment in the Early Twentieth Century', paper presented as part of the History of the Health Sciences Lecture Series, Claude Moore Health Sciences Library, University of Virginia, Charlottesville, Virginia, 16 March 2005.

22 Boardman, 'Rapidly Developing Cataract'.

23 For example, A.R. Vanore, 'Dinitrophenol – A Warning', *Medical Times and Long Island Medical Journal*, 62 (1934): 350; J.M. Strang and F.A. Evans, 'An Evaluation of Dinitrophenol as an Aid in Weight Reduction', *J. Am. Med. Assoc.*, 104 (1935): 1963; W.Z. Rundles, 'Dinitrophenol Cataract', *J. Michigan State Med. Soc.*, 34 (1935): 777; S.W. Imerman and C.P. Imerman, 'Dinitrophenol Poisoning', *J. Am. Med. Assoc.*, 106 (1936): 1088.

24 'Dinitrophenol Not Acceptable for N.N.R.', 31–33.

illustrate the commercial messages the public received about this drug and a close analogue, and how the federal agencies attempted to regulate the drug, its labeling, its advertising, and its distribution.

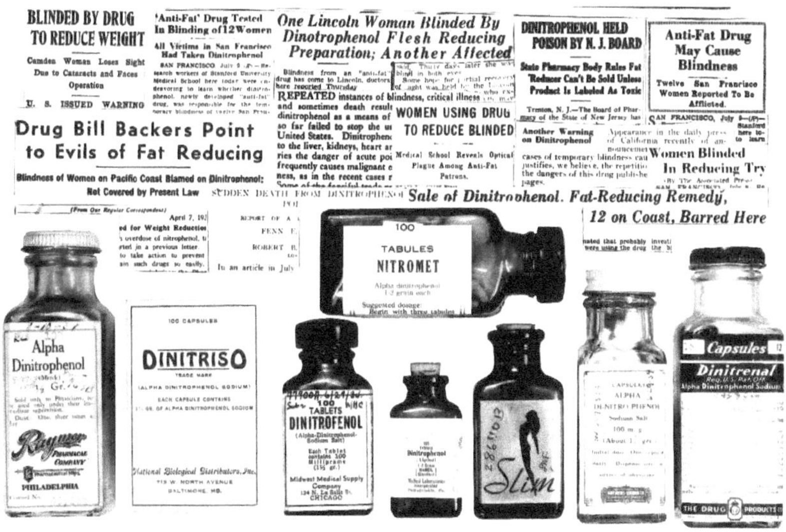

Figure 10.2: A selection of over-the-counter dinitrophenol preparations.
Source: FDA History Office

Redusols was a nationally advertised, mail-order product that employed splashy newspaper notices with slim, attractive, and barely attired female models juxtaposed to promises of quick and safe reduction. The monthly advertising budget was about $4000, and it seemed to pay off; the company received 75 to 100 requests daily for the treatment, which cost $3.00 per order for a 30-day supply.[25] Testimonials were important as well, one of which came from John J. Lyons, Secretary of the State of New York from

25 Post Office Department, Fraud Order no. 8248, 11 January 1936, 18-page typescript, p. 3, folder 22, box 603, AMA Archives, and 'Dilex-Redusols: a Dinitrophenol Nostrum declared fraudulent and debarred from the United States mails', *J. Am. Med. Assoc.*, 106 (1936): 1587.

1921 to 1922, who reputedly was able to take off 18 pounds in five weeks with Redusols. Interestingly, the same ads pointed out that a 15-pound reduction in one month was harmful for anyone.[26] Redusols notices deprecated thyroid as harmful, even advocating that it should be available only through a physician's prescription. Yet the active ingredient of Redusols was not divulged.[27]

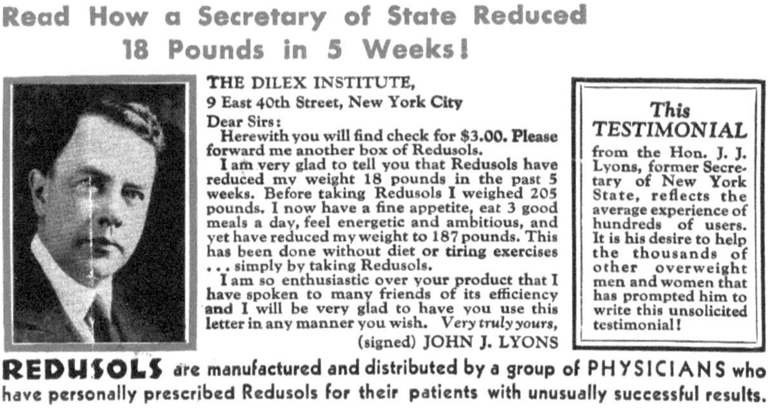

Figure 10.3: Portion of an advertisement for Redusols.
Source: Historical Health Fraud and Alternative Medicine Collection, American Medical Association.

26 'Guaranteed Weight Reduction', newspaper clipping, attached to A.E. Backman to A.J. Cramp, 5 June 1935, folder 1, box 604, AMA Archives; 'The Political Graveyard: Index to Politicians, Lyons to Lyversa', 10 March 2005, http://politicalgraveyard.com/bio/lyons-lytle.html.; and 'Reduced 37 Pound with Dilex-Redusols Writes Mrs. H.H. Langley', newspaper clipping, attached to Mrs. M. Keating to American Medical Association, 13 November 1935, folder 1, box 604, AMA Archives.

27 'Here's Good News for Men and Women Wishing to Grow Thin!' circular, stamped 8 July 1935, folder 1, box 604, AMA Archives. Many letters questioning the composition of Redusols can be seen in folder 1, box 604, AMA Archives, such as Mrs. Joseph McGraw to American Medical Association, 15 December 1934.

The Dilex Institute of New York, founded in 1934 and whose inventory largely consisted of feminine hygiene and contraceptive products, distributed the drug. It reassured the public on the same page of one circular that Redusols would 'positively have no harmful effects', that it had 'no ill effects whatsoever', that 'you are safe with Dilex Redusols', that it could be taken 'without any possibility of injury to the system'.[28] And as a reminder of how badly the customer needed Redusols, Dilex conveniently included in its circulars a table of appropriate weights as a function of height and age for both women and men.[29]

The Post Office Department, experienced in fighting those who would use the mail to perpetrate fraud on the population, as James Harvey Young has shown,[30] filed charges again Dilex and its officers in November 1935.[31] The fraud order cited the frequent and hyperbolic statements of safety in the ads and the claim that only alteration of the recommended dosage would result in harm. In addition, the firm maintained that Redusols would remove the cause of obesity, such that the weight removed would remain lost even after ending the treatment. This, the Post Office pointed out, constituted an implied cure of obesity. That Dilex officer Madeline McCarthy had testified that she lost about 40 pounds with Redusols to a specific weight, yet appeared at the hearing rather heavy and refusing to be weighed, certainly did not help the Dilex case. The fraud order was issued in January 1935, and thereafter any mail intended for Dilex was returned to the sender.[32]

28 'Here's Good News'.

29 'Directions for Using Dilex Redusols', circular, folder 1, box 604, AMA Archives, and 'Here's Good News'.

30 For example, see J.H. Young, *The Toadstool Millionaires: a social history of patent medicines in America before Federal Regulation* (Princeton: Princeton University Press, 1961); idem, *The Medical Messiahs: a social history of health quackery in twentieth-century America* (Princeton: Princeton University Press, 1967).

31 Post Office Department, Fraud Order 8248, pp. 3–4, 15.

32 Post Office Department, Fraud Order 8248, pp. 6–10, 16–17; F.W. Norris to A.J. Cramp, 10 December 1935, folder 1, box 604; Bureau of Investigation, American Medical Association to F.W. Norris, 14 December 1935, folder 1, box 604, AMA Archives; and 'Dilex-Redusols'.

The Post Office Department received some assistance from the Food and Drug Administration in the chemical analysis of Redusols. The two agencies, in fact, had worked together frequently on fraud cases, particularly when it involved products that were beyond the reach of the 1906 Food and Drugs Act.[33] The modern Food and Drug Administration began with the passage of this law, which prohibited the adulteration and misbranding of food and drugs in interstate commerce. The law required labeling of certain hazardous ingredients, established the United States Pharmacopoeia and the National Formulary as the official standards for drugs, and led food and drug inspectors to visit manufacturing facilities, warehouses, and other food and drug processing establishments on a regular basis to check for compliance with the law. But there was no provision for premarket clearance for pharmaceuticals.

That lacuna was made tragically apparent in 1937, when a firm introduced a formulation of sulfanilamide prepared with diethylene glycol. A search of the literature on this substance – much less rudimentary toxicological screening – would have revealed the poisonous nature of this solvent. However, the firm merely examined its Elixir Sulfanilamide for color and flavour. Over one hundred died after taking the Elixir, further deaths prevented in large part by aggressive action by the FDA to remove the product from the market. Interestingly, the legal basis for the agency's action was not because of the health hazard posed by this preparation, but rather because it was misbranded (an 'elixir' must contain alcohol, which was absent from this product). Public outrage helped push a moribund bill through Congress that comprehensively reexamined consumer product regulation in the US, and the 1938 Food, Drug, and Cosmetic Act righted many of the shortcomings of the 1906 law. Among other provisions, it required manufacturers to establish a drug's safety before it could be marketed.[34]

33 John Swann, 'The Formation and Early Growth of the Drug Laboratory of the USDA Bureau of Chemistry', presented at the Edelstein Award Symposium, Annual Meeting, History of Chemistry Division, American Chemical Society, Boston, Massachusetts, 20 August 2002, 21-page typescript, p. 12.

34 A wide body of literature discusses the history of drug regulation in the US, but for a brief overview and references to some of this literature, see, e. g., J.P. Swann, 'Pharmaceutical

The drug provisions of the 1906 act affected neither medical devices nor cosmetics, and the clinical search for effective reducing agents such as dinitrophenol notwithstanding, many believed obesity was not an actual disease and, concomitantly, treatments were regarded as cosmetics rather than drugs from a legal standpoint. When the FDA issued a public warning against dinitrophenol-containing reducing nostrums in July 1934, Chief Chemist Walter Campbell lamented that 'all that the Food and Drug Administration can do is to warn the public that these compounds are dangerous, and then spell out the specific hazards of self-medication with this powerful drug'.[35] Continuing concern for this problem led the agency to institute a survey of reducing remedies containing dinitrophenol by early 1935, and in August FDA announced a new regulatory approach. Since the 1906 act was 'practically inoperative against this public health hazard ... the only application of the law to these products is through some misstatement of fact or some false and fraudulent curative claim in the labeling. In any event, the law can be invoked only when the product has been transported across a State line'.[36] In actuality, it was not long before FDA seized and destroyed a dinitrophenol-containing product.

Regulation Before and After the Food, Drug, and Cosmetic Act', in I.R. Berry (ed.), *The Pharmaceutical Regulatory Process* (New York: Marcel Dekker, 2005), pp. 1–46.

35 'Dinitrophenol Dangerous in "Reducing" Nostrums', press release, US Department of Agriculture, 18 July 1934, folder 22, box 603, AMA Archives.

36 'Weight Reducer Causes Blindness: Government again warns public', press release, US Department of Agriculture, 22 August 1935, folder 22, box 603, AMA Archives, and T.F. Pappe to Chief, Eastern District, 22 April 1935, folder 28597-B, accession 59A2098, box 1006, FDA Seizure Case Files, History of Medicine Division, National Library of Medicine, Bethesda, Maryland (hereafter FDA Seizure Files).

Figure 10.4: Advertisement for Formula 281.
Source: Historical Health Fraud and Alternative Medicine Collection, American Medical Association.

In April 1936 the agency met with representatives of Isabella Laboratories, manufacturers of another direct-to-consumer dinitrophenol preparation, to alert them of the dangers of this active ingredient.[37] Isabella must have listened carefully, because they removed it in favour of another ingredient. Isabella had been founded in February 1934 by Chicago advertising executive Harry Gorov. Gorov no doubt called upon his background to come up with the preparation's name, Formula 281, meaning a formula for the 'too weighty one'.[38] His circulars reminded customers that 'Formula 281 is

37 P.B. Dunbar to Chief, Central District, 12 February 1937, folder 12804-C, box 1248, accession 59A2098, FDA Seizure Files.

38 Exhibit A In Re: Isabella Laboratories, 24 February 1937, attached to H. Runkel, 'Summary and Recommendation', 24 February 1937, folder 12804-C, box 1248, accession 59A2098, FDA Seizure Files.

absolutely safe. It must be safe or prominent national magazines would not accept our advertising.'[39]

After meeting with Food and Drug, Gorov's medical consultant recommended replacing dinitrophenol with one-half grain dinitrocresol, its chemical cousin, based on both a recent article that appeared in *The Lancet* on the use of dinitrocresol in obesity and the consultant's use of the drug in his own practice. In addition, Isabella added one-eighth grain of phenolphthalein 'which would physic a person freely if taken in excessive doses'. The preparation was reborn as Improved Formula 281. Alarmed that dinitrocresol was potentially an even more dangerous drug, FDA collected a sample of 281 in an Ohio pharmacy in November 1936. This switch from dinitrophenol to dinitrocresol showed a 'wanton and reckless disregard' for the lives of the public, according to Paul Dunbar, deputy to the FDA chief chemist, and the firm's circulars and especially its testimonials would form the heart of the case against Formula 281.[40]

The government's action against Isabella Laboratories centered on a misbranding charge, based on Formula 281's alleged false and fraudulent claims that it was a scientifically correct fat reducer and that it purported to be safe beyond any question or doubt. The issue of whether or not Formula 281 was a cosmetic, the argument continued, was settled when Isabella labeled 281 with detailed testimonies from patients who claimed to have derived specific curative or therapeutic effects from this substance, such as the woman from Grand Rapids, Michigan, who supposedly said that she was '... having wonderful results from your preparation. Walking more and feeling better than I have for a good many years', or the judge

39 See, e.g., 'Read These Letters!' (reproduced testimonial letters on one sheet), stamped 27 May 1935, folder 1, box 604, AMA Archives; 'Facts for Actual Letters in Our Files', stamped 10 September 1935, *ibid.* (includes flyer, 'Read How You Can Reduce Fat – Quickly and Safely with Formula 281', which has quote); and 'Reduce Fat Safely', circular, stamped 6 May 1937, ibid.

40 Exhibit A In Re: Isabella Laboratories (first quote); A.E. Paul to Chief, Cincinnati Section, 6 November 1936, folder 12804-C, box 1248, accession 59A2098, FDA Seizure Files; C.K. Glycart, analyst's report, sample 12804-C, 8 January 1937, ibid.; P.B. Dunbar to Chief, Central District, 12 February 1937, ibid. (third quote); Runkel, 'Summary and Recommendation'; A.E. Paul to Chief, Administration, 10 March 1937, ibid.; and P.B. Dunbar to Chief, Central District, 16 April 1937, ibid. (second quote).

from Chicago who was quoted saying: 'have lost 18 pounds. Never felt better in my life'. FDA argued that, unlike other preparations, when 281 put these statements in the labeling, it conveyed to others a sense that they, too, could restore their health by taking this substance to treat overweight.[41] Isabella did not contest the charge, so it remained unknown if FDA's arguments would have held up in court. Gorov chose not to find out, and paid the fine of $50.00 in accordance with the law. That marked the end of the interstate sale of Formula 281, but Harry Gorov's troubles were only going to get worse.

In April 1937 FDA alerted the Chief Inspector at the Post Office Department of its dealings with Isabella Laboratories, including 281's longstanding safety problems and inflated therapeutic claims. The Post Office agreed that it was worth pursuing a fraud investigation against Formula 281, and by the end of that year a fraud case was in preparation.[42] The Federal Trade Commission, using for the first time since its establishment in 1914 its new powers to protect consumers as well as businesses from unfair trade practices under the 1938 Wheeler-Lea Act, also took action against Isabella. The retail outlet for Formula 281, the Hartman drugstore chain in Chicago, also became part of FTC's investigation. On September 16, 1938, the US District Court in Chicago enjoined Isabella Laboratories and the Hartman stores from advertising Formula 281.[43]

Soon under its new law and authorities of 1938, FDA was operating under a policy that officially ended dinitrophenol as a treatment for

41 Dunbar to Chief, Central District, 16 April 1937; 'Charge Sheet', attached to W.G. Campbell to The Solicitor, 20 April 1937, folder 12804-C, box 1248, accession 59A2098, FDA Seizure Files; and Attorney for the United States, Northern District of Illinois, United States of America v. Harry Gorov, trading as Isabella Laboratories, F & D No. 39730, 3 May 1937, passim, folder 12804-C, box 1248, accession 59A2098, FDA Seizure Files.

42 J.J. Durrett to Chief Inspector, Post Office Department, 6 April 1937, folder 12804-C, box 1248, accession 59A2098, FDA Seizure Files; K.P. Aldrich to J.J. Durrett, 10 April 1937, ibid.; and Dunbar to Chief, Central District, 6 November 1937.

43 Federal Trade Commission, press releases, 8 September 1938 and 16 September 1938, and 1 October 1938, folder 22, box 603, AMA Archives; 'Reducing Drug's Sale Restrained as Peril to Eyes', Chicago Tribune, 9 September 1938, newspaper clipping, folder 2, box 604, AMA Archives; and An act to amend the act creating the Federal Trade Commission, 75th Cong., 3d sess., 21 March 1938, 52 US Stat. 111.

obesity, regardless of where it might be obtained. The Food, Drug, and Cosmetic Act deemed misbranded any drug that was dangerous to health when used according to the labeling. The agency developed mandatory and detailed warning statements for use with many drugs or drug classes. But within weeks of that act, FDA promulgated a policy to deal with so-called dangerous drugs, that is, those that would pose hazards when used indiscriminately, or if their use required special knowledge vis-à-vis the disease being treated, such as sulfanilamide. Such pharmaceuticals would have to be labeled by the manufacturer with a de facto prescription legend. However, according to the agency's regulatory policy, there were other drugs that simply could not be used safely to treat any disease even under the strictest controls possible. Dinitrophenol fell into such a category. The agency could not identify any traffic in this drug at first, but Eastman Kodak, a bulk supplier of the drug since the early 1930s, was nearly prosecuted in 1941. The firm had shipped dinitrophenol to manufacturers under the label, 'not for drug use', and some of the product briefly found its way into the drug marketplace.[44]

Many sought means to control the use of dinitrophenol – and dinitrocresol – during these years. To be sure, these were not the only dangerous pharmaceuticals on the market, and efforts had been underway to change the drug law even before dinitrophenol was on the market. Still, dinitrophenol was a poster drug in FDA's traveling show to highlight the shortcomings of the 1906 law that one reporter dubbed, the 'chamber of horrors'.[45] Californian Edna Gibbs, who lost her sight while taking the drug under a physician's care, carried forth a poignant fight to prevent her tragedy from happening again.[46] State Boards of Health and of Pharmacy in Louisiana,

44 J.P. Swann, 'FDA and the Practice of Pharmacy: prescription drug regulation before the Durham-Humphrey Amendment of 1951', *Pharmacy in History*, 36 (1994): 59–60; P.B. Dunbar to R.O. Butterfield, 25 October 1941, folder '511.02 – 511.04–06', box 405, General Subject Files, 1938–1974, FDA Records, RG 88, National Archives, College Park, Maryland; and *Federal Food, Drug and Cosmetic Law: Administrative Reports, 1907–1949* (Chicago: Commerce Clearing House, 1951), pp. 929 and 971.

45 R. de Forest Lamb, *American Chamber of Horrors: the truth about food and drugs* (New York: Farrar & Rinehart, 1936), pp. 94–97.

46 'Sacramento Woman, Blinded by Reducing Drug, Fights for New Protective Law', newspaper clipping, folder 2, box 604, AMA Archives.

Colorado, and elsewhere pushed to have the drug restricted to prescription only status.[47] Cities like San Francisco and Toledo, Ohio, banned its sale, and victims filed lawsuits against physicians and pharmacists responsible for its sale.[48] And this certainly was not a uniquely American problem, either. British officials proposed its placement on the Poisons List, thus restricting dinitrophenol to distribution only through a physician's orders, Germany issued an official warning against its use, and Canada severely restricted its movement.[49]

In a sense, the story of dinitrophenol is somewhat like the story of cloning today, or really any case where advances in research and technology proceed ahead of our ability to come to grips with them legally, ethically, or otherwise. The dinitrophenol researchers emphasized forcefully that the drug should be used only under close supervision, but one must wonder exactly to whom they were talking. Could they reasonably have expected a curtain of regulation to spring forth to keep this drug out of the hands of the public? It was a national issue, and thus a federal solution was needed. Ironically, that solution was triggered, not by a powerful chemical that was quickly losing legitimacy in regular medicine while posing a genuine hazard to the public health through self-medication, but rather by a wayward preparation of the new wonder drug that had achieved virtual universal admiration, sulfanilamide. America seemed to respond then, as now, to the acute problems much more than the chronic ones.

47 'Anti-Fat Drug Is Treacherous Asserts Visitor', *Rocky Mountain News*, n.d., newspaper clipping, folder 2, box 604, AMA Archives, and 'Louisiana Board of Health Issues Warning on Alpha Dinitrophenol in Treatment of Obesity', *Food Drug Rev.*, 18 (1934): 165.

48 'At Whose Risk?', *San Diego Daily Union* (12 July 1935), newspaper clipping, folder 22, box 603, AMA Archives; 'Coroner Will Ask City Ban on Pill' [*Toledo Times*?], 26 April 1936, attached to Bureau of Investigation, AMA to William M. Shapiro, 11 May 1936, folder 2, box 604, AMA Archives; and Joseph Rubinstein to Bureau of Investigation, AMA, 11 August 1936, folder 2, box 604, AMA Archives; Mark Butler to Bureau of Investigation, AMA, 26 January 1939, folder 2, box 604, AMA Archives; 'Dinitrophenol: Blindness Attributed to Use of Drug', *J. Am. Med. Assoc.*, 119 (1942): 903; and 'Dinitrophenol: Blindness Attributed to Use of Drug', *J. Am. Med. Assoc.*, 121 (1943): 975.

49 'The Danger of Drugs Used for Weight Reduction', *J. Am. Med. Assoc.* (5 May 1934), p. 1508, and 'Dinitrophenol Not Acceptable', 33.

DOMINIQUE A. TOBBELL

11 Charitable Innovations: the political economy of thalassemia research and drug development in the United States, 1960–2000

In a 1998 editorial of the *Journal of Laboratory and Clinical Medicine*, British hematologist, Victor Hoffbrand, attested to

> the difficulties for clinical scientists in developing orally active chelators without the support on one the hand of the pharmaceutical industry when the products are nonproprietary or on the other of a government funding agency that might regard drug development to be of low scientific merit and the province of drug firms, not of university-employed academics.

The consequence, as Hoffbrand continued, has been that 'Much of the research in this particular field has been funded by charities such as the Cooley's Anemia Foundation in the United States or the Thalassemia Society in Great Britain'.[1]

As this quotation indicates, researchers investigating the genetic blood disorder, thalassemia, have faced a significant funding problem. Since the 1970s, the field has been characterized by a lack of corporate interest in the development of new drugs (known as iron chelators) to treat the toxic accumulation of iron caused by the life-long bi-weekly or monthly transfusions on which all thalassemia patients are dependent. In the early 1960s the Swiss pharmaceutical company CIBA discovered, developed, and marketed the only iron chelator available to thalassemia patients in the US until late

1 A.V. Hoffbrand, 'Iron chelation therapy: the need for orally active drugs', *Journal of Laboratory and Clinical Medicine*, 131 (1998): 290–291 (291).

in 2005.[2] However, desferrioxamine (or Desferal as it is marketed) is a less-than-ideal chelator, first because it cannot be administered orally and thus must be given to patients by painful injection, and second because the drug is very expensive.

After a gap in the 1970s, during which their interest in the field slackened, Ciba (since 1970 Ciba-Geigy, after the merger between the two companies), established an iron chelator program in the 1980s.[3] However, it is clear from the patent record, the published biomedical literature, and the experiences of researchers in the field, that the company's investment in chelator development was very limited, at least until the early 1990s. Rather than venturing into the research and development of new classes of molecules better suited to oral delivery, the company focused its research activities to modifying desferrioxamine and its derivatives.

In light of Desferal's limitations, thalassemia researchers have considered the development of orally active chelators a research priority since the 1970s. However, pharmaceutical firms saw little market potential in new iron chelators because the majority of the world's three million thalassemia patients lived outside of their major markets, the United States and Western Europe. Thalassemia researchers were thus forced to look to other institutions for research support. As a result, the history of oral iron chelator development from the 1970s onwards stands in contrast to corporate-centered accounts of drug development, which characterize academic researchers as the producers of basic scientific knowledge and industrial researchers as the innovators. Rather, this history reveals a political economy of public sector drug development in which academic laboratories have been the

2 On 9 November 2005, the United States Food and Drug Administration approved Novartis's Exjade (deferasirox, ICL670), the first oral iron chelator available to US patients. http://www.fda.gov/bbs/topics/NEWS/2005/NEW01258.html (accessed 7 March 2006). Novartis, formed by a merger of Ciba-Geigy and Sandoz in 1996, had a long history in the field of iron chelation therapy.

3 Several of the people I interviewed commented that the company established an iron chelator program after, and thus in response to, the US government's increasing investment in the search for and development of new chelators in the late 1970s and 1980s, a subject which I will discuss later in this paper.

primary sites of innovation and the NIH and disease-based organizations have been the major sponsors of that innovation.[4]

Furthermore, the geographic concentration of thalassemia patients has contributed another important feature to the research economy. Major centers of thalassemia research and treatment developed in cities where elite biomedical research institutions and large populations of immigrants (and thus thalassemia patients) from Southern Europe, Asia, and the Middle East converged. As such, Britain and the United States have been the key sites of thalassemia research and chelator development. However, the comparatively small population of thalassemia patients in these countries encouraged American and British drug developers to collaborate with clinical investigators in countries with larger populations of thalassemia patients, particularly those in Italy, Greece, and Israel, in order to fully research and test their new compounds. As a result, collaborative international networks of academic researchers have dominated the field of chelator development since the early 1960s.

Based on interviews with thalassemia patients and their families, representatives from disease-based organizations and the National Institutes of Health, and thalassemia researchers, along with the published patent and medical literature, this paper describes that political economy, examining how it emerged, how it was sustained, and with what consequence for the

4 While scholars of drug and vaccine development have long-recognized the importance of academic research in the discovery and development of new therapeutic agents, few historical studies have been made on the subject. Exceptions include : J. Liebenau, 'Public Health and the Production and Use of Diphtheria Antitoxin in Philadelphia', *Bulletin of the History of Medicine*, 61 (1987): 216–236; M. Bliss, *The Discovery of Insulin* (Basingstoke: Macmillan, 1987), L. Galambos and J. Sewell, *Networks of Innovation: vaccine development at Merck, Sharp & Dohme, and Mulford, 1895–1995* (Cambridge: Cambridge University Press, 1995); L. Marks, *Sexual Chemistry: a history of the contraceptive pill* (New Haven: Yale University Press, 2001); N. Rasmussen, 'Steroids in Arms: Science, Government, Industry, and the Hormones of the Adrenal Cortex in the United States, 1930–1950', *Medical History*, 46 (2002): 299–324, and idem., 'Making the First Anti-Depressant: amphetamine in American medicine, 1929–1950', *Journal of the History of Medicine and Allied Sciences*, 61: (2006) 288–323, idem, 'On Speed: the many lives of amphetamine' (New York City: New York University Press, 2008).

nature of pharmaceutical innovation. This paper, then, adds to the growing literature on the role of patient activists and disease-based organizations in the second half of the twentieth century in securing greater government investment in and research attention to traditionally marginalized, and not-so-marginalized, diseases.[5] At the same time, the history of iron chelator development expands the rather limited literature on pharmaceutical innovation.[6] In particular, by examining drug development as it occurs in the context of limited resources, the history of iron chelator development highlights the ways in which the innovation process is shaped by the political economy.

5 On patient activism and AIDS research, see S. Epstein, *Impure Science: AIDS, activism, and the politics of knowledge* (Berkeley: University of California Press, 1996). On patient activism and breast cancer research, see B.H. Lerner, *The Breast Cancer Wars: hope, fear, and the pursuit of a cure in twentieth-century America* (New York: Oxford University Press, 2001). On the influence of patient activism on drug regulation see A. Daemmrich, *Pharmacopolitics: drug regulation in the United States and Germany* (Chapel Hill: University of North Carolina Press, 2004). On the role of disease-based organizations in securing greater research funding for rheumatoid arthritis in Britain, see D. Cantor, 'The aches of industry: philanthropy and rheumatism in inter-war Britain', in J. Barry and C. Jones (eds), *Medicine and Charity before the Welfare State* (New York: Routledge, 1991), pp. 225–245. On the role of disease-based organizations in raising public awareness, increasing research in and federal funding of Tourette syndrome, see H.I. Kushner, *A Cursing Brain? The histories of Tourette syndrome* (Cambridge, Mass.: Harvard University Press, 1999), and of multiple sclerosis, see C. Talley, 'The combined efforts of community and science: American culture, patient activism, and the multiple sclerosis movement in the United States', in R.M. Packard, P.J. Brown, R.L. Berkelman, and H. Frumkin (eds), *Emerging Illnesses and Society: negotiating the public health agenda* (Baltimore: Johns Hopkins University Press, 2004), pp. 39–70.

6 Examples of this historical literature include; R. Vos, *Drugs Looking for Diseases: innovative drug research and the development of beta blockers and the calcium antagonists* (Boston: Kluwater Academic Publishers, 1992); K. Wailoo (ed.), *Drawing Blood: technology and disease identity in twentieth-century America* (Baltimore: the Johns Hopkins University Press, 1997), pp. 99–133; Marks, *Sexual Chemistry*; Rasmussen, 'On Speed'; A. Tone, *The Age of Anxiety: a history of America's turbulent affair with tranquilizers* (New York: Basic Books, 2009); J. A. Greene, *Prescribing by Numbers: drugs and the definition of disease* (Baltimore: Johns Hopkins University Press, 2007); and V. Quirke, *Collaboration in the Pharmaceutical Industry: changing relationships in Britain and France, 1935–1965* (New York: Routledge, 2007) .

Blood, Clockwork Pumps, and the Limits of Therapeutic Innovation

Thalassemia is a collection of genetic blood disorders in which sufferers are unable to produce normal hemoglobin molecules. One of the most severe forms of the disease, thalassemia major (or Cooley's Anemia as it is also known) is prevalent among individuals of Mediterranean descent such as Greeks, Italians, and Cypriots and in people from the Middle East, South Asia, and Southeast Asia. Until the 1970s, most American patients were of Italian and Greek descent. Over the past three decades, however, the increased immigration of people from the Middle East and Asia has had a corresponding effect on the thalassemia population in North America. In Toronto, for example, a city with one of the largest thalassemia populations in North America, the proportion of patients of Greek and Italian heritage declined from 65.5 per cent in the 1970s to 21 per cent in the 1990s. During the same period, the proportion of thalassemia patients of Middle Eastern or Asian ethnicity rose from 34.5 per cent to 79 per cent.[7]

In order to survive beyond their fifth birthday, patients must receive regular blood transfusions that provide them with a temporary supply of normal hemoglobin every two- to three-weeks. In the mid-1950s, physicians discovered, however, that repeated transfusions led, over time, to a toxic build up of iron in the heart and other major organs of thalassemic patients, leading ultimately to cardiac failure. Patients would die before they reached their twentieth birthdays. In light of this, French, German, and American workers engaged in blood-disorder research, began in the late 1950s to search for compounds, termed iron chelators, that could remove this accumulated iron from the thalassemic's body.

7 For a review of the demographic changes in the thalassemia population in North America over the past forty years, see E.P. Vichinsky, E.A. MacKlin, J.S. Waye, F. Lorey and N.F. Olivieri, 'Changes in the epidemiology of thalassemia in North America: a new minority disease', *Pediatrics*, 116 (2005): 818–825.

In 1960, researchers from the Swiss pharmaceutical company CIBA and the Swiss Federal Institute of Technology, discovered, while looking for new antibiotic compounds, a potential new iron donor, ferrioxamine B. This hydroxamic acid, a metabolite of the bacterium *Streptococcus pilus*, was just one of numerous iron-binding agents isolated from microorganisms and studied by researchers in the US and Europe during the mid-twentieth century. The Swiss researchers, along with others in the field of microbial biochemistry were interested in these microbial chelates because their high specificity for iron and the ubiquity of iron in the environment meant they could have useful applications in medicine, industry, and agriculture.

Following the Swiss team's investigation of ferrioxamine B they found that removal of the iron group from the molecule transformed the resultant compound, desferrioxamine B, into a highly specific chelator of iron ions. This, together with its low toxicity and rapid removal from the body made desferrioxamine B a good candidate for use in chelation therapy.[8]

Though desferrioxamine B was used initially to treat acute iron poisoning, British clinicians in Oxford and London began testing desferrioxamine B in patients suffering from a variety of iron-overload disorders. In 1968, following several encouraging reports from British researchers, CIBA began marketing desferrioxamine as Desferal in both Europe and the United States. By the late 1970s, desferrioxamine had become the standard treatment for thalassemia patients undergoing hypertransfusion, extending patients' lives well into their twenties and thirties. Thus, desferrioxamine has effectively transformed thalassemia from an acute, lethal disease to one that is life-long and chronic. Many of the first patients treated with Desferal are surviving into their forties.

However, because desferrioxamine is inactive when administered orally, the drug can only be given by injection. In addition, desferrioxamine has only a short half-life in the body. Thus, for the drug to remove sufficient amounts of iron from the body it must be administered to patients

8 Von H. Bickel, E. Gäumann, W. Keller-Schierlein, V. Prelog, E. Vischer, A. Wettstein and H. Zähner, 'Über eisenhaltige Wachstumsfaktoren, die Sideramine, und ihre Antagonisten, die eisenhaltigen Antibiotika Sidermycine', *Experientia*, 16 (1960): 128–168.

subcutaneously over the course of 9–12 hours a day by way of a portable infusion pump, six out of seven nights a week their entire lives. This requires patients (or in the case of younger patients, parents) to inject themselves with a needle each night in order to attach the pump, a procedure that is physically painful. According to one patient:

> You had all kinds of lumps. And you're small and you don't have much stomach area to work with, so it was a little difficult. And then they wanted you to do it seven days a week. You really needed to have a break – mentally, physically – you looked forward to having that one or two nights off.[9]

Desferal therapy is arduous, inconvenient, and emotionally painful, as another patient recalled:

> [M]y biggest problem always was Desferal. I just could never ever, ever, ever, ever deal with it. Psychologically, physically, mentally, spiritually ... psychologically it was just, like, it was a highly sick kind of thing – having to find your own time, having to stop your activities in order to do that, having to carry all that. If I didn't have it [the pump attached] I would hear it, I would feel it ... It just does something to you when you have to put in so many hours ... I had been tortured physically ... the pump would be every day and for so many hours and the fact that nobody knew about it. I just couldn't share it.[10]

As these quotations reveal, thalassemia patients contend with the physical and emotional scars inflicted by the syringe pump. Because of the Desferal regime, many children with thalassemia are not able to attend sleepovers at their friends' houses, thalassemics of all ages are often forced to miss work or school because of sickness or hospital visits, and patients have confronted the stigma that is associated with any blood-linked disorder. As one patient recalled: 'Of course I was teased ... second through eighth grade I was know as "Contamination"'.[11] Another patient, who spent her

9 T.P., interview with author, 2 February 2006.
10 I.L., interview with author, 9 January 2006.
11 C.J.L., interview with author, 5 February 2006.

childhood in Greece, described the impact this stigma can have on the emotional development of Greek patients:

> There has not been a lot of acceptance [in Greece]. A lot of people have been discriminated against. A lot of people are living in fear that they will be discriminated against – that their child is going to be differentiated and discriminated. So that develops a sense of secretiveness and nobody will tell anyone. Even in my family people would not know ... I was doing a lot of hiding. So in a way you become a seeker and you become someone you are not and then you try to realize who you are when you become an adult, since you've been hiding all those years ... that's why many Thalassemia patients – I don't want to generalize or categorize – but a lot of us are lost.[12]

The parents of thalassemic children also struggle with the implications of desferrioxamine therapy. Peter Chieco, a former president of the Cooley's Anemia Foundation and the father of a teenage daughter with thalassemia, explained,

> when my daughter was young it was very easy to control her chelation because my wife and I were the ones who were in control of the pump and the meds ... But as kids get older and bigger, you may literally have to sit on your child to get the syringe pump in.

For many parents, Chieco continued, 'they are simply unable – physically or mentally – to do it to their children'.[13]

Because of the nature of the treatment, compliance was and still is a significant problem. As 19 year-old thalassemic, Alicia Somma, testified to the US House Appropriations Committee in March 2004, 'It's just hard to convince someone that you have to do something so barbaric to yourself to save your own life'.[14] Just a few months earlier, one of Somma's

12 I.L., interview with author, 9 January 2006.

13 Peter Chieco, interview with author, 25 January 2006.

14 A. Somma and M. Giammalvo, 'Testimony Before House Appropriations Committee – 2004', Cooley's Anemia Foundation, http://www,thalassemia.org/sections.php?sec=4&tab=51&sub=62 (accessed 16 March 2005).

teenage friends had died because he'd repeatedly refused to undergo his nightly treatment.

In addition to the severe compliance problem, the high cost of desferrioxamine was also a significant limitation of the therapy. As a team of clinical researchers put it in 1977,

> The thalassemia syndromes are relatively rare in the United States and Western Europe, and this complicated and expensive form of treatment is not likely to become generally available to the countries where it is needed most. The vision of thousands of young children walking about clutching clockwork pumps is more of an Orwellian dream than a medical reality![15]

Indeed, the expense of producing desferrioxamine, reflected in its cost, proved particularly prohibitive to the majority of thalassemia patients living in underdeveloped countries in Southeast Asia and the Indian subcontinent. A year later, an editorial in the *Lancet*, captured the profundity of the situation:

> the price of the drug at £3.12 ($6) per g is sufficient to cause anxiety even in Britain, and is simply too much for some other communities. The annual cost of 2g desferrioxamine per day for all existing patients in Britain (300) would be nearly £¾ million; in Cyprus (500) £1 ¼ million; in Greece (2000) nearly £5 million, and in Italy (12,000) £28 million; owing to new births these costs would double in 10 years. Progress in treatment will cause much bitterness, and many families in the Middle East and the Far East will beggar themselves in attempting to obtain adequate treatment for their child. It is therefore of pressing importance to establish the minimum effective dose of desferrioxamine given subcutaneously, to support the development of cheaper iron-chelating drugs, and to explore alternative approaches to these diseases.[16]

15 D.J. Weatherall and M.J. Pippard, et al., 'Iron loading and thalassemia – experimental success and practical realities', *New England Journal of Medicine*, 297 (1977): 445–456 (456).

16 Editorial, 'Desferrioxamine and Transfusional Iron Overload', *Lancet*, 1, no 8062 (1978): 479–480 (480).

In light of the serious cost and compliance issues associated with Desferal therapy, by the mid-1970s, several academic researchers (based primarily in the United States and Britain) were searching for alternative iron chelators that could be produced more cheaply and administered orally. The problem, as noted in an editorial that appeared in the *British Medical Journal* in 1978, was, however, 'to persuade drug companies to take up the challenge of finding a better way'.[17]

Economies of Scarcity: orphan disease and the limits of industrial development

> 'The hard fact of life is that development resources are always scarce and fully committed'.[18]

Despite the substantial number of thalassemia patients worldwide, there was little interest among pharmaceutical companies in developing new iron chelators once Desferal reached the marketplace.[19] The problem, as noted by the editors of a special volume of the journal, *Birth Defects* in 1976, was that '[a]lthough the number of [thalassemia] patients worldwide is quite large (estimated at 3 million), the number living in the United States and Europe is quite small and thus limits the amount of funds available for this disease'.[20] In other words, the actual market of patients able to afford treat-

17 Editorial, 'Intensive Iron-Chelation in Thalassemia', *British Medical Journal*, 2, no 6140 (1978): 782–783.

18 F.P. Hauck, 'Drug Development in Industry', in A.E. Martell, W.F. Anderson and D.G. Badman (eds), *Development of Iron Chelators for Clinical Use* (New York: Elsevier/North-Holland, 1981), pp. 235–244 (236–237).

19 This continued to be the case through the early 1990s, even after Desferal had proven to be a significant economic success for Ciba-Geigy. Alan Cohen, Interview with author, 9 November 2005.

20 A. Cerami, C.M. Peterson, and J.H. Graziano, 'Introduction', *Birth Defects. Iron Metabolism and Thalassemia*, 12 (1976): xiii.

ment was limited principally to the United States and Western Europe, where the patient population appeared to be around 15,000 and Ciba-Geigy already held the market with Desferal.

The lack of corporate interest in drug development is a problem shared by all who conduct research on orphan diseases, so-called in the United States because they affect less than 200,000 patients and because 'they traditionally have not had funding', a significant cadre of 'investigator[s], or agency dedicated to researching their prevention, diagnosis, and treatment'.[21] As Lisa Basara and Michael Montagne describe, the reason for industry's reluctance is purely economic: '[t]hese drugs have low commercial value because very few people need them; their ability to recoup research and manufacturing costs is minimal, and their ability to generate substantial profits is virtually nil'.[22] As a result, responsibility for developing potentially life-saving drugs to treat these rare disorders has fallen to academic researchers, as a thalassemia researcher noted in a 1983 editorial in the *New England Journal of Medicine*:

> It is essential ... that we [academic researchers] continue to try to develop an efficient oral chelating agent; to date the response of the pharmaceutical industry to this problem has been disappointing.[23]

Yet, in the 1970s and 1980s, thalassemia researchers did not find it easy to secure the resources necessary to develop new iron chelators. As retired director of the National Institute of Diabetes, Digestive, and Kidney Disease's (NIDDK) hematology program, David Badman recalled, inves-

21 L.R. Basara and M. Montagne, *Searching for Magic Bullets: orphan drugs, consumer activism, and pharmaceutical development* (New York: Pharmaceutical Products Press, 1994), p. 127. For the history and an analysis of the influence of the Orphan Drug Act, which was signed into law by President Reagan in 1983, on the development of drugs for rare diseases, see P.S. Arno and K.B.M. Davis, 'Rare Diseases, Drug Development, and AIDS: the impact of the Orphan Drug Act', *Milbank Quarterly*, 73 (1995): 231–251.

22 Basara and Montague, *Searching for Magic Bullets*, p. 128.

23 D.J. Weatherall and M.J. Pippard, et al., 'Editorial Retrospective. Iron Loading in Thalassemia – Five Years with the Pump', *New England Journal of Medicine*, 308 (1983): 456–458.

tigators submitting grant applications to government agencies 'often had trouble securing funding for drug development work as it was seen by many reviewers as too mundane'.[24] This was the case especially for researchers seeking support for the preclinical testing of potential therapeutic agents. As Anthony Cerami, a thalassemia researcher, observed in 1981:

> Too often, review committees are quick to say that this [preclinical testing] is not 'real science' or the studies do not warrant the expense and cut the budget severely. The FDA, on the other hand, is increasingly insistent on further pharmacological and toxicological studies before proceeding to human trials. This places the investigator in the difficult situation of having very limited to non-existent funds to perform the work demanded by the regulatory agency.[25]

And yet if academic researchers had safety and toxicity data for a compound, drug companies were more inclined to pick up the drug for subsequent development and marketing. As an industry executive explained to an audience of thalassemia researchers in 1980, drug firms often looked to academic institutions and government agencies for potential new therapeutic agents, however, 'products from outside sources must earn their way into the [company's drug development] system by convincing us that they offer greater and/or more immediately realizable market potential than ongoing projects'.[26] The executive continued, 'it is easiest to make a case for developing compounds that have at least been in early human trials. Concepts with limited animal results are much more difficult to justify'.[27]

The challenge for thalassemia researchers then was to overcome the reticence of grant review boards toward drug development work and secure public sector support for the development of new iron chelators. To this end, the disease-based organization, the Cooley's Anemia Foundation, was instrumental.

24 David G. Badman, interview with author, 18 January 2006.

25 Anthony Cerami, 'The Future of New Iron Chelator Drugs', in A.E. Martell, W.F. Anderson and D.G. Badman (eds), *Development of Iron Chelators for Clinical Use* (New York: Elsevier/North-Holland, 1981), pp. 197–200.

26 Hauck, 'Drug Development in Industry', p. 237.

27 Ibid.

Building a Charitable Base for Innovation: ethnicity and parent advocacy

The Cooley's Anemia Foundation, created by several parents of thalassemic children in New York City in 1954, has, since the 1960s, been at the center of efforts by thalassemia researchers to develop alternative iron chelators. The Foundation was established initially to fund-raise, organize blood drives, and raise awareness about the disease. However, concerned by the lack of understanding about the disease in high-risk communities, limited treatment options, and the poor life expectancy of thalassemic children, the Foundation began its own research program in the early 1960s.

Recognizing early on that it would never have the money to support an entire research laboratory, the Foundation instituted a fellowship program in which they awarded two years of funding to individual researchers who were working on projects related to thalassemia. 'Because most members of the Foundation did not have any medical background and because they were emotionally involved',[28] the Foundation relied on a permanent advisory panel composed of lead investigators in thalassemia research to initiate requests for research proposals and to review and rank all funding requests received by the Foundation. Although the Foundation's executive board had the final say on which projects recommended by the Medical Advisory Board's would receive funding, this outsourcing of scientific decision-making to a body of qualified experts reflects the Foundation's understanding of and appreciation for the scientific process of peer review. Researchers applying to the Foundation's fellowship program competed for the organization's research dollars in the same way that they competed for NIH money, their research proposals subject to critical peer-review. In this way, the Cooley's Anemia Foundation presented itself not only as a source of financial support for thalassemia researchers but, more significantly, as a legitimate participant in the political economy of scientific research.

28 Chieco interview.

Data on the fellowships awarded during the 1960s is unavailable, but from 1973 through the mid-1980s, the Foundation awarded, on average, four two-year fellowships a year, initially at the level of $5,000 a year. As the Foundation's membership base grew and with it the organization's fund-raising capacities, both the size of the fellowship and the number awarded expanded. By the late 1980s, the organization was awarding between 10 and 15 two-year fellowships each year. Today, the Foundation's fellows receive closer to $40,000 a year toward their research.[29] Thirty-one of the fellowships awarded since 1972 have been specifically for research into the development of new chelators.

Despite the subsequent growth of their fellowship program, in the 1960s the CAF's executive board members were well aware of the limits of a charity-based research program. As former president of the Foundation, Nunzio Cazzetta recounted: 'We realized ... that we were a poor man's foundation. We were never going to raise millions and millions of dollars. It was not going to happen. We're [both the disease and the Foundation] unknown. To this day we're unknown. We're not cancer; we're not heart; we're not kidney; we're not liver'.[30] As such, the Cooley's Anemia Foundation was also committed to raising the level of government funding of thalassemia research. To this end, and as part of a broader effort to raise awareness about the disease, the Foundation launched a legislative program in the late 1960s.

Through its local chapters, the CAF searched for members of Congress – both at the state and federal level – who might be responsive to the thalassemia community's requests for greater government investment in the disease. Because, in the 1960s and 1970s, thalassemia was mostly concentrated in Italian-American and Greek-American communities, the CAF used ethnicity in the early years of their legislative program to galvanize support for thalassemia among state and federal legislators.[31] 'As a rule',

29 Spreadsheet of 'CAF Fellowships as of 6/30/73 – 2006', given to author by Gina Cioffi, Executive Director, Cooley's Anemia Foundation, February 2006.

30 Nunzio Cazzetta, interview with author, 9 February 2006.

31 The ethnic picture of thalassemia has changed significantly over the course of the past 40 years, reflecting in large part changes in patterns of immigration in the United States.

Cazzetta explained 'we focused on our own ethnic group because we felt it was almost an obligation on their part to take a look at us'.[32]

Ethnicity similarly served as a way of increasing the Foundation's fund-raising capacities and facilitating its public education and outreach efforts. The Foundation forged relations with ethnic-based charities such as the Order Sons of Italy of America and (though to a more limited extent) the American Hellenic Education Progressive Association,[33] utilizing the networks of these charities to raise both awareness of thalassemia and money for the Foundation's programs among the Italian- and Greek-American communities served by the charities.

As testimony to the efficacy of the Foundation's ethnicity-based legislative strategy, in the early 1970s the Foundation's Connecticut chapter identified two Italian-American members of Congress who were responsive to the Foundation's calls for legislative support. Representatives Robert Giaimo and Ellen Grasso subsequently introduced a bill that called for greater government funding of thalassemia research, medical care, and public education. The CAF called on its members to write letters to their members of congress, urging passage of the bill. The Foundation's president, Carmine Geoni together with Nunzio Cazzetta, who was at that time head of the legislative program, visited the NIH and congressional staffers in Washington D.C., to advocate for support of the bill. In August 1972, their efforts paid off as President Nixon signed the National Cooley's Anemia Control Act (PL 92–414) into law.

The Act, which followed the passage of similar national control acts for sickle cell anemia, hemophilia, and cancer, authorized 'annual funding (for three years) for support of ... (1) screening, treatment and counseling

---

Now, thalassemia has a relatively high prevalence in individuals of South- and South-East Asian and Middle Eastern descent. See H.A. Pearson, A.R. Cohen, P.J. Giardina, an dH.H. Kazazian, 'The changing profile of homozygous beta-thalassemia: demography, ethnicity, and age distribution of current North American patients and changes in two decades', *Pediatrics*, 97 (1996): 352–356.

32 Cazzetta interview.

33 According to Nunzio Cazzetta, American Hellenic Education Progressive Association established its own foundation committed to community outreach and education and research.

programs ($1 million); (2) research ($1.7 million); and (3) development and dissemination of information and educational materials ($1 million)'.[34] An Interagency committee formed after the Act's passage to determine the distribution of responsibilities charged the NIH with implementing the research portion of the law.

Despite passage of the Act, it took several years before that authorization translated into appropriated dollars in the NIH's budget. It was, as Nunzio Cazzetta described it, 'a lesson in Government 101'.[35] Although the Act had authorized $1.7 million dollars to be spent on a research program for thalassemia, no money had actually been appropriated. Rather, Congress expected those funds to be drawn from the general appropriations for research activities at the NIH. However, the NIH's budget was subject to competition from many different research groups and in fact, as the agency reported to Congress on the Control Act in 1974, the NIH was already spending some $2 million on basic research programs related to thalassemia. The NIH was thus meeting 'its "legal" requirements by listing in one place all the individual grants already supporting Cooley's anemia'.[36]

Frustrated, the CAF sought the advice of government affairs consultant, John Grupenhoff, who recommended they begin advocating with congressional and NIH staff members to secure appropriations specifically for thalassemia research.[37] Heeding Grupenhoff's advice, the Foundation has, since 1972, sent representatives to Washington D.C. at least once a year to advocate with the National Institute of Diabetes, Digestive, and Kidney Diseases (which had responsibility for research related to hemoglobin synthesis and iron metabolism), the National Heart, Lung, and

34 The Cooley's Anemia Newsletter, 'Cooley's Anemia Act – Report by Dr. French Anderson', 4 (Fall 1974), p. 4.

35 Cazzetta interview. As one government staffer explained to Cazzetta, 'you have the license to drive but you don't have the money to buy the gas'.

36 W.F. Anderson, 'Development of Iron Chelators – Chairman's Opening Remarks', in W.F. Anderson and M.C. Hiller (eds), *Proceedings of a Symposium: Development of Iron Chelators for Clinical Use* (Bethesda: US Department of Health, Education, and Welfare, 1976), p. 2.

37 John Grupenhoff, interview with author, January 27, 2006.

Blood Institute, and congressional staffers to get legislative language put into NIH appropriations bills asking for initiatives specifically for thalassemia research.[38] (More recently, the Foundation also began to petition the National Institute of Child Health and Human Development). As part of this strategy, the Foundation has also invited patients – often children and adolescents – to testify in front of the health appropriations committees of both the House and Senate in Congress each year about the importance of continued government support of thalassemia. And, through its publications and chapter meetings, the Foundation has called on its members to 'contact your Congressmen and Senators ... and get them moving, put pressure on them ...' to push for greater funding of thalassemia research.[39]

One product of the Foundation's legislative program (and a direct result of the National Cooley's Anemia Control Act) was the establishment, by the National Institute of Arthritis, Metabolic and Digestive Diseases in 1973, of a contract program to synthesize, isolate and test new iron chelator compounds. The iron chelator program was operated in a manner similar to the technology development projects funded by other government agencies in the second half of the twentieth century. Academic institutions competed for the government contracts to isolate and synthesize new chelating agents, to screen and run preliminary tests of those new compounds, and, beginning in 1983, to conduct preclinical studies on potential lead candidates.

38 Until 1972, the National Institute of Diabetes, Digestive, and Kidney Disorders was named the National Institute of Arthritis and Metabolic Diseases. In 1972, the Institute was renamed the National Institute of Arthritis, Metabolic and Digestive Diseases, followed by the National Institute of Arthritis, Diabetes, and Digestive and Kidney Diseases in 1981. The Institute received its current designation in 1986. The National Heart, Lung, and Blood Institute was named the National Heart and Lung Institute until June 1976 when hematologists, blood disorder researchers, and blood disease-based organizations included the Cooley's Anemia Foundation lobbied to include 'Blood' in the Institute's name.

39 The Cooley's Anemia Newsletter, 'Cooley's Anemia Act – Report by Dr. French Anderson', 4 (Fall 1974), p. 4.

Over the course of the 1970s and through the 1980s, the Institute awarded five contracts for the synthesis of new chelators to researchers at Duke University, Research Triangle Institute, North Carolina, Texas A&M, the University of Mississippi, and West Virginia University. While each of the contractors pursued a different approach to the development of new chelators, they relied on one of two research strategies. The first, pursued by researchers in the Department of Microbiology at the University of Mississippi Medical School, involved the isolation of naturally occurring chelating agents from microorganisms. The second, pursued separately by researchers at the Research Triangle Institute, Texas A&M, West Virginia University, and Duke University, involved the chemical synthesis of chelators based on the properties of naturally occurring and effective synthetic chelators.

The Institute also awarded two contracts – one to Rockefeller University and the other to the Mason Research Institute in Worcester, Massachusetts – for the screening of potential new chelators. The two research groups developed different screening assays using rodent models of the iron overloaded state, and proceeded to test the iron-binding capacities, efficacy, and toxicity of all potential products isolated and synthesized by the contract and by non-contract researchers. And in 1984, the Institute awarded a contract to SRI International, an independent, nonprofit research institute, to conduct animal toxicity studies of lead candidates produced by both contract and non-contract researchers.

Unlike the size of the contracts awarded for defense-related development, the chelator contracts were relatively small. The synthesis and screening program, which lasted about ten years, cost the Institute an average of $200,000 per year, while the preclinical toxicity contracts (which continue today) typically cost around $600,000 per year (although some years little work was done while the contractors were between compounds).[40]

The nature of the NIH budget, however, meant that the NIDDK program competed with other projects and other Institutes for the finite amount of government money appropriated by Congress for biomedical

40 Badman interview.

research each year. As such, the extent of the funding available for the iron chelator program depended in large part on the effectiveness of the CAF's legislative campaign. As David Badman recounted,

> every Institute [of the NIH] has its own constituent of patient- and disease-based organizations and as such, every Institute is being lobbied by those constituents and each of those Institutes in turn support the President's request to Congress to award them the appropriations for each specific disease ... the NIDDK has always had to justify the continued funding of the chelator program.

Fortunately the Cooley's Anemia Foundation has been very effective in petitioning Congress, as Badman continued, 'I doubt we could have done what we have without the CAF'.[41]

In addition to the NIDDK contract program, between 1974 and 2005, the combined institutes of the NIH (principally the NIDDK and the NHLBI) funded at least forty different research projects related to the isolation, synthesis, screening, preclinical and clinical testing of iron chelators, with several projects receiving continuous NIH funding for more than ten years.[42] By the mid-1970s, then, due in large measure to the activities of the Cooley's Anemia Foundation, there were two main funding options for researchers seeking to develop iron chelators in the United States: the NIH and specifically the NHLBI and the NIDDK – either through its contract or grants systems, and the Cooley's Anemia Foundation. By underwriting the cost of the preclinical stages of pharmaceutical innovation, the NIH

41 Badman interview.

42 Researchers who have received more than ten years of NIH funding through research grants include Kenneth Raymond, University of California, Berkeley (1980–2005), Colin Pitt of Research Triangle Institute (1974–1985), John B. Neilands, University of California, Berkeley (1974–1983), Arthur E. Martell, Texas A&M University (1978–1993), Robert Grady, Rockefeller University (1977–2005) and Weill Medical College of Cornell University (1980–2005), and Raymond J. Bergeron, University of Florida, Gainesville (1982–2005). Information on federally funded biomedical research projects since 1972 can be found using the NIH's CRISP database at http://crisp.cit.nih.gov/. Search terms used by the author were: chelator, thalassemia, iron, and the names of individual thalassemia researchers identified in the published medical literature.

and the Cooley's Anemia Foundation were instrumental in facilitating the development of new iron chelators by academic researchers.

To be sure, the academic development of iron chelators has not been a solely American enterprise. Several European researchers based at academic and clinical centers in Switzerland, Israel, Greece, Italy, and Britain, often working in collaboration with American researchers and supported by CAF fellowships and NIH research grants, have contributed significantly to the innovation process. In particular, a similar but distinct political economy of public sector chelator development emerged in Britain over the course of the 1970s and 1980s.

In 1978, working in the lab of medicinal chemist, Robert Hider at the University of Essex, George Kontoghiorghes completed his doctoral dissertation on the development of hydroxypyridones for the treatment of iron-overload in thalassemia. As a Greek Cypriot who had grown up in Cyprus in the 1960s and had lost two cousins to thalassemia and seen numerous other young Cypriots die of the disease, Kontoghiorghes was committed to pursuing research on thalassemia.[43]

The hydroxypyridones are a class of synthetic compounds that contain a hydroxamate group similar to that in desferrioxamine, but which are orally active. Kontoghiorghes' dissertation work and his subsequent post-doctoral work at University College Hospital in London, under the supervision of Hider and Ernest Huehns, was funded by the publicly-owned National Research Development Corporation and its successor, the British Technology Group. The NRDC was created by the post-war British government in 1949 as a means of translating the nation's scientific expertise and technological inventiveness into international commercial competitiveness. Until 1985, when the Thatcher government relinquished its monopoly rights, the NRDC and its successor, the BTG, were granted first refusal on the commercialization of all government-funded research in universities.[44]

43 M. Shuchman, *The Drug Trial: Nancy Olivieri and the science scandal that rocked the Hospital for Sick Children* (Toronto: Random House Canada, 2005).

44 For a history of the NRDC see, S.T. Keith, 'Inventions, patents and commercial development from governmentally financed research in Great Britain: the origins of the National Research Development Corporation', *Minerva*, 19 (1981): 92–122.

By the mid-1980s, Kontoghiorghes and his colleagues had identified L1 (also known as deferiprone) as the most promising of the hydroxypyridone compounds. However, as Miriam Shuchman has recently described, following Kontoghiorghes' increasing discontent regarding the ties of the University College Hospital lab to BTG and the company's desire to delay publication of the lab's results because of its commercial interests, Kontoghiorghes left Hider and Huehns' lab in 1984 and moved to The Royal Free Medical School in London where he continued to develop L1. From this point on, and continuing through the 1990s, Kontoghiorghes' work on L1 – including preclinical and early clinical studies – was funded entirely by the disease-based organization, the United Kingdom Thalassemia Society; the charity's contributions totaling around £500,000.[45] (British-Cypriot parents of a thalassemia child founded the UK Thalassemia Society in 1976). The drug development work of the British scientists reveals that the development and early testing of promising new chelators was, as in the United States, dependent upon the support of a charitable association and two publicly owned technology transfer companies, NRDC and BTG.

What influence then has this political economy of public sector drug development, in the United States in particular, had on the nature of chelator innovation?

45 The total figure cited is £512, 275. United Kingdom Thalassemia Society, http://www.ukts.org/pages/research.htm (accessed 6 March 2006).

Making Orphan Diseases into 'Feasible Industrial Markets'[46]

> In the history of chelation it's taken thirty years to come up with an oral chelator ... In thirty years – millions of dollars poured into it – you've got to be kidding me![47]

As Nunzio Cazzetta asserted, the rate of innovation in the field of iron chelators has been frustratingly slow – for patients, parents, and researchers alike. It is only in the last few years that compounds identified as potential new chelators by academic researchers in the late 1970s and early 1980s have begun to enter clinical trials. In spite of the substantial investments of the NIH and the Cooley's Anemia Foundation and their counterparts in Britain, and despite the innovative work of academic researchers, clinical development of promising chelating agents has still been hindered by the absence of significant corporate investment in the field. One way of unpacking this relationship between political economy and pharmaceutical innovation is to examine how researchers have made decisions about prioritizing and managing the limited resources available to them, and in particular, how they have determined which of their innovations has the best potential for success and which are most likely to fail.

Between 1974 and 1990, the team of contractors at Rockefeller University, led by Anthony Cerami and later by Robert Grady at Cornell University Medical Center, and subsequently by Mason Research Institute, screened as many as 250 molecules isolated and synthesized by the Institute's contractors and by other NIH-funded researchers.[48] Over the course of these 16 years, through the work of its contractors, the NIDDK identified

46 Hauck, 'Drug Development in Industry', p. 244.

47 Cazzetta interview.

48 Members of the New York team also received funding during this time from NIH research grants and research fellowships from the Cooley's Anemia Foundation. For example, Anthony Cerami received NIH research grants from 1976–1981, Robert Grady from 1977–2005, and Margaret W. Hilgartner from 1989–1991. Data obtained from NIH Crisp database, see n. 34.

five potential lead candidates for development.[49] In each case, the contractors identified the potential chelator using the Rockefeller-devised screening program, following which they conducted preliminary toxicity studies in animal models of the disease. In instances where the preclinical results were favourable, the group pursued limited clinical testing in a handful of thalassemia patients. During this time, the New York group discounted four of the potential oral chelators, after sub-optimal preclinical or clinical testing data for a compound forced them to consolidate their resources and focus on potentially more promising alternatives.

In 1974, for example, the team identified 2,3-dihydroxybenzoic acid as a potential candidate drug. Despite initially encouraging clinical results, by the end of a year-long double-blind study in 15 thalassemia patients conducted in 1977, the results were decidedly discouraging. Though 2,3-DHB showed no signs of toxicity when given for a year, the team was unable to document any indication that the drug reduced patients' accumulation of iron. As the group reported that year in the *New England Journal of Medicine*, 'Because of the greater efficacy of slow infusions of deferoxamine chelating iron, the clinic has shifted its emphasis toward further evaluation of that compound'.[50] Although the group ceased development work on the compound they were reluctant to entirely condemn 2,3-DHB, encouraging other researchers to pursue its development. The group continued, 'in view of the minimum toxicity of 2,3-DHB and the short duration of this trial, further work appears warranted to define its role in the treatment of iron

49 The NIDDK was given responsibility for identifying these compounds as the Institute not only worked with the contractors but also made all the decisions regarding lead compounds (from personal communication with David Badman). The five potential lead compounds were rhodotorulic acid, cholylhydroxamic acid, 2,3-dihydroxybenzoic acid, N,N'-bis(2-hydroxybenzyl) ethylenediamine-N, and ethylenediamine-N,N'-bis(o-hydroxyphenylglycine). See, for example, A. Cerami, R.W. Grady, C.M. Peterson, and K.K. Bhargava, 'The Status of New Iron Chelators', *Annals of the New York Academy of Sciences*, 344 (1980): 425–435.

50 C.M. Peterson, R.W. Grady, et al., "2,3-DHB Ineffective in Treatment of Iron Overload', *New England Journal of Medicine*, 297 (1977): 1404.

overload, and other investigators may wish to evaluate the drug'.[51] However, no other researchers picked up 2,3-DHB for further analysis.

The New York team's decision to turn their attention to other, more promising chelators indicates a critical limitation of research and development conducted in academic centers with government- and charity-funds: because these resources are limited, researchers are often forced to drop potentially promising lines of research. This has had implications for how researchers have characterized the therapeutic potential of their innovations. The histories of three subsequent lead candidates developed in the United States and Britain also reveals that 'success' and 'failure' have to be understood in the social and economic context of pharmaceutical innovation (see Table 11.1). Furthermore, their histories point to a significant shift in the political economy of chelator development that took place during the 1990s.

By 1990, the participants at the Sixth Annual Cooley's Anemia Symposium in New York had identified four orally active drugs as potential candidates for commercial development.[52] One of these compounds, HBED (and its derivative, dimethyl-HBED), had been synthesized by two of the NIDDK's contractors at Research Triangle Institute in 1978 and subsequently identified as a promising chelating agent by Grady's team at Cornell University Medical Center.[53] Based on preclinical rodent studies conducted by Grady and his collaborators, 'there was reason to believe that the HBED compounds would be orally active in humans'.[54] However, just as 2,3-DHB had shown great promise in preclinical studies in the late 1970s and subsequently failed to deliver on that promise in the clinic, HBED

51 Ibid.

52 See, for example, S. Piomelli, 'Oral Iron Chelators for the Clinical Management of Iron Overload. Current Hopes and Problems', in *Sixth Cooley's Anemia Symposium. Annals of the New York Academy of Sciences*, 612 (1990): 311–314.

53 C.G. Pitt and G. Gupta, 'The Design and Synthesis of Chelating Agents for the Treatment of Iron Overload in Cooley's Anemia', in W.F. Anderson and M.C. Hiller, *Proceedings of a Symposium*, pp. 137–168; R.W. Grady and C. Hershko, 'HBED: A Potential Oral Iron Chelator', in *Sixth Cooley's Anemia Symposium. Annals of the New York Academy of Sciences*, 612 (1990): 361–368.

54 Raymond J. Bergeron, Interview with author, 16 February 2006.

and its derivative, dimethyl-HBED, proved similarly disappointing in the early clinical studies.

Another of the candidates highlighted at the 1990 conference was desferrithiocin (DFT) and its derivatives. In 1980, Swiss researchers discovered and isolated the compound from a strain of *Streptomyces antibioticus*. Over the following decade, researchers at Ciba-Geigy, in collaboration with academic researchers in the United States, Britain, and Australia, began investigating the clinical potential of DFT in limited animal studies.[55] Although these studies were initially encouraging, more extensive testing in animal models of iron overload conducted shortly after the 1990 conference revealed that the compound caused significant damage to the kidneys, leading to the death of some of the animals tested. Ciba-Geigy subsequently dropped DFT from its chelator program.[56] However, despite these disappointing studies, unlike the earlier drug candidate, 2,3-DHB, neither DFT nor HBED were dropped as potential lead candidates by their academic sponsors. Rather, one of Ciba-Geigy's academic collaborators, Raymond Bergeron at the University of Florida (whose work has been supported by NIH research grants since 1982), continued to study DFT and its derivatives. Over the course of the 1990s, researchers in Bergeron's lab identified the molecular components of DFT responsible for its toxicity and subsequently synthesized several different compounds based upon the essential components of the DFT molecule but without the toxicity.[57]

Around the same time that Bergeron's lab had begun investigating the derivatives of DFT, David Badman of the NIDDK asked Bergeron

55 As noted in this paper's introduction, Ciba-Geigy did maintain a limited chelator program from the mid-1980s. However, the consensus among those I spoke with is that the company maintained its interest in the field because of the investment being made by the NIH.

56 Raymond Bergeron speculated that, in addition to the toxicity problems associated with DFT, the company might have dropped the compound because management of the iron chelator program changed hands around the same time. Raymond J. Bergeron, interview with author, 16 February 2006.

57 For example, see Raymond J. Bergeron, Charles Z. Liu, James S. McManis, Michael X.B. Xia, Samuel E. Algee, and Jan Wiegand, 'The Desferrithiocin Pharmacophore', *Journal of Medicinal Chemistry*, 37 (1994): 1411–1417.

to conduct further animal studies of HBED.[58] While the results were 'profoundly disappointing', Bergeron's lab continued studying the iron chelation properties of several HBED salts and in 1999 determined that the monosodium salt of HBED, if given subcutaneously or intravenously (the compound is inactive when administered orally), was significantly more effective at removing iron from non-iron overloaded rats and iron-loaded *Cebus apella* monkeys than Desferal.[59]

Thus, almost ten years after both HBED and DFT had tested poorly in preclinical and clinical studies, academic researchers still viewed these compounds (or at least their derivatives) as potential candidates for commercial development. Pharmacological reasons may in part explain why HBED and DFT continued in the innovation pipeline where a decade earlier 2,3-DHB had failed. However, several social and economic factors were at least equally as important.

First, by the time HBED and DFT tested poorly in preclinical studies, the NIH, the CAF, and thalassemia researchers had committed significant money, manpower, and time to their development. In purely economic terms, the size of both the CAF fellowships and NIH research grants had grown significantly between the mid-1970s and the early 1990s. Furthermore, the NIDDK had launched the preclinical testing phase of its contract program in 1984, several years after 2,3-DHB had been discarded as a potential drug candidate. Thus by the 1990s more money and more manpower were available to continue the development of HBED and DFT even after they had failed in the clinical and preclinical arena, respectively. Indeed, by 1990, both compounds had been in the innovation pipeline for more than a decade compared to the three years of research and testing put into 2,3-DHB. Researchers had more safety and efficacy data on HBED and DFT than they had had for 2,3-DHB because of the resource investment already made on the compounds; they also had a greater economic stake in the drugs' continued development.

58 Bergeron interview.

59 Raymond J. Bergeron, Jan Wiegand, and Gary M. Brittenham, 'HBED: the continuing development of a potential alternative to Deferoxamine for iron-chelating therapy', *Blood*, 93 (1999): 370–375.

Drug Name	Chemical or Alternative Names	Trade Names	Route of Administration	Main Developer(s)	Funding	Company
2,3-DHB	2,3-dihydroxybenzoic acid		Oral	Grady et al	NIH, CAF	
HBED	N,N'-bis(2-hydroxybenzyl) ethylenediamine-N		Oral	Grady et al	NIH, CAF	
Dimethyl HBED			Oral	Grady et al	NIH, CAF	
Sodium HBED			Intravenous or Subcutaneous	Bergeron et al	NIH, CAF	SunPharm GelTex
DFT	Desferrithiocin		Oral	Bergeron et al	NIH, CAF, Ciba-Geigy	GelTex, Genzyme
L1	Deferiprone	Ferriprox	Oral	Kontoghiorghes	NRDC, BTG, UKTS	Apotex
ICL670	Deferasirox	Exjade	Oral	Novartis	Novartis	Novartis

Table 11.1: Leading candidates for chelator development, 1977–2005.

A second factor helping to sustain continued investment in these compounds was the increased demand from patients for an orally active chelator. Although clinicians had recognized by the late 1970s that compliance posed a serious problem for patients undergoing Desferal therapy, many believed that patients would ultimately submit to the therapy if they were fully aware of the fatal consequences of noncompliance.[60] As Cazzetta recounted, through the 1980s there was a general feeling among researchers and clinicians of 'why bother, we have something and it works'.[61] Cazzetta continued, 'It took us [the CAF] forever to convince them that you could say comply all you want ... [but] you cannot have a human being [comply with the painful and arduous Desferal therapy] twenty-four/seven ... It's unrealistic'.[62] The efforts of the Cooley's Anemia Foundation to change researchers' perceptions of the therapeutic problem also came during a period of patient activism: in the wake of the women's health movement, which forced changes in the structure and demography of clinical trials, and during the wave of challenges to the AIDS drug trial process.[63]

In addition, because patients were not being optimally chelated, clinicians were beginning to see increasing numbers of patients die prematurely from iron overload in their twenties.[64] Thus, by the early 1990s thalassemia researchers had recognized the severity of the compliance issue and were more responsive to patients' demands for an oral chelator.

A third factor that contributed to the continued development of HBED and DFT by academic researchers was the growing interest of drug

60 Cazzetta interview.

61 Ibid.

62 Ibid.

63 On patient activism and AIDS research, see Steven Epstein, Impure Science. On the activism of female breast cancer patients on breast cancer research and treatment in the height of the women's health movement, see Barron H. Lerner, The Breast Cancer Wars. And for an analysis of the recent US health research policies that mandate the inclusion of women and racial and ethnic minority groups as experimental subjects in biomedical research, see Steven Epstein, 'Bodily Differences and Collective Identities: the politics of gender and race in biomedical research in the United States', *Body & Society*, 10 (2004): 183–203.

64 Chieco interview.

firms in oral chelators. For example, in 1999 Raymond Bergeron secured a patent on the sodium salt of HBED for use in the treatment of iron overload. The University of Florida, which held the patent, licensed the discovery to SunPharm, a small, Florida-based biotechnology company established in 1990 'specifically to move technologies created in [Bergeron's] lab to the clinic'.[65] The license was subsequently sold to GelTex when the firm bought SunPharm later that year.[66] However, because sodium HBED was not an orally effective iron chelator, GelTex put further development of the drug on the backburner. Bergeron explained, 'At that time, everyone wanted an *oral* chelator – and it actually looked like we might have one coming soon ... Because of this, the interest level in HBED waned'.[67] Apparently, when GelTex considered sodium HBED competing with other orally active drugs currently in the clinic, the company saw the drug as a potential economic failure and refocused its development resources to compounds that were orally active. Despite this setback, sodium HBED remains in the pipeline and Bergeron is optimistic that either Genzyme (or a smaller company) will soon renew its clinical interest in the compound.[68]

The histories of HBED and DFT highlight a critical shift in the political economy, from public-funded academic innovation without corporate-backing to the entrance of small start-up firms willing to invest in the clinical development of new chelators. Indeed, Bergeron was also able to secure corporate sponsorship for the DFT compounds produced and patented by his lab. In 2005, the University of Florida licensed the DFT

65 Raymond J. Bergeron, Interview with author, 16 December 2005.

66 Ibid.

67 Raymond J. Bergeron, Interview with author, 16 February 2006.

68 Bergeron is optimistic because the economic picture for NaHBED has changed since the arrival of an orally active chelator on the market. Unlike Novartis's ICL670, Desferal, and the other chelators currently being tested in the clinic, animal studies suggest that NaHBED can be given safely to patients in the high doses needed to remove iron quickly enough from severely iron overloaded hearts (these patients have a very high risk of cardiac failure and death). This represents a potential clinical niche for NaHBED as the current treatment, high intravenous doses of Desferal, increases the patient's risk of hypotensive events, which in turn can lead to the patient's death.

compounds to Genzyme, which had acquired GelTex in 2001.[69] The history of a third lead candidate points to another new entrant in the political economy: the generic manufacturing firm.

George Kontoghiorghes, one of the British researchers responsible for developing the hydroxypyridones during the 1980s and 1990s, also managed to secure corporate investment in his lead candidate, L1. Ciba-Geigy expressed initial interest in the drug in 1990 and 1991, sponsoring extensive preclinical studies on the compound. However, after Kontoghiorghes's former collaborators began raising questions about the potential toxicity of the compound, the company dropped the product. Two years later, in part due to the campaigning of the Cooley's Anemia Foundation and members of its Medical Advisory Board, the Canadian firm, Apotex, which develops and manufactures generic pharmaceuticals, took over commercial development of the drug. By the mid-1990s, L1 became the first oral chelator shown to be effective in clinical studies and in 1998 Apotex began marketing the drug in Europe, under the brand name Deferiprone. As of 2006, the company had yet to apply for FDA approval for Deferiprone.

The history of L1's clinical development is extremely contentious. Published accounts (such as Miriam Shuchman's book) highlight the personality clashes, professional conflict, and questionable medical ethics that characterized this stage of the drug's development.[70] However, this history, together with that of HBED and DFT, reveals that the political economy

69 One of these compounds is now being tested in the clinic by Genzyme following the encouraging results of preclinical studies conducted by the NIDDK contractor, SRI International. I suspect that Bergeron's success in securing corporate sponsorship for chelators produced in his lab is due, in large part, to the social and economic capital Bergeron has with these firms. During the 1990s, Bergeron helped SunPharm introduce a new anticancer and a new AIDS drug in the clinic. Currently, Genzyme has another of Bergeron's compounds undergoing clinical trials.

70 For the fullest account of the controversial development of L1 see Miriam Shuchman, *The Drug Trial*. A selection of the extensive number of biomedical articles, editorials, and reviews published on the subject include: David G. Nathan and David J. Weatherall, 'Academic Freedom in Clinical Research', *New England Journal of Medicine*, 347 (2002): 1368–1371; Elaine Gibson, Françoise Baylis, and Steven Lewis, 'Dances with the Pharmaceutical Industry', *Canadian Medical Association Journal*, 166 (2002): 448–450; C. David Naylor, 'The Deferiprone Controversy: time to move on', *Canadian Medical*

of chelator development had shifted significantly by the early 1990s: while reluctant to invest in chelation therapy in the 1980s, despite the therapeutic promise of academic innovations, a handful of pharmaceutical companies were expressing new interest in the field. The crucial point here is that these companies were start-up biotechnology and generic manufacturing firms, not large research-based pharmaceutical companies. Their entrance was paved by two pieces of legislation in the early 1980s.

The 1980 Bayh-Dole Act enabled universities and small businesses to patent discoveries that resulted from government-funded research (such as the NIDDK's Iron Chelator Program) and grant exclusive licenses to pharmaceutical and biotechnology firms. This Act made it possible for academic researchers such as Robert Grady and Raymond Bergeron to begin licensing their innovations to interested firms.

In 1983, in an effort to stimulate innovation, President Reagan signed the Orphan Drug Act into law. The Act promised tax credits and government grants to companies that developed, tested, and marketed orphan drugs. And, after Congress amended the Act the following year, granted seven years marketing exclusivity to the first firm to complete all the testing requirements and file for a New Drug Application with the FDA. Certainly, Apotex, SunPharm, GelTex, and Genzyme were beneficiaries of these incentives when they picked up L1, sodium HBED, and DFT from their academic innovators. However, while the Bayh-Dole and Orphan Drug Acts opened the door for corporate investment in the chelator field, it took a decade for the first firm to walk through it. The published medical literature and the experiences of those working in the field may indicate several reasons why this happened.

First, academic researchers such as George Kontoghiorghes, Robert Grady, and Raymond Bergeron had, by the mid-1990s, accumulated relatively extensive efficacy and toxicity data on some of their lead compounds; just the kind of data that pharmaceutical companies needed (but refused to pay for). Second, chelator developers were beginning to identify other diseases with substantially larger patient populations (such as diabetes,

---

Association Journal, 166 (2002): 452–453; and Julian Savulescu, 'Thalassemia Major: the murky story of Deferiprone', *British Medical Journal*, 328 (2004): 358–359.

cancer, malaria, and HIV) that might respond well to treatment by iron chelators, thereby expanding the potential marketplace for iron chelators.[71] And third, companies such as Ciba-Geigy and Apotex were under increasing pressure from thalassemia researchers, clinicians, and the Cooley's Anemia Foundation, to begin investing in chelator development.

Ultimately, the arrival of L1 in the clinics of Europe and the United States in the mid-1990s permanently transformed the political economy of chelator development. Not only did L1 potentially offer thalassemics an alternative to Desferal and the syringe pump, it also provided Ciba-Geigy with an economic incentive to accelerate its own chelator program. Specifically, L1 placed Ciba-Geigy's market position under threat for the first time in thirty years. The threat was only temporary. In November 2005, Novartis (formerly Ciba-Geigy) reclaimed that position when it received FDA approval for its own oral chelator, ICL670 (which is marketed in the United States as Exjade). Though no evidence is available to explicate the relationship between what seems to be Novartis' rapid development of an oral chelator and the arrival of L1 in the marketplace, Nunzio Cazzetta (along with others whom I have spoken with), speculates, 'Novartis pushed quickly when they thought they might have a competitor'.[72]

71 For example, see David G. Badman, Raymond J. Bergeron, and Gary M. Brittenham (eds), *Iron Chelators. New Development Strategies* (Ponte Vedra, Florida: The Saratoga Group, 1998). In particular, for rationale behind chelation therapy for diabetes see, Mingwei Qian and John W. Eaton, 'Transition Metal Chelators: possible new therapies for diabetic complications', 399–414; for chelation therapy for HIV see, Johan R. Boelaert and Jacques Piette, 'Iron Overload: exacerbation of fungal infections by siderophores and iron's role in HIV infection', 415–426, and for chelation therapy for malaria see, Mark Loyevsky, Victor R. Gordeuk, and Gunter Weiss, 'Iron Chelation as a Therapeutic Strategy for the Treatment of Malaria', 427–454. On the rationale behind using chelation therapy in anti-cancer treatment see, Des R. Richardson, 'Analogues of Pyridoxal Isonicotinoyl Hydrazone (PIH) as Potential Iron Chelators for the Treatment of Neoplasia', *Leukemia and Lymphoma*, 31 (1998): 47–60; and G. Darnell and Des R. Richardson, 'The Potential of Iron Chelators of the Pyridoxal Isonicotinoyl Hydrazone Class as Effective Antiproliferative Agents III: the effect of the ligands on molecular targets involved in proliferation', *Blood*, 94 (1999): 781–792.

72 Cazzetta interview. For a review of Novartis's chelator program and their efforts to develop an orally active chelator, see N. Hanspeter et al., 'Development of Tridentate

Conclusion

The history of the development of iron chelators teaches us several things about drug development and biomedical research in the second half of the twentieth century. First, it shows us precisely how a disease-based organization helped guide the course of research for a specific disease. It shows us how they lobbied, who they lobbied, and what exactly they lobbied for.[73] Since the 1970s, the Cooley's Anemia Foundation's broad-based strategy of identifying the key actors in the biomedical research economy – legislators, their staffers, NIH researchers *and* drug developers – has proven remarkably effective at securing greater funding for thalassemia research and in pushing forward chelator development. By organizing patients to testify every year in front of Congress the Foundation has, according to government affairs consultant John Grupenhoff, been able to 'convey the seriousness of the disease … [and] put a human face on the requests for more research money'.[74] For researchers like Raymond Bergeron the effect has been equally powerful:

> [Having] a mother come up to me after giving a presentation, asking me 'When can I treat my child with this?'… gives you a whole new perspective on your work. And it certainly puts a fire under you to get things done. And this is a truly marvelous thing. The CAF should take as much credit as the scientists and physicians do in getting things done. They really have greased a lot of wheels.[75]

Yet the experiences of the Cooley's Anemia Foundation also reveal how contested the lobbying environment is for groups interested in shaping the political economy of biomedical research. For instance, lobbying Congress to get $3 million authorized for spending on thalassemia research,

Iron Chelators: From Desferrithiocin to ICL670', *Current Medicinal Chemistry*, 10 (2003): 1065–1076.

73 Because representatives from the CAF were not paid for the legislative work they did, it is actually more accurate to characterize their legislative activities as advocacy work. John Grupenhoff, interview with author, October 1, 2006.

74 John Grupenhoff, interview with author, 27 January 2006.

75 Bergeron interview, 16 December 2005.

prevention, treatment, and education means very little, in practice, if no money has been *appropriated*. As such, disease-based organizations like the Cooley's Anemia Foundation must continually advocate with Congress and the NIH to ensure *sustained* government investment in the research field.

Second, the history of chelator development highlights the critical role of the NIH in securing not only the financial but also manpower resources needed to advance drug research. With the research behind much drug development seen as too mundane by many scientists, the NIDDK's contract program offered an incentive to researchers to work in the field. In addition to materially supporting the work of the contractors, the program also provided researchers working outside the program with an opportunity to have their new innovations tested by someone else and have the cost of that testing underwritten by the government. Not incidentally, the National Cancer Institute established a similar – although much larger – program in the 1950s to develop new anticancer agents. That the NIDDK program generated at least two drugs now being tested in the clinic suggests that the government contract program serves as a model for the development of other orphan drugs when drug companies are apathetic.

Third, we currently know very little about how pharmaceutical innovation happens. In the current litigious climate it is increasingly difficult for historians to gain access to drug company records for post-World War Two drug developments. At the same time, for proprietary reasons, industrial researchers have generally been reluctant to publish their development work. In contrast, the history of public-sector drug development provides an important window into the processes of pharmaceutical innovation. For example, in the field of chelator development, an analysis of the published medical literature and published data on NIH funding, together with conversations with the academic developers and their sponsors, has shown the ways in which researchers make decisions about how to prioritize and manage the resources available to them.

Yet, while this history reveals the critical role of public sector organizations such as disease-based organizations and the NIH in forging a political economy that is supportive and indeed enabling of orphan drug development, it also illustrates the limits of that support. Although the NIH and Cooley's Anemia Foundation were able to underwrite the cost of the preclinical development of potential new chelators, the cost of bringing a new chelator to market exceeded their financial support. The emergence of start-up biotechnology and generic manufacturing firms willing to pick up promising innovations from academic researchers and introduce them into the clinic thus proved crucial. Not only did these companies make possible the continued development of academic innovations, the presence of these new competitors in the political economy also encouraged large firms such as Novartis to take seriously the challenge of developing a new oral chelator. As a result, the history of chelator development illustrates the important role of start-up biotechnology companies in the marketplace of orphan drugs.

Acknowledgements

I would like to thank Ruth Schwartz Cowan, Matthew Tontonoz, Jeremy Vetter, and Roberta Bivins for their critical insights into various drafts of this article. I owe thanks, in particular, to Ruth Schwartz Cowan for introducing me to the subject of iron chelation therapy. Earlier versions of this work were presented at Perspectives on Twentieth Century Pharmaceuticals (July 2005), and at the History of Biomedical Research Interest Group of the National Institutes of Health (March 2006). I thank the audiences of both meetings for their valuable comments. I am indebted to the generosity of all those I interviewed as research for this article: David Badman, Raymond Bergeron, Nunzio Cazzetta, Peter Chieco, Gina Cioffi, Alan Cohen, John Grupenhoff, and the patients who openly shared with me their experiences of living with thalassemia. Finally, many thanks go to Anne Waginger for transcribing those interviews.

HEATHER MUNRO PRESCOTT

12 Imperfect Methods and Uncertain Choices: the regulation and marketing of emergency contraception in the United States

Introduction

In a 1990 *New York Times* Sunday Magazine column, science and health reporter Philip J. Hilts declared: 'Years of litigation and agitation have left America in the dark ages of contraception'. Since the introduction of the contraceptive pill in the early 1960s, Hilt argued, the United States had 'fallen farther behind Europe in the birth-control choices available', as the number of US pharmaceutical companies involved in contraceptive research and development fell from nine in 1980 to only one, Ortho Pharmaceutical, in 1990. Hilts blamed the dearth of birth control innovation on an 'unwitting coalition of longtime adversaries – manufacturing companies and courtroom litigators; feminists, right-to-life groups and religious activists, including the Catholic Church – who over the last two decades, working independently, created what is now described as a birth-control backlash'.[1]

The term 'unwitting coalition', masks the multiplicity of responses to various birth control options developed since the 1960s. As Adele Clarke and Theresa Montini argue, contraceptives are examples of 'contested technologies' which are 'constructed within as well as disseminated through extant, contentious arenas composed of heterogeneous actors committed to action on the core issue'. Using RU-486 (mifepristone) as an example,

1 P.J. Hilts, 'Birth-Control Backlash', *New York Times* (16 December 1990), SM41.

Clarke and Montini offer an 'arena analysis' of the 'various actors, including scientists, pharmaceutical companies, medical groups, antiabortion groups, women's health movement groups, and others who have produced situated knowledges' of the drug.[2]

Likewise, this article will explore the multiple constituencies involved in the development and marketing of emergency contraceptives since the 1960s. Because various forms of emergency contraception have existed since the 1960s, this technology has a significantly longer history than does mifepristone. Therefore, it is possible to trace not only the 'heterogeneous constructions' of this technology by various actors, but also how the attitudes of some key participants involved in this story have evolved over time.

After-the-Fact Contraception

The first observations that steroids could be used as postcoital methods of birth control were made in the mid-1920s by A.S. Parkes and C.W. Bellerby, British physiologists who found that injection of estrogen compounds into rats and mice within two days of copulation caused reappearance of the estrus cycle and pregnancy interruption. Although Parkes and Bellerby did not speculate on the implications of their research for human contraception, they did suggest that estrin injections could be used to treat amenorrhea and other menstrual disorders. During the mid-1960s, birth control pioneer Gregory Pincus drew upon these earlier studies and demonstrated similar findings in laboratory animals.[3] Soon after, Ary A. Haspels, Professor of

2 A. Clarke and T. Montini, 'The Many Faces of RU486: tales of situated knowledges and technological contestations', *Science, Technology, and Human Values*, 18 (1993): 42–78.

3 G. Pincus, 'Hormonal Steroids and Preimplantation Stages', in G.E.W. Wolstenholme and M. O'Connor (eds), *Ciba Foundation Symposium: preimplantation stages of pregnancy* (Boston, MA: Little, Brown, and Company, 1965), pp. 378–396. For more on the history of Pincus's work, see L. Marks, *Sexual Chemistry: a history of the pill* (New Haven, CT:

Obstetrics and Gynecology at the University of Utrecht in the Netherlands, and Yale gynecology researchers John McLean Morris and Gertrude van Wagenen made the first reports of the use of this technology in humans. In both cases, requests for human use came from rape counseling centers. In 1964, Dutch rape counselors asked Haspels to treat a 13-year-old girl who had been gang-raped by three men.[4]

Morris and van Wagenen, who had published their research on post-coital contraception in the macaque monkey in the late 1950s, received similar requests from rape crisis workers at Yale New Haven hospital, starting in 1963. Although they noted in their article, 'in spite of confidence that success in the macaque should be paralleled by success in man, initial human experimentation was undertaken with some trepidation', partly because the impact on humans was unknown, but also because disseminating any form of birth control was illegal in the state of Connecticut until 1965. Their first cases, then, were rape victims, who received 50mg of diethylstilbestrol injected 4 to 6 days after exposure. Morris and van Wagenen acknowledged that 'the chance of pregnancy following rape is uncertain', none of their small sample of patients became pregnant, and 'side effects, when they occurred, were those usually associated with administration of estrogen', chiefly nausea and breast soreness that ceased after medication was stopped. The researchers then solicited a 'limited number of courageous volunteers', consisting of married graduate students and faculty wives, who 'furnished an opportunity for further and more adequate observation'. These preliminary trials also resulted in no pregnancies, but given the small number of subjects, they concluded: 'these results should be viewed with caution'.

In his discussion of the paper, Somer Sturgis, Professor of Gynecology at Harvard Medical School, suggested that Morris and Van Wagenen's

---

Yale University Press, 2001); A. Tone, *Devices and Desires: a history of contraceptives in America* (New York: Hill and Wang, 2001); E. Watkins, *On the Pill: a social history of oral contraceptives, 1950–1970* (Baltimore, MD: Johns Hopkins University Press, 1998).

4 A.A. Haspels, 'The "morning-after pill" – a preliminary report', *International Planned Parenthood Medical Bulletin*, 3 (1969): 6; Haspels, 'Emergency Contraception: a review', *Contraception*, 50 (1994): 101–108.

work 'may be the most important advance in antifertility research in the human being since the pathfinding work of Rock and Pincus more than a decade ago'. Indeed, he found it even more significant since the problem of overpopulation in 'deprived areas of the world', could not be solved by the standard birth control regimen of daily pills for 24 days of the menstrual cycle. 'Yet the coital act itself can be universally accepted as a signal to take medication to prevent conception by the most primitive of peoples', Sturgis argued. 'This is the ultimate goal of these preliminary studies'.[5]

Sturgis's remarks reflect what Johanna Schoen calls 'the complex set of factors that motivated health and welfare professionals' involvement in the delivery of birth control' to the developing world and indigent populations in the United States. On the one hand, many of these professionals were 'genuinely concerned with maternal and child health and hoped to improve women's access to health and contraceptive services'. On the other hand, they also 'hoped that limiting the reproduction of the poor would provide a scientific solution to poverty and poor health and would improve the quality of the race'. They also argued: 'The distribution of contraceptives among the poor would save taxpayers money by controlling state expenses for social services'. Although by the 1960s some health professionals were beginning to express the view that even the poor had a right to reproductive control, there was also a persistent belief that poor, uneducated women were unreliable contraceptors. Consequently, these professionals tended to favour technologies such as the IUD, that did not require client cooperation and could be under the control of physicians rather than patients.[6]

Sturgis's comments, therefore, suggest that some birth control professionals considered emergency contraceptives to be another way to handle the 'non-compliance' issue. Dr Alan Guttmacher, President of Planned Parenthood World Federation, told *New York Times* personal health col-

5 J.M. Morris and Gertrude van Wagenen, 'Compounds Interfering with Ovum Implantation and Development. 3. The role of estrogens', *American Journal of Obstetrics and Gynecology*, 96 (15 November 1966): 804–815.

6 J. Schoen, *Choice and Coercion: birth control, sterilization, and abortion in public health and welfare* (Chapel Hill: University of North Carolina Press, 2005), pp. 22–23.

umnist Jane Brody that the potential for 'after-the-fact' contraception was 'tremendous', especially for young women who had infrequent intercourse. 'From a physician's point of view', said Guttmacher, 'such intermittent medication is more desirable than the oral contraceptives in current use ... because it is less likely to give undesirable effects' that lead patients to discontinue use. In the same article, Morris underscored the significance of the pill for the young rape victims he saw from the impoverished neighborhoods surrounding New Haven Hospital: 'It means that the girl who has been raped and has a good chance of becoming pregnant doesn't have to worry', he said. 'Neither does the girl who was caught off guard and did not take prior precautions ... If all else fails, girls can wake up in the morning and know that there is available retroactive contraception'.[7]

Morality Based on Fear is No Morality at All

This concern about the consequences of contraceptive negligence are also apparent in discussions of unmarried college women. Commenting on Morris and Van Wagenen's work, Dr Charles E. McLennan, Professor of Obstetrics and Gynecology at Stanford University, flippantly suggested that 'the supply of experimental subjects might become enormous if campus vending machines could dispense little "six-packs" of stilbestrol tablets, each packet promising a refund if the user would simply report results promptly to a central registry in New Haven'. McLennan concluded that the US Food and Drug Administration 'would veto such a straightforward approach to science'.

McLennan's tongue-in-cheek remarks about campus vending machines alluded to what some referred to as a 'contraceptive revolution' on North American college campuses that paralleled the sexual revolution of the 1960s. According to a report published by the Group for the Advancement

7 J.E. Brody, 'Human Tests Considered', *New York Times* (1 May 1966): 1.

of Psychiatry, *Sex and the College Student* (1965), 'college students are asserting with increasing militancy their right to privacy and self-determination in matters pertaining to their own sexual activity. There is consensus that the double standard is increasingly being discarded and that college girls more often seek sexual experience during the college years than did their mothers'. The report observed that students were especially vigilant in their demands for contraception: 'Many college students view contraceptive information as a right that is due them', and were well informed about the various methods that were available, including the latest invention, the contraceptive pill.[8]

The notion of a 'right to privacy' in sexual matters alluded to the landmark US Supreme Court decision *Griswold v. Connecticut* (1965). Prior to that decision, Connecticut and other states with large Catholic populations forbad distributing contraceptives, even to married women. In 1961, Estelle Griswold, Executive Director of the Planned Parenthood League of Connecticut, and other birth control advocates in the state set up a clinic in New Haven in order to test state law. Griswold and other clinic staff were arrested, but appealed their case all the way to the US Supreme Court, who ruled: 'Connecticut's birth-control law unconstitutionally intrudes upon the right of marital privacy'.[9] It was not until 1972 that the Court extended the same right of privacy to the unmarried.[10] Until then, most states were either silent on the issue of whether the unwed had the same privacy rights, or explicitly outlawed prescribing or distributing contraceptives to unmarried individuals as was the case in Massachusetts and Wisconsin.[11] Consequently, an ACHA survey in 1970 indicated that the number of institutions offering contraceptive services had increased

8 Group for the Advancement of Psychiatry, *Sex and the College Student*, G.A.P Report Number 60 (New York, 1965), pp. 10, 44.

9 *Griswold v. Connecticut*, 381 US 479 (1965). For more on the evolution of the concept of the right of privacy in American jurisprudence, see J.W. Johnson, *Griswold v. Connecticut: Birth Control and the Constitutional Right of Privacy* (Lawrence: University of Kansas Press, 2005).

10 *Eisenstadt v. Baird*, 405 US 438 (1972).

11 H.F. Pilpel and N.F. Wechsler, 'Birth Control, Teenagers, and the Law', *Family Planning Perspectives*, 1 (Spring 1969): 29–35.

to 118, but this represented a small percentage of the over 2,500 colleges and universities in the US at that time. Although the reasons for these statistics were not stated, it possible that fear of prosecution played a role in the dearth of birth control services offered at college and university health centers.[12]

Yale started offering contraceptive and sex counseling services to unmarried undergraduates shortly after going coed in 1969, even though the *Griswold* decision only applied to married women. 'When Yale first started admitting women undergraduates', said Yale gynecologist Dr Philip Sarrel, 'the officials had fantasies that all the young girls descending on campus would get pregnant'. To alleviate these fears, Yale hired Sarrel and his wife Lorna, a psychiatric social worker, to set up a sex counseling service that provided not only contraceptives but sex education and relationship advice as well.[13] Sarrel urged his colleagues at other institutions not to 'hide behind the law ... If it is good medical practice to prevent a girl from becoming pregnant', and he believed this was the case for many young, unmarried women, 'then the law should be circumvented so that the girls does not go to a criminal abortionist'.[14] Marguerite Shearer, staff gynecologist at the University of Michigan, likewise told the *Detroit Free Press* that she would continue to prescribe contraceptives to female students, regardless of marital status, adding, 'Morality based on fear is no morality at all'.[15]

Students who could not find birth control on campus resorted to services in the local community. At the University of North Carolina, Chapel Hill, students sought the advice of Dr Takey Crist, a gynecologist at North Carolina Memorial Hospital and a member of the faculty in the

12 P.M. Sarrel and L.J. Sarrel, 'Birth Control Services and Sex Counseling at Yale', *Family Planning Perspectives*, 3 (July 1971): 33.

13 Pamela Swift, 'Student Sex Clinics', *Philadelphia Sunday Bulletin*, 29 August 1971, clipping from Radcliffe Archives News Office, RG X, Series 7, Box 27, Folder 406, Schlesinger Library, Radcliffe Institute for Advanced Study, Cambridge, Massachusetts.

14 'Contraceptive Counseling, Classes on Sex Seen Needed on Campus', *Ob. Gyn. News* (1 May 1970): 6.

15 Walker Lundy, 'Never Have So Many Been So Unabashed: U-M Meets Sex Problem Head On', *Detroit Free Press* (8 April 1969), courtesy of the Bentley Historical Library, University of Michigan, Ann Arbor.

UNC School of Medicine Department of Obstetrics and Gynecology. In 1969, Crist and his colleagues in the medical school conducted a survey of the contraceptive practices of 600 women at UNC. Crist reported: 'While the modern coed may now feel more free to indulge in sexual relations, her ignorance of contraception is appalling'. While some women failed to use birth control because they did not anticipate having sex, or did not want to 'ruin the moment', by the far the most common reason 'was that the girl did not know where to get such information regarding contraception'. Crist attributed this ignorance to the fact that 'contraceptives and related sexual information are not generally available to unmarried students at the student health service of this university. Most coeds seek help from OB residents or physicians in the adjacent community'.[16] He observed that while 'college students have begun to ask questions and take action, the adult world has allowed them to search for themselves and without any direction. How can we help our young people assume the responsibility that goes along with their desire for sexual freedom if we fail to listen and to attempt to guide them with understanding?' Crist called on universities and colleges to meet this demand for knowledge, 'for to abdicate this responsibility would leave instruction and consideration to the commercial interest of our society'.[17]

During the mid to late 1960s, students around the country led campaigns demanding that birth control services be available in student health centers. In response to these student movements, Planned-Parenthood-World Population initiated a Program of Student Community Action in 1969, appointing twenty-five year old Dan Pellegrom to direct the program. Pellegrom was a recent graduate of Columbia University, where he had served as Student Council President during the infamous student

16 T. Crist, 'Contraceptive Practices Among College Women', clipping, ca. 1971, Takey Crist Papers, Special Collections, Duke University (hereafter referred to as Crist Papers), Box 5.

17 Crist, 'Something is Happening on our College Campuses', *North Carolina Journal of Medicine,* 33 (1972): 42–44, Crist Papers, Box 5.

strike of 1968.[18] Planned Parenthood leaders hoped that a college program would 'acquaint college students with the latest contraceptive techniques and devices and provide them with information about availability and accessibility' and 'provide a vehicle' for establishing contraceptive services on college campuses. The College Program also drew upon broader concerns about the 'population explosion', although they intended to expand population concerns beyond the traditional focus on impoverished minorities in the United States and the developing world. Rather, they hoped to 'motivate college students to have smaller families through the voluntary use of the latest, most effective, and safest contraceptive methods'.[19]

One of these new and effective methods was the 'morning-after pill', which proved especially popular during an era when abortion was illegal in many areas of the country. According to a report in *Medical World News* in 1969, even college physicians who 'were loath only three years ago to prescribe oral contraceptives' were now 'doling out big doses' of DES 'as a "morning-after" preventive measure for girls unprepared for the night before'. The report noted that this 'enthusiasm' for the morning-after pill was spreading even though the US Food and Drug Administration (FDA) had not approved DES for the purposes of emergency contraception.[20] In the United States, the FDA is the main federal agency overseeing the safety of most food substances, human and animal drugs, medical devices, and other therapeutic substances of biological origin. In order for a drug to be distributed in the United States, pharmaceutical manufacturers must submit a New Drug Application (NDA) to the FDA. Yet the agency cannot regulate medical practice, and it is common for physicians, based on scientific evidence and professional standards of care, to prescribe a medication 'off-label', i.e. for diseases or conditions other than that for which it

18 Program Description, Planned Parenthood College Programs, Planned Parenthood Federation of America Papers, Series II, Sophia Smith Collection, Smith College, Northampton, MA (hereafter referred to as PPFA II Papers), Box 56, Folder 11.

19 Samuel Taylor to Douglas Stewart, 27 March 1969, PPFA II Papers, Box 56, Folder 1; 'A Program of Student Community Action (A working document) for Planned Parenthood-World Population', June 1969, PPFA II Papers, Box 56, Folder 1.

20 'Morning After Pill Gets Around', *Medical World News* (5 December 1969): 13.

originally received FDA approval. Although manufacturers cannot market or advertise uses of a drug that are not FDA approved, the FDA has no oversight over how a physician uses a drug already on the market. In the case of DES, an FDA official declared: 'Our authority is over the manufacturer and promoter. We can only advise doctors'. Yet, he also noted, 'the safety and efficacy of diethylstilbestrol as a morning-after contraceptive has not been established. What's needed is a well-monitored scientific investigation'.[21]

The first to meet this call for a systematic study was University of Michigan Health Service senior physician Lucile Kirtland Kuchera, who published her work with students in the *Journal of the American Medical Association* in 1971. The article described a study of 1,000 patients given DES as a postcoital contraceptive at the University Health Service during 1966 and 1967. Kuchera gave 'great credit' to the work of Morris and van Wagenen whose 'work on effective postcoital antifertility agent' had inspired her to begin administering the treatment to her patients who had come to her for help when exposed to unprotected intercourse because of rape, neglect to use a contraceptive, or failure of prophylaxis such a broken condom. 'Formerly in this clinic we had only told the patient to wait for her next menses – not very consoling advice to a woman who does not want to be pregnant'. Kuchera reported that the pill was 100 per cent effective, and that 'no serious side effect was known in any of the 1,000 cases while taking diethylstilbestrol or afterward'. Some patients reported nausea or headache, but 45 per cent 'had virtually no side reactions'.[22] When asked by *New York Times* personal health columnist Jane Brody why she published the study, Kuchera said she wanted not only to confirm the results of other physicians, but also 'to acquaint the general practitioner, who sees many of the women who might benefit from the drug, with the treatment schedule'.[23] Unfortunately, Kuchera's work would soon become the focal

21 'Morning After Pill Gets Around', 14.

22 Lucile Kirtland Kuchera, 'Postcoital Contraception with Diethlystilbestrol', *JAMA*, 218 (25 October 1971): 562.

23 J.E. Brody, 'Success Reported for Next Day Pill', *New York Times* (26 October 1971): 19.

point of a controversy regarding the use of postcoital contraception on college campuses.

Morning-After or Mourning-After?

In April of 1970, cancer researchers in Boston published the first reports of vaginal cancer in the daughters of women who had consumed DES while pregnant.[24] Although there was no established link between the use of DES as a postcoital contraceptive and vaginal cancer, when Kuchera's work reached the national media, Belita Cowan, a Masters student in English at Michigan, began an investigation into the use of DES as a 'morning-after' pill. Cowan worked part-time at the University Hospital where physicians routinely gave the drug to female students and other women in the community who wanted to prevent unwanted pregnancies.[25] Cowan found that Kuchera's results did not fit with her experience working with some gynecology patients in the University Hospital. According to Cowan, some women who had received the drug experienced serious side effects such as severe headaches, nausea, and vomiting. Cowan was also horrified that doctors continued to administer DES as postcoital contraception even after the cancer reports began to appear. Although there was no link between use of DES as an emergency contraceptive and vaginal cancer in women who took it, Cowan and her peers were sufficiently alarmed to begin investigating the matter. In 1971, Cowan joined with women who had been patients at the University of Michigan Health Service to form Advocates for Medical Information (AMI), the forerunner of the National Women's Health Network. Hoping to inform women at both Michigan

24 A.L Herbst and R.E. Scully, 'Adenocarcinoma of the Vagina in Adolescence', *Cancer*, 25 (April 1970): 745–757.

25 S. Morgen, *Into Our Own Hands: the Women's Health Movement in the United States, 1969–1990* (New Brunswick, NJ: Rutgers University Press, 2002), p. 10.

and other colleges, AMI first warned of the dangers of postcoital use of DES in an article entitled 'Cancer and the Morning-After Pill (will you be Mourning-After)' in their feminist newspaper, *her-self* in 1972.[26] Using a grant from the Student Government Association, Cowan conducted her own survey of the morning-after pill in 1974, interviewing 65 of the more than 1,400 students who had received the pill at the Health Service by the end of 1971. According to the students surveyed, gynecologists at the health center did not inform them about the dangers of DES, nor were they informed that they were part of an experimental study.[27]

In 1972, Cowan's findings and similar reports from other campuses prompted the Health Research Group of Ralph Nader's consumer watchdog organization the Public Citizen to allege that 'many university health centers' were prescribing a carcinogenic drug 'without warning coeds of the risk to themselves and their possible offspring'. The group charged that college women served as 'guinea pigs' for this new drug, while doctors failed to provide 'even the most rudimentary observance of professional standards and informed consent'.[28]

This exposé prompted Kuchera to do a follow-up study for the journal *Contraception*, in which she interviewed an additional 1,410 women who received the morning-after pill between Fall 1967 and April 1971. 'In patient interest and for scientific validity, it was important to make an effort to complete the follow-up of women who had not been contacted', she wrote. She added that at the time of their initial visits for emergency contraception, she told patients DES 'was not a new drug but this was a new use and a follow-up was expected. The possibility of failure was explained, and it was only given to consenting women'. Kuchera added that

26 K. Weiss, 'Cancer and the Morning-After Pill (will you be Mourning-After?)', *her-self*, 1 (September 1972): 1.

27 B.H. Cowan, 'Special Report: The "Morning-After" Pill', *her-self* (Sept/Oct 1974): 6–7.

28 'Birth Curb Pill is Termed Risky', *New York Times* (12 December 1972); 'Coeds Said Getting Cancer-Linked Contraceptive', *Durham Morning Herald* (December 12, 1972), clipping, Crist Papers, Box 15. See also 'K.U. Clinic Dispenses Morning-After Birth Pill', *Kansas City Times* (21 December 1972), p. 6A (courtesy of John Swann, FDA History Office).

she explained the possible side effects of the drug, and 'emphasized that this was an emergency type of treatment, and should a continuing need for contraception exist, other means should be sought'.[29]

The Issue of Choice

The DES controversy did not stop the demand for the drug, however, even after the *Roe v. Wade* Supreme Court decision in 1972 made abortion more widely available in the United States. Workers at rape crisis centers acknowledged that while the use of DES was controversial, 'some patients ask for it even after they have been warned of its possible consequences'.[30] Physicians at college health centers reported that they still wrote hundreds of prescriptions for postcoital pills per year. A survey of 42 health institutions, including the student health services of 14 large universities, outpatient gynecology services of 14 major hospitals, conducted by the Cancer Control Bureau of the New York State Health Department, indicated that 12 prescribed DES to women directly and 11 others referred them to a physician who would prescribe it. The study estimated that 450 women were treated or referred in New York State alone during 1973, with the heaviest use at college and university health centers.[31]

Because of this ongoing demand for and use of the drug, in 1973 the FDA approved DES 'as an emergency treatment only, and not as a routine method of birth control'. Like other oral contraceptives, the FDA insisted that each prescription of DES include a package insert warning

29 Kuchera, 'Postcoital Contraception with Diethylstilbestrol – Updated', *Contraception*, 10 (July 1974): 47–54.

30 M. Henry, 'DES Dispute: Contradictory Points of View Swirl Over Estrogen's Use as Rape Treatment', *Florida Times-Union* (21 September 1982), A-1, A-3, clipping, National Women's Health Network (NWHN) Papers, Sophia Smith Collection, Smith College, Box 7.

31 'Expert on Cancer Wary of New Pill', *New York Times* (23 February 1973): 7.

of the potential dangers of the drug. The FDA noted: 'there is at present no evidence that the restricted use of diethylstilbestrol for postcoital contraception carries a significant carcinogenic risk either to the mother or the fetus'. The only documented association between DES and cancer in women who took the drug were those who 'underwent prolonged treatment (five years or more) with DES for gonadal dysgenesis (incomplete development of the ovaries) due to chromosome abnormalities'. Although the FDA did not consider this 'applicable to the short course of therapy involved in postcoital contraception', because of these cases and 'the numerous side effects (especially those related to thromboembolism) associated with high doses of estrogens', the agency insisted that the leaflet warn the patient that the drug was 'for emergencies only and should not be used repeatedly'. Instead, if patients found themselves using the drug more than once, they should consult their physicians for 'an adequate method of routine contraception'.[32]

Critics remained skeptical that such caveats would resolve the ongoing debates over the risks and benefits of the drug. As Jane Brody wrote in a *New York Times* column, the package warning that DES be used only as an emergency treatment and not a regular form of birth control had 'already become a new focus for controversy'. An FDA spokesman told her that the agency was unable to prevent the use of the drug for an unapproved purpose or to enforce restrictions on the drug, leaving it to individual doctors to determine what constituted emergency use. 'For us to do so would be to interfere with the practice of medicine', said the spokesman. The FDA also could not prevent frequent use of DES, since a licensed physician could write numerous prescriptions for the same drug. Anita Johnson of the Health Research Group sarcastically remarked that the FDA's warnings were useless: '"Emergency" has no meaning at all', she scoffed, and 'far more women will now take it with fewer safeguards'.[33] In her testimony before the Senate Subcommittee on Health in 1975, Cowan

32 'FDA Approves Marketing of DES as "Morning-After" Pill but Limits its Use to "Emergency Treatment Only"', *Family Planning Perspectives*, 7 (March/April 1975): 57.

33 Jane E. Brody, 'The Birth Pill Dispute', *New York Times* (23 February 1973): 6.

stated that even emergency use of DES for rape victims was inappropriate. 'What FDA fails to realize is that the emergency aspect of treating rape victims has to do with the trauma of the assault', not the possibility of pregnancy. 'By its failure to regulate the use of DES' Cowan argued: 'the FDA has created a drug abuse problem'. She claimed that less than 5 out of 100 unprotected sexual encounters resulted in pregnancy, which meant that 95 per cent of women given DES did not need it. 'In dispensing DES to rape victims, we are unnecessarily exposing the majority of them to carcinogenic risks'.[34] Greenwald, director of the Cancer Control Bureau in the New York State Health Department, likewise argued, 'it seems likely that widespread use of DES as a morning-after pill will result in some daughters born with this vaginal cancer risk, either because of an excessive time lapse between conception and DES prophylaxis, or because of other forms of treatment failure'. He concluded it was 'unfortunate' the FDA had taken 'an official action that will encourage widespread use' before comprehensive research on drug safety could be done.[35]

Although college health experts argued that claims about the dangers of DES as an emergency contraceptive were 'based on a relatively limited sample and without any scientific evidence', the controversy had succeeded in making everyone in college health 'much more cautious in their use of DES'.[36] Instead, researchers both in the US and abroad looked for other methods of emergency contraception. In 1972, Haspels demonstrated that ethinyl estradiol, a compound made of a mixture of natural estrogens, was just as effective as an emergency contraception as DES.[37] The following year, Morris and van Wagenen published a similar study, adding that estradiol caused fewer side effects than did DES.[38] In 1977, Canadian gynecologist

34 'Testimony of Belita H. Cowano the Use of Diethylstilbestrol as a "Morning-After" Pill', Senate Subcommittee on Health, 27 February 1975, NWHN Papers, Box 7.

35 'Expert on Cancer Wary of New Pill', *New York Times* (23 February 1973): 7.

36 James Dilley to Gene Vadies, 25 January 1973, PPFA II Papers, Box 56, Folder 2.

37 Haspels, 'Post-coital Oestrogen in Large Doses', *International Planned Parenthood Federation*, 6 (1972): 3–4.

38 Morris and van Wagenen, 'Interception: The Use of Postovulatory Estrogens to Prevent Implantation', *American Journal of Obstetrics and Gynecology*, 115 (1 January 1973): 101–106.

Albert Yuzpe showed that a dose of 4 tablets of the standard birth control pill Ovral administered over 12-hour period was just as effective as DES, without the risks.[39] Due to the safety concerns about DES, by the late 1970s the 'Yuzpe method' had become the method of choice for postcoital contraception.[40]

The National Women's Health Network, however, continued to oppose any form of emergency contraception, arguing that all estrogens – including birth control pills and hormone replacement – were unsafe and unreliable. In her testimony before the FDA in April 1980, Cowan stated that studies on the safety of other estrogen regimes as emergency contraceptives had not been done and until they could be assured that there were no long-term risks, she and the Network would continue to oppose these drugs. Instead she proposed that doctors wait to see what happened, and then if the woman had a late or missed period perform a menstrual extraction.[41] Women's health activists did not agree on this issue, however. Kristin Luker, for example, said that while she shared her colleagues' skepticism about technological quick fixes, she also argued that the key issue was choice. She warned that if feminist health activists were not tolerant of a 'wide range of personal choices', including such potentially dangerous measures as hormonal contraception, then these options may not be available to future generations of women.[42]

39 A.A. Yuzpe and W.J. Lancee, 'Ethinylestradiol and dl-norgestrel as a Postcoital Contraceptive', *Fertility and Sterility*, 28 (September 1977): 932–936.

40 B. Pillsbury, F. Coeytaux and A. Johnson, *From Secret to Shelf: how collaboration is bringing emergency contraception to women* (Pacific Institute for Women's Health, 1999).

41 Presentation by Belita Cowan to the Commission of the US Food and Drug Administration, April 20, 1980; Meeting of the Fertility and Maternal Health Drugs Advisory Committee, FDA, 11 April 1980, NWHN Papers, Box 7.

42 Kristen Luker's reply to Belita Cowan's essay: 'Ethical Problems in Government-Funded Research', in H.B. Holmes, B.B. Hoskins, and M. Gross (eds), *Birth Control and Controlling Birth* (Clifton, NJ: Human Press, Inc., 1980), p. 82.

An Underground Technology

Luker's words proved prophetic: opposition from feminist health activists and consumer groups and escalating lawsuits from daughters of women who had consumed DES during pregnancy made pharmaceutical companies unwilling to file a New Drug Application to the FDA for approval of the Yuzpe method. Paradoxically, this meant that the only federally approved method of emergency contraception was DES. By the early 1980s, only one manufacturer in the United States – Eli Lilly and Co. – still made the compound, and even they warned doctors that they should not use DES as an emergency contraceptive. 'From our standpoint, the drug is not effective for postcoital contraception', stated Lilly spokesman Jean Palmer. The sole manufacturer of Ovral, Wyeth Ayerst, believed that the cost of obtaining FDA approval for use of their product as an emergency contraceptive was too high, the risk of product liability too high, and the potential profitability of four pills packaged for one time use too low to make it worth their effort. Concerns about litigation also prevented the company from advertising that Ovral could be used for this purpose or listing this indication on the package insert. According to officials at the FDA, the lack of a dedicated product for the Yuzpe method meant that the FDA had little regulatory oversight over this regimen. 'When the drug is on the market, the doctor can prescribe it for whatever he thinks is appropriate', said FDA spokeswoman Faye Peterson. 'It's not illegal ... there's nothing we can do about it'.[43] This placed the FDA in a 'Catch-22' – without a new drug application, they had no way to ensure that drugs used 'off-label' were done so safely and that accurate patient information was included in the packaging.[44]

With no dedicated product on the market, it was difficult for population groups, emergency medicine experts, and other proponents of emergency contraception to make doctors and patients aware of this technology.

43 Henry, 'DES Dispute', p. A-3.

44 Meeting of the Fertility and Maternal Health Drugs Advisory Committee, FDA, 11 April 1980, NWHN Papers, Box 7.

Descriptions of emergency contraception appeared in a number of journal articles, gynecology technical bulletins, contraceptive textbooks, and popular works on birth control. Yet, surveys of both students and physicians throughout the 1980s and 1990s indicated that few knew about or employed this technology. Indeed, one study by Lee H. Schilling, staff gynecologist at the Student Health Service at California State University at Fresno, observed in 1979 that 'in the minds of many, DES, the "morning-after pill", and postcoital contraception (PCC) are synonymous', even though most physicians by that point had switched to the Yuzpe method. Because of 'perceived dangers' about the use of estrogens, 'postcoital contraception is a poorly understood and vastly underutilized tool'.[45]

In addition, concerns about lawsuits led some family planning services to refuse administering any form of emergency contraception to their clients. Indeed, because Planned Parenthood's insurer would not cover them for this procedure, Eve Paul, Director of Legal Services of PPFA forbad affiliates from prescribing postcoital pills of all kinds to their patients and advised instead they should be referred to outside practices. Nevertheless, internal correspondence in the PPFA records suggests that individual affiliates acted on their own, and some continued to administer the Yuzpe method.[46] Yet for many practitioners and patients alike, emergency contraception remained 'the nation's best kept secret'.[47]

45 L.H. Schilling, 'An Alternative to the Use of High-Dose Estrogens of Postcoital Contraception', *Journal of the American College Health Association*, 27 (1979): 247–249. See also Schilling, 'Awareness of the Existence of Postcoital Contraception among Students who have had a Therapeutic Abortion', *Journal of American College Health*, 32 (1984): 244–246.

46 E.W. Paul, Director of Legal Services, to Mr. Jack Shettle, Brooks, Shettle & Garman, 20 September 1978; Eve W. Paul to Winston Gaye, 27 July 1978, PPFA Papers, Smith College, Series II, Box 68, Folder 61.

47 R.A. Hatcher, J. Trussell, F. Stewart, S. Howells, C.R. Russell, and D. Kowal, *Emergency Contraception: the nation's best-kept secret* (Decataur, GA: Bridging the Gap Communications, 1995).

A Call to Action

During the early 1990s, various groups dedicated to reproductive rights set out to bring emergency contraception 'out of the closet'. Among the most prominent activists in this area was James Trussell, a Professor of Economics and director of the Office of Population Research at Princeton University, who harnessed the new technology of the World Wide Web by creating a website devoted to distributing information about this technology.[48] Other activists also began a movement to bring a dedicated product to market. Even the National Women's Health Network became convinced by the early 1990s that the Yuzpe method was safe and effective. In a position paper entitled 'Postcoital Contraception: Time for Cautious Approval' (1993) the Network stated that their new position on this issue was 'prompted by dialogue in a coalition of women's health groups concerning two recent contraceptive developments'. The first of these was the development of new methods of emergency contraception – including RU-486 and the Copper-T IUD. The second was discussion about the possibility of making oral contraceptives available without a prescription. Furthermore, 'sufficient evidence on safety and effectiveness of the Ovral postcoital method is available for the Network to cautiously support its use'. Finally, increasing restrictions on abortion and access to federally-funded birth control under Presidents Ronald Reagan and George H.W. Bush made the Network convinced that they needed to help ensure that women had access to this technology when all else failed. 'As long as condoms break, we get caught in the moment, and women are raped, unintended pregnancies will occur', they wrote.[49]

According to the World Health Organization Department of Reproductive Health and Research, many other family planning experts and women's health activists believed that the availability of a dedicated

48 See http://ec.princeton.edu/.

49 National Women's Health Network Position Paper, 'Postcoital Contraception: Time for Cautious Approval', 1993, NWHN Papers, Box 7.

product commercially marketed for the purpose of emergency contraception would 'help legitimize the method, improve safety and reliability, and allow the method to be publicized'. In response to these efforts, the Population Council of the Rockefeller Foundation funded a conference on emergency contraception in Bellagio, Italy in April 1995. The conference brought together representatives from various family planning organizations around the world, who developed a *Consensus Statement on Emergency Contraception*, calling on family planning providers to learn about this technology and ensure that women everywhere have access to this technology.[50] The Bellagio conference also led to the creation of an International Consortium on Emergency Contraception, consisting of seven organizations involved in health and family planning: the Concept Foundation (Bangkok), International Planned Parenthood Federation (London), Pacific Institute for Women's Health (Los Angeles), Pathfinder International (Boston), Population Council (New York), Program for Appropriate Technology in Health (Seattle), and the Programme. Members of the Consortium believed that without a commercial product, packaged and labeled for emergency contraceptive use, they could never realize the full potential of the technology for preventing unwanted pregnancies. The Consortium also claimed that a dedicated emergency contraceptive would not be commercially viable without the active support of health care providers, family planning organizations, and women's health activists.[51] To facilitate this process, the WHO Expert Committee on the Use of Essential Drugs agreed to include the Yuzpe regimen in the WHO Model List of Essential Drugs.[52]

In the United States, reproductive rights activists pressured the FDA to increase awareness of emergency contraception among American consumers. President Clinton's administration proved much more favourable to expanding the availability of birth control options. In November of 1994,

50 'Consensus Statement on Emergency Contraception', *Contraception*, 52 (1995): 211–213.

51 'Expanding Family Planning Options', *WHO Annual Technical Report* (1995), http://www.who.int/reproductive-health/publications/HRP_ATRs/1995/execsum.html

52 'Improving methods of emergency contraception', *Progress in Reproductive Health Research*, 51 (1999), http://www.who.int/reproductivehealth/hrp/progress/51/news51_1.en.html

the Center for Reproductive Law and Policy filed a citizen petition with the FDA asking that the agency require manufacturers of certain birth control pills to revise their packaging information to include the use of such products as emergency contraceptives. FDA Commissioner David Kessler, a Clinton appointee, asked the Advisory Committee for Reproductive Health Drugs to review the safety and effectiveness of certain oral contraceptives for emergency contraception. The Advisory Committee met in 1996 and unanimously confirmed that certain regimens were effective. In March of 1997, the FDA took the unprecedented step of publishing a notice in the *Federal Register* requesting that manufacturers submit supplemental new drug applications for emergency use of oral contraceptives, thereby putting 'an official stamp of approval' on the Yuzpe regimen.[53]

An article from *Fortune* magazine was blunter: 'We were annoyed as hell', says a former FDA official. 'All they had to do was take something they were already making, put it in another package, and market it. They wouldn't do it' because of concerns about possible lawsuits, as well as fears about boycotts by consumers opposed to what they believed to be an 'abortion drug'.[54]

The first to answer the FDA's 'call to action' was Roderick MacKenzie, a contraceptive chemical engineer and former president of Ortho Pharmaceutical who had founded the company GynoPharma to reintroduce the Copper-7 IUD to the US. MacKenzie heaped scorn on pharmaceutical companies who perpetuated the myth that feminist activism and a litigious climate were to blame for 'all the problems facing women and their contraceptive choices in the United States'. The truth, he said, was that the cost of litigation was 'very small' when 'measured as a percentage of sales'. Rather, pharmaceutical companies were more interested in developing products for ulcers, cardiovascular disease, and other drugs that did not carry with them 'a sea of bad publicity endangering other drugs'.[55] After selling the

53 'Birth Control, The Morning After', *New York Times* (5 July 1996): A22; D.A. Kessler, Commissioner of FDA, 'Prescription Drug Products: certain combined oral contraceptives for use as postcoital emergency contraceptives', *Federal Register*, 62 (25 February 1997): 8610–8612.

54 Peter Keating, 'Where Drug Firms Fear to Tread', *Fortune* (26 October 1998): 48 (1).

55 Hilts, 'Birth Control Backlash', 72.

rights to the IUD to Johnson and Johnson in 1995, 'I was wondering what I was going to do next', Mackenzie recalled. 'So it was fortuitous that the FDA came to me and asked me to help make emergency contraceptives happen'.[56] In response to the FDA's appeal, MacKenzie started the company Gynetics, and submitted a new drug application Preven, essentially a repackaging of the Yuzpe method, to the FDA in 1997, which approved the product in September 1998.[57] Shortly thereafter, Gynetics launched a direct-to-consumer marketing campaign via radio advertisements in major metropolitan areas and print advertisements in women's magazines. 'During the first year on the market, promotion of Preven caused awareness of emergency contraception to increase from 11 per cent to 15 per cent', said Sherry Bump, executive director of marketing for Gynetics. Bump estimated that it would 'take $40 to $50 million to move women's awareness levels from 15 per cent to 50 per cent'. In order to meet this commitment, Gynetics entered into a partnership with Barr Laboratories, and 'enlisted the help of MTV's "Loveline" co-host Drew Pinsky to speak with the media and the public on the importance of emergency contraception'.[58] Gynetics also donated 25,000 Preven kits to Planned Parenthood clinics to distribute to low-income women. Mackenzie stated: 'We are proud to support Planned Parenthood's initiative to educate women about emergency contraception and to provide the organization with 25,000 PREVEN Kits so that emergency contraception can be made available to any woman who needs it, regardless of her financial status'.[59]

The story of Plan B represents a somewhat different approach to development and distribution. Plan B differs from Preven in that it contains only progestin, while Preven contains both progestin and estrogen. International studies had shown progestin-only regimes were more effective and had fewer side effects than did combined progestin and estrogen

56 Keating, 'Where Drug Firms Fear to Tread'.

57 FDA, CDER NDA 20–946.

58 'Companies commit to emergency contraception – Have you?', *Contraceptive Technology Update*, 20 (December 1999): 137.

59 Feminist Daily Newswire, 17 March 1999, http://www.feminist.org/news/newsbyte/uswirestory.asp?id=2186 (accessed 24 March 2005).

pills, but MacKenzie believed it would be easier and faster to get the Yuzpe method through the FDA first. Sharon Camp, former coordinator of the International Consortium on Emergency Contraception, now CEO of the Alan Guttmacher Institute, disagreed and started her own company, the Women's Capital Corporation, to test and create a progestin-only product. Together with the WHO, the WCC organized and supported a large-scale clinical trial to compare levonorgestrel, given as two 0.75 mg doses 12 hours apart, with the Yuzpe regimen for emergency contraception. 21 research centers in 14 countries (Australia, Canada, China, Georgia, Hungary, India, Mongolia, New Zealand, Nigeria, Panama, Slovenia, Sweden, the United Kingdom and the USA) took part in the study that was completed in 1997 and published in the *Lancet* the following year.[60]

In September 1999, the FDA approved the American version of the product, which WCC called Plan B. Rather than direct-to-consumer marketing and distribution to major pharmacies, however, the WCC chose to initially market and distribute Plan B to facilities that prescribed and distributed contraceptives on-site – e.g. college and university health services, Planned Parenthood, rape crisis enters, and federally-funded birth control clinics.[61] 'We are an outgrowth of a private/public consortium, so we have a very strong public-sector focus', said Camp.[62] The WCC also had a very small marketing budget (only about $300,000) and had to target it carefully. In 2002, the company enlisted DDB advertising company in Seattle to launch its first major advertisement campaign, which was directed at sexually active college women. On Valentine's Day of that year, 30 college newspapers around the country featured ads and inserts with such slogans as 'Accidents happen. That's why there's morning-after contraception', and 'So many men. So many reasons to have backup contraception'.[63]

60 'Improving methods of emergency contraception', *Progress in Reproductive Health Research*, 51 (1999), http://www.who.int/reproductivehealth/hrp/progress/51/news51_1.en.html

61 Interview with Sharon Camp, 17 March 2005.

62 'Emergency contraception use up – new ECP arrives', *Contraceptive Technology Update*, 20 (September 1999): 108.

63 S.J. Heim, 'DDB pushes morning-after pill: with limited funds, plan B marketer focuses on college women', *ADWEEK Western Edition*, 52 (11 February 2002): 5.

Conclusion

Despite the success of dedicated products such as Plan B, the earlier controversy over DES continues to haunt current debates about emergency contraception in the United States, as opponents of the technology have misappropriated the vocabulary of an earlier generation of feminist health activists. For example, the Population Research Institute, a conservative group that has been instrumental in convincing the current Bush administration to halt family planning funding to developing countries, declares on its website that it is 'dedicated to stopping human rights abuses committed in the name of *family planning*' [emphasis in the original].[64] In a recent position paper opposing making emergency contraception available over-the-counter, the Population Research Institute, in addition to raising moral objections to emergency contraception, draws quite explicitly on earlier tensions between the women's health movement and traditional population control organizations to condemn emergency contraception, along with other forms of 'family planning', as a form of 'genocide' directed at impoverished individuals within the United States and in developing nations. They also use phrases such as 'women as guinea pigs' and 'reckless experimentation on women' to describe studies on the use and efficacy of emergency contraception.[65]

By raising this point, I do not intend to cast blame on the early feminist health care pioneers. They provided an invaluable service to women health care consumers by revealing the sexism and abuses of the health care establishment, critiques that have greatly affected the health care professions and empowered women as health care consumers. More importantly, there have been significant changes in the perspectives of feminist health advocates since the early 1970s. Nowhere is this historical shift more apparent

64 http://www.pop.org/

65 Population Research Institute, *Under the Table: why the US Food and Drug Administration should not approve the over-the-counter distribution of morning after pills* (Front Royal, VA: Population Research Institute, 2004).

than in the changing position of the National Women's Health Network (NWHN) on emergency contraception. Initially a vehement opponent of the 'morning-after pill' and other products containing estrogen in the early 1970s, by the early 1990s the organization had joined broader efforts to develop a dedicated emergency contraceptive product. Emergency contraception came to represent a 'middle ground' between an older liberal feminist position that tended to support technological innovations such as the contraceptive pill, and a more radical feminist position that criticized the use of hormonal contraception but was otherwise in favour of reproductive rights. Therefore, the story of emergency contraception represents a rapprochement – an intentional coalition if you will – between mainstream population control groups, representatives from the pharmaceutical industry, and members of feminist women's health advocacy groups. Like the AIDs activists described by Arthur Daemmrich, this coalition was instrumental in creating a regulatory and political environment in the United States that was favourable towards developing emergency contraceptive technology and bringing it to market.[66]

66 A.A. Daemmrich, *Pharmacopolitics: drug regulation in the United States and Germany* (Chapel Hill: University of North Carolina Press, 2004).

PART 5

From Pharma to Biotech

BEAT BÄCHI

13 Producing Ascorbic Acid: how microbes transformed the pharmaceutical company Hoffmann-la Roche, 1933–1953[1]

Introduction: Tadeus Reichstein, Hoffmann-la Roche and ascorbic acid

Tadeus Reichstein's[2] synthesis of l-ascorbic acid (vitamin C) in 1933 was the first total synthesis of a vitamin. The start of vitamin C manufacturing was marked by the industrial production of 50 kilogram of ascorbic acid in 1934 by the Swiss pharmaceutical company Hoffmann-la Roche (hereafter Roche). For this, Roche used a slightly modified version of Reichstein's purely chemical synthesis. Reichstein had had to rework his original method of synthesis because it had been too complicated to use for large-scale production. This had led him to integrate a fermentation process in the chemical synthesis in order to reduce the number of intermediary stages from ten to six. However, to make Reichstein's second, hybrid laboratory synthesis competitive on a large scale, both he and Roche's chemical engineers had

1 This paper is part of my dissertation concerning the history of synthetic vitamin C, published under the title: 'Vitamin C für alle! Pharmazeutische Produktion, Vermarktung und Gesundheitspolitik' (Zürich: Chronos, 2009). For stimulating discussions and criticisms I would like to thank Alexander Bieri, Michael Bürgi, Thomas Casutt, David Gugerli, Bruno Halm, Hyo Yoon Kang, Alison Kraft, Viviane Quirke, Christina Ratmoko, Carsten Reinhardt, Tobias Straumann and Hermann Wichers.

2 See http://nobelprize.org/medicine/laureates/1950/reichstein-bio.html [cited 14 February 2006]. See also Rothschild, *'Tadeus Reichstein'*. The bequest of Tadeus Reichstein (1897–1996) is archived at Staatsarchiv Basel-Stadt.

had to overcome many problems, as it was the first time that they had had to manipulate micro-organisms for pharmaceutical production.

In this paper, I will analyse the emergence of Roche's world leading position in the production of vitamins through the history of the industrial synthesis of l-ascorbic acid. Using ascorbic acid manufacture as a 'tracer' helps to shed light on how industrial practices impose their own rationale on modes of research and production. I will therefore explore how ascorbic acid synthesis contributed to Roche's transformation into an important player in the area of biotechnology. At first sight, this transformation may appear to be accidental. But since the economically competitive synthesis of ascorbic acid included fermentation technology, the manufacture of this vitamin compelled Roche to work for the first time not only with living organisms, but also with scientists and research institutions in related scientific disciplines. Hence, the central questions addressed in this paper are: how did Tadeus Reichstein become a lead user[3] of microbiological techniques? How and why did Roche take up fermentation technology? And, resulting from this: which resistance needed to be overcome for biosynthetic methods to be developed at Roche? To answer these questions, I will follow Michel Morange's lead, and focus on the trajectory of ascorbic acid production.[4] In the context of the long-term history of biotechnology, this involves following the microbes from academic and industrial laboratories into the head offices and production plants of Roche.

To clarify, the advent of the 'new' biotechnology, i.e. recombinant DNA-based and/or tissue culture-based processes, is not tackled in this paper. There is no straightforward line between 'classical' biotechnology, meaning the use of living organisms in a production process, and the 'new' biotechnology. Nevertheless, within the pharmaceutical industry, the use of classical biotechnology helped to pave the way for the 'new' biotechnology

3 For an interesting application of Eric von Hippel's concept of lead user to organic chemistry, see C. Reinhardt, 'A Lead User of Instruments in Science. John D. Roberts and the adaptation of nuclear magnetic resonance to organic chemistry, 1955–1975', *Isis*, 97 (2006): 205–236.

4 M. Morange, *A History of Molecular Biology* (Cambridge MA: Harvard University Press, 1998), pp. 5–6.

in the late 1960s. To shed light on how the integration of biosynthesis into Roche's innovative strategies prepared the firm for the 'new' biotechnology, I will first examine how Roche entered the vitamin business.

Ascorbic Acid and the Long-Term History of Biotechnology

In recent years, the history of 'biotechnology, especially the introduction of molecular biological methods into the pharmaceutical industry'[5] has attracted much interest from historians of science.[6] In contrast, its

5 R. Bud, 'Molecular biology and the long-term history of biotechnology', in Arnold Thackray (ed.), *Private Science: biotechnology and the rise of the molecular sciences* (Philadelphia: University of Pennsylvania Press, 1998), pp. 3–19 (4, 7 and 16). For different uses of the notion 'biotechnology' see also Bud, 'Biotechnology in the twentieth century,' *Social Studies of Science*, 21 (1991): 415–457 (416, 446). The word 'biotechnology', although often referred to as being new, had already been coined in 1917. Karl Ereky called his fattened pigs 'biotechnological working machines'. Thus the concept entered the major German dictionaries as early as the 1920s, and from this one can conclude that biotechnology has a long (conceptual) tradition.

6 The history of biotechnology and molecular biology has been the subject of a huge amount of historical literature. For my case study, research on the long-term history of biotechnology was of special interest. See P.G. Abir-Am, 'The Molecular Transformation of Twentieth-Century Biology', in John Krige and Dominique Pestre (eds), *Science in the Twentieth Century* (Amsterdam: Harwood, 1997), pp. 495–524; idem, 'Molecular Biology and its Recent Historiography: a transnational quest for the "Big Picture"', *History of Science*, xliv (2006): 95–118; R. Bud, *The Uses of Life: a history of biotechnology* (Cambridge: Cambridge University Press, 1993); idem, 'Molecular biology'; S. de Chadarevian and Harmke Kamminga (eds), *Molecularizing Biology and Medicine: new practices and alliances, 1910s–1970s* (Amsterdam: Harwood Academic Publishers, 1998); S. de Chadarevian and Bruno Strasser (eds), 'Molecular Biology in Postwar Europe', Special issue of *Studies in the History and Philosophy of Biological and Biomedical Sciences*, 33C (2002); J.-P. Gaudillière, *Inventer la biomédecine: la france, l'amérique et la production des savoirs du vivant (1945–1965)* (Paris: la Découverte, 2002); L.E. Kay, *The Molecular Vision of Life: Caltech, the Rockefeller Foundation, and the rise of the new biology* (New York: Oxford University Press, 1993); L. Marschall, 'Industrielle Biotechnologie im 20. Jahrhundert. Technologische Alternative oder Nischentechnologie?', *Technikgeschichte*, 66

prehistory, i.e. the way in which biosynthesis as a form of classical biotechnology prepared pharmaceutical companies for the era of new biotechnology by integrating 'the two cultures' of chemistry and biology,[7] has received far less attention. Vitamins have not yet moved into the focus of historians of biotechnology, with the exception of Luitgard Marschall, who has studied the (re)introduction of fermentation technologies in the pharmaceutical industry since the 1880s. In particular, she has examined the innovative processes that were directly concerned with ascorbic acid production at Merck during the 1930s. To be more accurate, Marschall has analysed how Merck mastered the biosynthetic step involved in the so-called Reichstein procedure to transform sorbitol into sorbose. She has come to the conclusion that fermentation was a conservative innovation and a niche technology in the pharmaceutical industry, which tended to use traditional working steps and already existing equipment.[8]

(1999): 277–293; idem, *Im Schatten der chemischen Synthese. Industrielle Biotechnologie in Deutschland (1900–1970)* (Frankfurt/New York: Campus, 2000); Max Planck Institute for the History of Science (ed.), *History and Epistemology of Molecular Biology and Beyond: problems and perspectives* (Berlin: Max Planck Institute for the History of Science, Preprint 310, 2006); Morange, *A History of Molecular Biology*; V. Quirke, 'Making *British* cortisone: Glaxo and the development of corticosteroid drugs in the UK in the 1950s and 1960s', *Studies in History and Philosophy of Biology and Biomedical Sciences*, 36 (2005): 645–674; idem, *Collaboration in the Pharmaceutical Industry: changing relations in Britain and France in the twentieth century* (London/New York: Routledge, 2008); N. Rasmussen, 'Of "Small Men", Big Science and Bigger Business: the Second World War and biomedical research in the United States', *Minerva*, 40 (2002): 115–146; Thackray, *Private Science*.

7 A. Kornberg, 'The Two Cultures: chemistry and biology', *Biochemistry*, 26 (1987): 6888–6891, and Quirke, 'Making British cortisone', p. 673.

8 Marschall, *Im Schatten der chemischen Synthese*, especially pp. 324–332. For the history of fermentation technologies at the end of the nineteenth and the beginning of the twentieth century see ibid., pp. 25–87. Harmke Kamminga and Christiane Sinding have addressed the issue of the relationship between vitamins and molecularization, while Knut Sogner has analysed the connections between vitamins and fermentation. See H. Kamminga, 'Vitamins and the Dynamics of Molecularization: biochemistry, policy and industry in Britain, 1914–1939', in de Chadarevian and Kamminga (eds), *Molecularizing Biology and Medicine*, pp. 83–105; C. Sinding, *Le Clinicien et le chercheur: des maladies de carence à la médicine moléculaire* (Paris: Presses Universitaires de France, 1991) ; K. Sogner, 'The Double Meaning of Vitamins', *Scandinavian Journal of History*, 22 (1997) : 187–198

As Marschall's work demonstrates, a cultural history of biotechnology can only be written by examining the practical developments, such as biosynthetic methods, achieved by pharmaceutical companies. Of course, this approach requires the study of sources not usually consulted by traditional – mainly conceptual – historians of biotechnology. Personal communications between inventors and companies, as well as archival material on day-to-day production problems, enrich our understanding of the long-term history of biotechnology within companies like Roche, by allowing us to focus on practices, material objects, and modes of production. Such a perspective implies a criticism of the conceptual historiography of biotechnology, which has often concluded that everybody has always had high hopes for biotechnology – even before the advent of the 'new' biotechnology.[9] In contrast to the published sources that are usually consulted, these unpublished documents bring to light problems, conflicts, and resistances, at both individual and institutional levels. Furthermore, they allow a shift in focus from the heroes of academic disciplines, to 'small' players, such as chemical engineers and industrial chemists. Their critical role would otherwise remain unknown, as they did not generally publish their work.

(188–189). Important studies concerning the cultural history of vitamins include R.D. Apple, *Vitamania: vitamins in American culture* (New Brunswick, New Jersey: Rutgers University Press, 1996); M. Bürgi, *Die Anfänge der industriellen Vitaminproduktion. Fotografien aus dem Historischen Archiv Roche* (Basel: Historisches Archiv Roche, 2004); K.J. Carpenter, *The History of Scurvy and Vitamin C* (Cambridge: Cambridge University Press, 1986); S.M. Horrocks, 'The Business of Vitamins: nutrition science and the food industry in inter-war Britain', in H. Kamminga and A. Cunningham (eds), *The Science and Culture of Nutrition, 1840–1940* (Amsterdam: Rodopi, 1995), pp. 235–258; H.J. Teuteberg, 'The discovery of vitamins: laboratory research, reception, industrial production', in A. Fenton (ed.), *Order and Disorder: the health implications of eating and drinking in the nineteenth and twentieth centuries* (East Lothian: Tuckwell Press, 2000), pp. 253–280; P. Werner (ed.), *Vitamine als Mythos. Dokumente zur Geschichte der Vitaminforschung* (Berlin: Akademie Verlag, 1998).

9 See Bud, 'Biotechnology in the twentieth century', p. 421: 'From a practical view, it seemed as if biology would replicate the forms and potential of chemistry as the scientific basis for the solution of technical problems'.

The Political Economy of Ascorbic Acid

At the end of the 1920s, Roche did not have a stake in the vitamin business. This was all the more surprising since the commercial promise of vitamins had been discussed for the first time as early as 1919.[10] But this subject ceased to be on Roche's research agenda for about ten years. Furthermore, the people who were advocating vitamin development, especially the head of scientific research at Roche, Markus Guggenheim, were subjected to mockery within the company.[11] Therefore, to explain how and why Roche entered the vitamin business one has to bear in mind the global political and economic context of the 1920s and 1930s. Roche had been an important producer of alkaloids, and the International Drug Convention (signed by Switzerland in 1924–1925)[12] created a major impetus for companies to explore new fields for research. Even before 'Black Friday' had taken place at the New York Stock Exchange on 25 October 1929, the convention had encouraged the Roche management to investigate potential new product groups.[13] Consequently, when Roche's head of research admonished the management once again in May 1929 – this time using biblical metaphors, predicting that the seven fat years might soon be over[14] – no one laughed at him anymore. In addition, the 1929 Nobel Prize for physiology or medi-

10 Historisches Archiv Roche (HAR), FE.0.9 – R–102166 b: Rapport No. 3747 von Dr Markus Guggenheim, Abteilung VI, an die Direktion. Basel, den 24. September 1919. Betr. Vitamine. S. 3.

11 HAR, PD.2.2.VIC - 103409 b: Aus der Geschichte der Vitamin C Produktion bei Roche (1976?), von Dr J.J. Fuchs, mit Begleitschreiben vom 22.9.1982. S. 2.

12 For example in 1921, Roche generated 8.5 million Swiss francs of its total turnover of 18.2 million Swiss francs with chemical products and alkaloids. H.C. Peyer, *Roche. Geschichte eines Unternehmens 1896–1996* (Basel: Editions Roche, 1996), p. 105.

13 At this point in time, Roche was still a relatively small enterprise with about 1300 employees in 15 countries, generating a turnover of 50.4 million Swiss francs in 1929 (36 million with pharmaceutical specialities and 14.4 million with chemical products and alkaloids). However, by 1953, it counted 7000 employees, generating a turnover of 419 million Swiss francs. See Peyer, *Roche*, pp. 92, 105.

14 HAR, FE.0.9 - R - 102166 h: Rapport No. 15900 von Dr Markus Guggenheim, Abteilung VI, an die Direktion. Basel, den 10. Mai 1929. S. 6f.

cine had been awarded to two researchers working in the field of vitamins. Thus, for the first time, Roche started taking this possibility seriously, but without coming to a decision.

Then, at the beginning of the 1930s, the situation worsened dramatically. Between 1931 and the end of 1932 the company had to reduce its staff in its headquarters in Basel by more than 20 per cent. By the end of 1932, there were some 197 people working at Roche Basel.[15] But it was not until 1933 that Roche was finally to reach a decision concerning vitamins. In that year, the coming into power of the Nazis in Germany, as well as the drop in value of the dollar in the United States, forced Roche to try its luck by taking up something new, whatever that might be.[16] This terrible uncertainty, which at once propelled, eased and hampered the re-orientation process, is the point of departure when Roche entered the field of vitamins – and that of biosynthesis. And this was a most unlikely departure from their traditional business.

In the same year, 1933, Tadeus Reichstein and his collaborators achieved (independently from Roche) a chemical synthesis of l-ascorbic acid in a laboratory of the Swiss Federal Institute of Technology (ETH Zurich). When the patent rights for the Reichstein procedure (owned by the small Swiss foodstuff company Haco, which funded Reichstein's work on the vitamin C synthesis) were offered to Roche in May 1933, Roche expressed interest, but did not buy the patent rights in the end. Roche's reservations against *synthetic* vitamins – produced by means of this first, purely chemical Reichstein synthesis – were not based on an aversion towards the 'synthetic paradigm'.[17] Since 1912, Roche had gradually integrated this paradigm,

15 HAR, FE.0.9 - 102226 f: Rapport No. 22652 von Dr H.M. Wüest, Abt. V, an die Direktion. Basel, den 8. Februar 1933, 2f.

16 HAR, FE.0.9 - 102226 f: Rapport No. 22667 von Dr H.M. Wüest, Abt. V, an die Direktion. Basel, den 11. August 1933, 1.

17 The 'synthetic paradigm' can be characterized as the belief that a total synthesis can be found for every chemical compound. On the 'extractive paradigm' and the 'synthetic paradigm' see V. Walsh, 'Paradigms in the Evolution of Life Sciences Research, and the Changing Structure of the Innovation Organization', in K. Grandin, N. Wormbs, A. Lundgren, and S. Widmalm (eds), *The Science-Industry Nexus: history, policy, implications* (Nobel Symposium 123) (Sagamore Beach: Science History Publications, 2004),

especially in connection with the production of analgesics and barbiturates, into its manufacturing practices. But rather it was the startling complexity of Reichstein's procedure that caused the company to back off. More than ten intermediary stages (estimated by Roche's chemical engineers as too many for an efficient industrial process) would have been necessary to produce vitamin C. Moreover the yield was quite low. Therefore, the extraction of vitamin C from *natural* sources seemed more rational at the time. Two important scientists in the field of vitamin C research, Paul Karrer and Albert von Szent-Györgyi, who later received the Nobel Prize, presented this view in person to the Roche authorities, whilst offering Roche a method for the extraction of vitamin C from red pepper.[18]

The Geneticist's 'Domestic Animal' and Ascorbic Acid Production

While Roche's research department was still evaluating the two possibilities for producing vitamin C on a large scale – extraction or synthesis – Tadeus Reichstein and his collaborators improved their process in the autumn of 1933. The main obstacle was the transformation of sorbitol into sorbose in order to achieve a simpler synthesis with an enhanced yield. Reichstein and his collaborators overcame this problem by introducing a traditional or classical biotechnological step: they used the bacteria named *Acetobacter suboxydans* to convert sorbitol into sorbose.

---

pp. 189–221 (190–194). Concerning the notion of 'Denkstil' and paradigms, see L. Fleck, *Entstehung und Entwicklung einer wissenschaftlichen Tatsache. Einführung in die Lehre vom Denkstil und Denkkollektiv* (Frankfurt a. M.: Suhrkamp, 1980 (1935)).

18 HAR, FE.0.9 - R - 102166 j: Rapport No. 24247 von Dr Markus Guggenheim, Abteilung VI, an die Direktion. Basel, 8. Mai 1933, 1–3. For Paul Karrer and Albert von Szent-Györgyi see http://nobelprize.org/chemistry/laureates/1937/karrer-bio.html and http://nobelprize.org/medicine/laureates/1937/szent-gyorgyi-bio.html [cited 14 February 2006].

What had led Reichstein to the idea of using microbes? A method for transforming sorbitol into sorbose by using micro-organisms had been found at the end of the nineteenth century by the French chemist and biologist Gabriel Bertrand (1867–1962). As demand for sorbose remained low, his procedure had not been elaborated on a large scale. In the autumn of 1933, after having consulted Bertrand's publications concerning the production of sorbose,[19] Reichstein ordered some *Acetobacter suboxydans* as specified by Bertrand from a number of 'Büros für Schimmelkulturen' (offices for mould cultures). However, those bacteria did not meet Reichstein's expectations. As an alternative, he endeavoured to capture some wild drosophila melanogaster, the 'geneticist's domestic animal',[20] which carries the bacteria in its intestines. Reichstein was following the advice of a friend, who was a bacteriologist working at the ETH in Zurich, and had told him to add some yeast broth to Bertrand's recipe. Fortunately, the autumn of 1933 was quite mild. Even though it was October, Reichstein was able to catch some drosophila on the windowsill of his laboratory located in a cellar of the institute for general and analytical chemistry at the ETH, and using these they succeeded in transforming sorbitol into sorbose.[21]

Thanks to the introduction of micro-organisms, the number of intermediary steps in the synthesis could be reduced from ten down to six. Furthermore, this new synthesis permitted the use of glucose as a cheap basic starting material. So, when Reichstein and his friend, the director of the Haco company and owner of Reichstein's patent rights, presented this new, biosynthetic method to Roche in the autumn of 1933, the Roche management and the research director considered this second synthesis

19 Bertrand, *Bull. Soc. Chim.* (3), 5 (1891): 554, and Bertrand, *Annales de chimie et de physique* (8), 3 (1904) : 230. See StABS, Mäppchen 'Hoffmann-La Roche, V. 1933–XI. 1935', Brief von T. Reichstein an die F. Hoffmann-La Roche, Abt. 6, ohne Datum. [Ende September, Anfang Oktober] 1933.

20 Roche Magazin Nr. 21, Mai 1984, 13.

21 Staatsarchiv Basel-Stadt (StABS), Abl. 1996/90, Wiss. Korr. 2; Korr. mit Akademie-Verlag, La Roche; Doss., Hoffmann-La Roche, V. 1933-XI. 1935: Brief von T. Reichstein an die F. Hoffmann-La Roche, Abteilung 6. Zürich, 5. Oktober 1933; StABS, Abl. 1997/42, A 1.2: Curriculum Vitae (engl.), Basel, May 31th, 1993, 6f.; Roche Magazin 11, Dezember 1980, 7 and 10; Roche Magazin 21, Mai 1984, 10–15.

of vitamin C to be more promising than the extraction of vitamin C from natural sources. Roche bought the patents for the Reichstein procedure in November 1933. Since the Reichstein procedure later turned out to be superior to other ascorbic acid syntheses, this decision would eventually lead to Roche's world leading position in the production of ascorbic acid – and Roche's first involvement in (traditional) biotechnology.

However, this is only the short version of the story. In practice, it took much longer for Roche to integrate chemistry and biology into their innovative process.[22] The use of bacteria caused unforeseen difficulties, and a certain amount of resistance had yet to be overcome. First of all, when Reichstein presented his new procedure for synthesising vitamin C to Roche's general director, he was told that this microbiological reaction using bacteria pleased neither him, nor Roche's industrial chemists. Reichstein replied that regardless of whether this step was viewed favourably or not, the bacteria would be the only laboratory workers able to transform sorbitol into sorbose with a yield of 90 per cent.[23]

Roche's reaction is typical of the pharmaceutical industry's resistance towards biosynthesis in the 1930s. If we look at enterprises manufacturing ascorbic acid in Germany, where Roche had to fight bitter patent rights disputes against Merck and I.G. Farbenindustrie, we can observe the same reluctance.[24] In the case of Merck, their chemists' scepticism towards the new, microbiological procedure decreased only slowly over time. This is remarkable because, having had to acquire the fermentation skills necessary for the production of lactic and butyric acid, Merck's chemists had gained a long experience in the area of classical biotechnology. Merck's bacteriology department had been founded in 1895 for the production

22 See A.D. Chandler, *Shaping the Industrial Century: the remarkable story of the evolution of the modern chemical and pharmaceutical industries* (Cambridge: Harvard University Press, 2005), p. 31; Kornberg, 'The Two Cultures'.

23 Roche Magazin Nr. 21, Mai 1984, 14.

24 For comparative analyses of pharmaceutical innovation in the twentieth century see S. Chauveau, 'Produits de santé, agents thérapeutiques et démocratie sanitaire', *Entreprises et Histoire*, 36 (2004): 5–11 ; see also Quirke, *Collaboration in the Pharmaceutical Industry*.

of vaccines. Therefore, in order to begin the fermentation of sorbitol into sorbose, Merck was able to use equipment that dated back to the early days of its bacteriological division, in particular vessels used for the production of tuberculin.[25]

In the case of I.G. Farbenindustrie,[26] it rejected the Reichstein process and held on to the so-called 'Helferich process', which was of a 'purely chemical nature'.[27] Furthermore, the I.G. officials flatly refused to have any bacteria in their production plants – although a number of biosynthetic processes were running within I.G. Farbenindustrie at the time, for example at Hoechst. But as the I.G. Farbenindustrie officials told Roche's management on more than one occasion, the presence of bacteria made them fear 'poisoning and degeneration'.[28] What is striking about this rhetoric, in which bacteria, microbes, and parasites seemed to be associated with Jews and, taken all together, were thought to threaten the 'Volkskörper', is that the objections against the microbiological step were always uttered after the official negotiations between Roche's management and I.G. Farbenindustrie had ended. For instance such statements concerning the 'fear of bacteria'[29] were made on the road to the train station, after the meetings.[30]

25 Marschall, *Im Schatten der chemischen Synthese*, pp. 313f., 316–318, 321, 324, 327–332.

26 On the controversial history of I.G. Farbenindustrie see P. Hayes, 'Zur umstrittenen Geschichte der I.G. Farbenindustrie AG', *Geschichte und Gesellschaft. Zeitschrift für Historische Sozialwissenschaft*, 18 (1992): 405–417.

27 HAR, PB.3 - 102556 a, N 419: Bericht No. 15 von Dr H.M. Wüest betr. Besprechung mit der I.G. über Ascorbinsäure. Baden-Baden, 9. Oktober 1936. Basel, den 12. Oktober 1936, 12.

28 Ibid., 12. See also HAR, PB.3 - 102556 b, N 420: Bericht No. 65 von Dr H.M. Wüest betr. Besprechung mit I.G. in Basel am 25. April 1940 betr. Ascorbinsäure-Lizenz. Ergänzung zu Bericht Dr Veiel No. 29, 2.

29 HAR, MV.0.2.1 - 102185 d: Ergänzung zu Direktionsrapport Dr Niederberger No. 23210, 16.4. 1935, 1–3.

30 HAR, PB.3 - 102556 b, N 420: Bericht No. 65 von Dr H.M. Wüest betr. Besprechung mit I.G. in Basel am 25. April 1940 betr. Ascorbinsäure-Lizenz. Ergänzung zu Bericht Dr Veiel No. 29, 2.

As far as I can see, of all the companies Roche was in negotiations with, Rhône-Poulenc[31] was the only one that did not 'fear' fermentation.[32]

But why did Roche at first consider this microbiological step so unfavourably? One reason is that the company had never produced vaccines, and although it employed bacteria for toxicological and diagnostic purposes in the 1930s, there was not a single bacteriologist working within Roche at the time. Thus, Roche's industrial chemists simply had no experience in handling bacteria. They would have preferred a purely chemical synthesis like most of the other processes Roche used at that time. Moreover, the philosophy of the total chemical synthesis was well suited to the 'Zeitgeist' (the spirit of the age).[33] So microbes were not always welcomed by the pharmaceutical industry, which long clung on to organic chemistry like a silver bullet to make magic bullets.

Microbes between 'Science' and 'Technology'

Biosynthetic ascorbic acid production emerged at the intersection between academia and industry. Strong ties between academic and corporate research were well established during the twentieth century in Switzerland. Furthermore, the cooperation between professors from the ETH and the chemical industry was a long-standing tradition rooted in the nineteenth

31 For a history of Rhône-Poulenc see P. Cayez, *Rhône-Poulenc 1895–1975* (Paris: Armand Colin-Masson, 1988).

32 HAR, PB.3 - 102556 b, N 420: Bericht No. 298 von Dr R.E. Gsell betr. Besprechung über Diacetongulonsäure mit Rhône-Poulenc am 2. X. 1940 in Genf. Hotel des Bergues, 14–16.30 Uhr. Basel, 3. Oktober 1940, 6.

33 On the 'philosophy of the eternal chemical synthesis', see Marschall, *Im Schatten der chemischen Synthese*, pp. 110–114; see also T. Lenoir, *Instituting Science: the cultural production of scientific disciplines* (Stanford, California: Stanford University Press, 1997), pp. 203–238.

century.[34] Thus, there was no such thing as 'pure science'.[35] Even in the industrial chemist's self-conception, the distinction between applied and pure science did not feature.[36] As Roche's chief chemist, the Austrian Franz Elger (1879–1941), once wrote, he saw himself as a 'scientifically orientated chemist', who regarded the transfer of knowledge and know-how beyond the 'factory walls' as one of the keys to success. He declared that Roche's cooperation with universities and academic researchers in Switzerland and abroad was important, not only for obtaining scientific support, but also because in certain cases an 'outsider was more likely to find the right thing'.[37] He was aware of the fact that 'thought styles' within a 'thought collective'[38] restricted the chemical engineer's horizons.

34 D. Gugerli, P. Kupper, and D. Speich (eds), *Die Zukunftsmaschine. Konjunkturen der ETH Zürich 1855 – 2005* (Zürich: Chronos, 2005), p. 200. The first co-innovation by Hoffmann-la Roche and ETH dates from 1904.

35 See J. Goodman, 'Can it Ever be Pure Science? Pharmaceuticals, the pharmaceutical industry and biomedical research in the twentieth century', in J.-P. Gaudillière and I. Löwy (eds), *The Invisible Industrialist: manufactures and the production of scientific knowledge* (Basingstoke: Macmillan, 1998), pp. 143–166. On the relationship between academic and industrial research see G. Bowker, 'Der Aufschwung der Industrieforschung', in M. Serres (ed.), *Elemente einer Geschichte der Wissenschaften* (Frankfurt am Main: Suhrkamp, 1995), 829–867; H. Etzkowitz and L. Leydesdorff, 'The Endless Transition: a "Triple Helix" of university-industry-government relations.' *Minerva*, 36 (1998): 203–208; Gaudillière and Löwy (eds), *The Invisible Industrialist*; K. Grandin, N. Wormbs, A. Lundgren, and S. Widmalm, *The Science-Industry Nexus*; C. Ratmoko, 'Hormone aus dem Industrielabor. Die Erforschung und Herstellung von Geschlechtshormonen bei der Ciba zwischen 1910 und 1940', *Schweizerische Zeitschrift für Geschichte*, 55 (2005): 84–94; J. Tanner, 'Medikamente aus dem Labor. Forschungspraxis, Unternehmensorganisation und Marktstrukturen in der chemisch-pharmazeutischen Industrie', in T. Busset et al.. (eds), *Chemie in der Schweiz. Geschichte der Forschung und der Industrie* (Basel: Christoph Merian Verlag, 1997), pp. 117–146.

36 For a history concerning the development of the chemical industry in Basel see T. Straumann, *Die Schöpfung im Reagenzglas. Eine Geschichte der Basler Chemie (1850–1920)* (Basel, Frankfurt a. M.: Helbling & Lichtenhahn, 1995). See also Tanner, 'Medikamente aus dem Labor'.

37 HAR, PE.1.E – 101913: Handschriftlicher Brief von Franz Elger an Emil C. Barell, 29. Juni 1920, Beilage: Exposé, 5.

38 On Denkstil and Denkkollektiv see Fleck, *Entstehung und Entwicklung*.

In practice, however, these interdependencies created some frictions. As Roche did not have enough experience to handle bacteria, their industrial chemists had to collaborate closely with Tadeus Reichstein. Since his methods were sometimes more those of a 'bricoleur'[39] (a tinkerer), Reichstein was therefore obliged to describe his – largely improvisational – experimental system in detailed letters to Roche.[40] This becomes obvious when one reads his descriptions of his experimental system: in front of almost every figure came the inevitable 'circa'. Or there were phrases like: 'if possible'. For example, after Reichstein had tried to ferment sorbitol into sorbose between the double panes of the windows in his laboratory, he noted: 'if possible exposed to daylight'.[41] Therefore, it is not surprising that repeated personal visits and talks in Reichstein's laboratory in Zurich were necessary.

Furthermore, when neither Roche's researchers nor Reichstein himself were able to carry out a particular test, they handed the task over to the bacteriological-hygiene institute at ETH. When Reichstein forwarded the institute's bill to Roche, the company reproached him for the exorbitant amount charged. Confronted with the allegation of excessive charges, Reichstein added a short postscript to his response:

> If I am allowed to make an additional remark, then I would propose that you should consider work that you allowed to be accomplished or facilitated by academia, firstly as a grant to academia or to science in general, whereby it is surely hoped on both sides that there might emerge a direct material benefit for you. As a direct claim ab initio this cannot be established. Academia is, in many cases, dependent on effective assistance by industry, and the latter can draw from that only indirect benefit.[42]

39 See F. Jacob, *Of Flies, Mice and Men* (Cambridge, Mass./London: Harvard University Press, 1998) ; C. Lévi-Strauss, *The Savage Mind* (London: Weidenfeld & Nicolson, 1966).

40 On the notion 'experimental system', see H.-J. Rheinberger, *Toward a History of Epistemic Things: synthesizing proteins in the test tube* (Stanford, California: Stanford University Press, 1997).

41 StABS, Abl. 1996/90, Wiss. Korr. 2; Korr. mit Akademie-Verlag, La Roche; Doss., Hoffmann-La Roche, V. 1933-XI. 1935: Brief (Durchschlag) von T. Reichstein an die F. Hoffmann-La Roche, Abteilung 6. Zürich, 3. November 1933.

42 StABS, Abl. 1996/90, Wiss. Korr. 2; Korr. mit Akademie-Verlag, La Roche; Doss., Hoffmann-la Roche, V. 1933-XI. 1935: Brief (Durchschlag) von Tadeus Reichstein an die F. Hoffmann-la Roche, Abt. 6. Zürich, 30. März 1935., 1f.

In this extract of Reichstein's letter to Roche, the latent tensions between academic and corporate research become apparent.

Although a clear-cut medical indication for ascorbic acid did not exist (with the exception of scurvy, which was too rare a disease to create a lucrative business), Roche's ascorbic acid sales grew steadily after 1934. This growing demand led to the scaling up of the production process. Such industrialization of the microbiological step involved in transforming sorbitol into sorbose created problems of a different kind to those experienced in a research laboratory. Roche's chemical engineers had to establish new procedures, which did not play any role in the laboratory process. From 1934 onwards, they tried to develop an entirely new way of fermentation in order to handle the necessary amount of bacteria, the so-called 'Tiefgärverfahren' (submerged process). To initiate this new fermentation method, Roche Basel contacted Roche Nutley (New Jersey, USA), where attempts to produce citric acid by 'fermentation in deep layers' were being conducted at the time. Unfortunately, the know-how generated at Nutley turned out not to be very useful in producing sorbose by means of deep fermentation, as was still being deplored by Roche's chemical engineers two years later in 1936.[43]

43 HAR, TI.0.2 - R - 102222 c: Rapport No. 25239 von Dr Rudolf C. Vetter, Abt. V, an die Direktion. Basel, den 4. Mai 1934, 2; HAR, TI.0.2 - R – 102222 e: Rapport No. 25286 von Dr Rudolf C. Vetter, Abt. V, an die Direktion. Basel, den 27. März 1936, 10.

Figure 13.1 shows the transfer of a sorbose-producing strain being carried out at the time of the closing of the frontier – on 23 August 1940 at the 'Grenzacherhorn' customs station.

Regarding manufacturing facilities, the deep-fermentation process required new equipment. Or, to put it correctly, it *would have* required new equipment. However, faced with the economic crisis of the 1930s, Roche's general director hesitated about investing in expensive facilities. Moreover, the commercial prospects of synthetic vitamin C were by no means clear. In this context, it would have appeared highly risky to install new facilities for the sole purpose of deep-fermentation. To minimize economic risk, Roche's management, therefore, took a path typical of the pharmaceutical industry at the time. In so far as possible, it re-used facilities that were already at hand, and bought only multifunctional equipment that could be employed for other processes as well. This observation supports Luitgard

Marschall's argument that biotechnology was a 'conservative innovation' and a 'niche technology' within the chemical industry.[44]

The strategies for minimizing economic risk were not the only constraints weighing upon Roche's chemists. The micro-organisms themselves remained a major difficulty. The more efficient microbial strain, called *xylinum* or *gluconicum*, which was found in 1934, could not be used for deep-fermentation. Whereas bacterium *xylinum* acted efficiently in surface culture, it did no work in submerged culture. Apparently, it could not work if the layer was more than three centimetres deep, which was far too shallow for commercial production.[45] Additionally, it was quite difficult to get hold of 'good' bacteria, because those captured in the wild were more efficient than those supplied by the different 'offices for mould cultures'. Therefore, Roche continued either to breed or to search for other, more efficient bacteria or, later on, fungi. After several experiments, in October 1936, Roche's industrial chemists returned to the bacterium *Acetobacter suboxydans*. It was suitable for the submerged process, but since *Acetobacter suboxydans* is an aerobic micro-organism, it was necessary to develop more sophisticated ways of aeration. This proved to be quite a complex task, because aeration increased the risk of infecting the lugs with contaminants in air.[46] Hence biosynthesis required incessant control by a trained employee, both day and night.[47] A porter or a night guard would not do, the chemists complained. Nevertheless, the general director tried to reduce the costs for this work, and a couple of incidents occurred because there had been no supervision of the fermentation process during night time.[48]

Another issue complicating the fermentation process were the personal conflicts that developed within Roche's fermentation laboratory, located

44 See Marschall, *Im Schatten der chemischen Synthese*, pp. 124 and 213.

45 HAR, TI.0.2 - R – 102222 e: Rapport No. 25281 von Dr Rudolf C. Vetter, Abt. V, an die Direktion. Basel, den 21. Januar 1936, 2.

46 HAR, TI.0.2 - R - 102222 e: Rapport No. 27910 von Dr Rudolf C. Vetter, Abt. V, an die Direktion. Basel, 18. September 1936.

47 HAR, TI.0.2 - R – 102222 e: Rapport No. 25281 von Dr Rudolf C. Vetter, Abt. V, an die Direktion. Basel, 21. Januar 1936, 10.

48 HAR, TI.0.2 - R - 102222 h: Rapport No. 31148 von Dr Rudolf C. Vetter, Abt. V, an die Direktion. Basel, 8. Mai 1942, 3.

in a cellar. The production of sorbose had a high status, and was referred to as a 'holy cow'.[49] However, some of the people who were put in charge of this task were described as quite eccentric. Problems arose between two of them in particular, Kurt Warnat, a German who was one of the chemists in charge of the fermentation process, and his supervisor, Franz Elger, when the Swiss authorities turned down Warnat's application to become a Swiss citizen.[50] Warnat suspected that Elger had distorted some statements he had made to him in confidence when reporting about him to the authorities. Warnat believed that Elger had told the Swiss naturalization officials that he was an 'ardent admirer of the Nazis', and wanted to become a 'member of the fifth column'. He thought that he had failed becoming naturalized as a result. Although Warnat's assumption was wrong, as he discovered only after Elger's death in 1942, this incident severely damaged the relationship between them.[51] Furthermore, Warnat vehemently disagreed with the other chemists about the relevant parameters for the most profitable fermentation process. Whereas he was focused on improving the yield, other Roche chemists put priority on speeding up the process.[52] This conflict does not only highlight the fact that Warnat was perhaps an awkward employee and a difficult collaborator, but also that there was not one best way to produce sorbose.

For all these reasons, the fermentation process became of prime importance. This is demonstrated in a vivid appeal made by Emil C. Barell,[53] Roche's general director, on the occasion of an internal meeting in April 1936:

49 HAR, PD.2.2.VIC – 101887: Dr W. Roeser. 'Meine Zeit als "Chef im Ascorbin". Die Ascorbinsäure-Produktion in Grenzach in der Zeit von 1960 bis 1972', undated, 3f.

50 Of Kurt Warnat it is only known that he received his PhD in 1923 from the University of Kiel. His thesis was entitled 'Über das Verhalten mehrwertiger Nitrile bei der Hoesch'schen Synthese'.

51 HAR, PE.1.W – 101580: Interne Mitteilung von Dr K. Warnat an Abteilung Dr Vetter. Basel, 7. Oktober 1947, 2.

52 HAR, TI.0.2 - R – 102222 e: Rapport No. 25281 von Dr Rudolf C. Vetter, Abt. V, an die Direktion. Basel, 21. Januar 1936, 1.

53 Emil C. Barell (1874–1953) was Roche's general director from 1920 till his death in 1953.

> Since the end of March, problems that endanger regular production of ascorbic acid have been occurring in the sorbose-factory at Basel. Dr Barell draws attention to the fact that resolving these problems is the most urgent and most important issue for the scientific and the manufacturing divisions at the moment. [...] If the production of ascorbic acid was to stagnate, this would not only be deleterious to the enterprise's morale, but also cause serious financial loss to the company. He therefore asks all attendees to do their level best – at the expense of their own comfort and their usual hours of work; and to propose immediately the means by which might be useful to overcome this crisis as soon as possible.[54]

What possible strategies could be used to master this crisis? First of all, Roche's management looked for additional bacteriological support. They no longer wished to cooperate with the bacteriological institutes in Zurich and Paris, whose work they described as 'absolutely useless'.[55] Other bacteriologists were described as either incapable, or already collaborating with rival pharmaceutical companies, thus making them unsuitable for collaboration with Roche. Since Roche was not willing to hire a fully trained bacteriologist, they contemplated hiring a 'young bacteriologist' and looking for another laboratory assistant with 'some bacteriological experience'.[56] However, Roche did not employ a trained bacteriologist until 1946. Therefore, the company's industrial chemists were compelled to draw upon their own experience of working with bacteria for diagnostic tests, an experience that went as far back as the 1920s, and adapt this experience to the production of vitamin C.

The problems, however, did not only arise because of a deficiency in (scientific) bacteriological support. It is important not to see technology simply as applied science. As an in-house Roche technical report from the production department concerning the production of ascorbic acid explained in 1936, the translation from 'scientific'[57] procedure to manufac-

54 HAR, TI.0.2 - R - 102222 e: Rapport No. 25288 von Dr Rudolf C. Vetter, Abt. V, an die Direktion. Basel, 24. April 1936, 1.

55 Ibid., 3f.

56 Ibid.

57 HAR, PB.3 - 100309 b: Rapport Nr. 23156 von Dr W. Lindenmaier und Dr H. Müller, Abt. F.B., an die Direktion. Basel, 17. August 1936, 5 and 7.

turing process often leads to extensive alterations of the initial 'scientific' instructions. Regarding the industrial production of sorbose, the Roche engineers complained that they had never before encountered so many obstacles in trying to transform a set of 'scientific' instructions into a viable manufacturing process. Consequently, the resulting large-scale production of synthetic vitamin C hardly had anything to do with the instructions Reichstein originally gave to Roche. Even the yields that were achieved in practice differed widely from those possible 'in theory'.[58] Despite these difficulties, a breakthrough in the deep-fermentation process was achieved in September/October 1936 after a large number of trials. As a result, Roche ordered a one-meter deep vessel with a capacity of 1500 litres (circa 2640 pints). To be on the safe side, surface cultures also continued to be used in order to produce sorbose.[59]

The Transferability of Microbes

Despite these successes, the first steps undertaken to introduce the Reichstein procedure at other Roche production plants, namely Grenzach (Germany) and Nutley (USA), illustrate how difficult it was to transform the fermentation technique into a standardized and transferable process. Roche Nutley did not master all the necessary production steps until 1939. These were still 'all batch operations with a great deal of manual handling. In later years we jokingly referred to this period as the years of pots and pans'.[60] To change this, and to rationalize the production steps, the head of industrial ascorbic acid production at Basel was sent to Nutley. Unfortunately, sources

58 All quotations derive from HAR, PB.3 - 100309 b: Rapport Nr. 23156 von Dr W. Lindenmaier und Dr H. Müller, Abt. F.B., an die Direktion. Basel, 17. August 1936, 27.

59 HAR, TI.0.2 - R - 102222 e: Rapport No. 27910 von Dr Rudolf C. Vetter, Abt. V, an die Direktion. Basel, 18. September 1936, 1–3.

60 HAR, PD.2.2.VIC – 101887: Vitamin C in USA. Abschrift ab Tonband. Dr J.J. Fuchs, December 1982, 1.

found in the Roche archive do not allow me to specify the most critical operations he observed there.

Things were quite similar at Roche Grenzach. The only difference was the geographical proximity between the different production plants, which facilitated the necessary exchanges. Therefore it was possible for the whole workforce of Roche Grenzach to receive training in the production plants at Roche Basel.[61] Moreover, when Roche Grenzach had problems in breeding their own, efficient strain of sorbose bacteria, fresh strains could be sent across from Basel. During the Second World War this transaction became rather difficult as Grenzach is situated on the other side of the Swiss-German border. The transport of sorbose-producing bacteria was allowed as an exception, and on condition that there were members of the Gestapo present. Furthermore, Roche Grenzach was informed that opening the containers would destroy the bacteria, which meant that the Gestapo had to be persuaded not to open the shipment.[62]

61 HAR, PE.1.S – 102025: Unterlagen Fritz Strub, Interview mit Fritz Strub vom 3.3.1987 / Dh, 3.

62 HAR, LG.DE - 101859 r: Aktennotiz über ein Telephongespräch mit Herrn Dr Hellmich am 13. August 1940, morgens. Anruf von Dr Hellmich. See also L. Straumann and D. Wildmann (eds), *Schweizer Chemieunternehmen im Dritten Reich. Veröffentlichungen der Unabhängigen Expertenkommission Schweiz – Zweiter Weltkrieg*, 7 (Zürich: Chronos, 2001), p. 220.

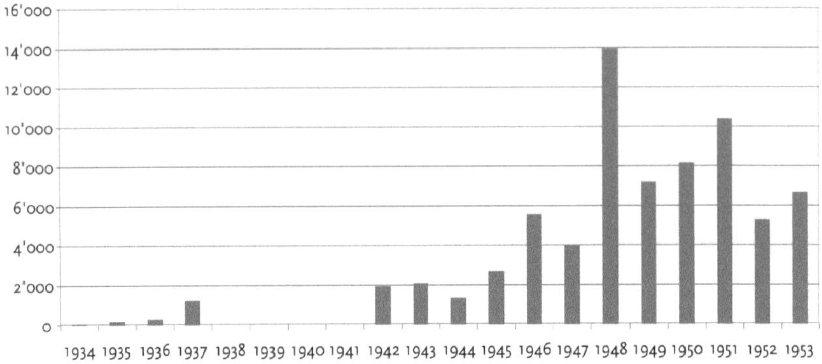

[Picture: HAR, LG.DE – 101696]

Figure 13.2: Ascorbic acid sales by Roche[63]

From Ascorbic Acid to Penicillin and beyond

The story of penicillin production provides a further example of the integration of multiple disciplines in biological research and development.[64] In this

63 HAR FR.2.3.6 104429 a-d; HAR PD.2.2–102638; HAR FE.0.8–103579a; HAR FO.3.4 – 102238; Roche Magazin 21, Mai 1984, 20f. For a discussion of the emergence of markets for ascorbic acid, see B. Bächi, '"Rein schweizerisches" Vitamin C aus Basel. Zur Kulturgeschichte einer soziotechnischen Innovation', *Basler Zeitschrift für Geschichte und Altertumskunde*, 105 (2005): 79–113 (100–113).

64 Concerning the history of penicillin production see J.-P. Gaudillière and B. Gausemeier, 'Molding National Research Systems: the introduction of penicillin to Germany and France', *Osiris*, 20 (2005): 180–202. See also Bud, *The Uses of Life*, pp. 103–107, and Gaudillière, *Inventer la biomédecine*, pp. 36–79.

respect, penicillin has been characterized as a 'paradigm for biotechnology'.[65] As Robert Bud has noted with regard to the history of penicillin production: '[I]t was the interface between chemical engineering and microbiology that seemed to have the most outstanding commercial implications on the front between biology and engineering'.[66] And Viviane Quirke has come to the conclusion that penicillin helped to resolve the dichotomy between companies specializing in synthetic organic chemicals and those specializing in biologicals.[67]

But how does the history of penicillin production relate to that of ascorbic acid? After the end of the Second World War, the production of ascorbic acid at Roche Basel came to a stop, because of overproduction and huge stocks stored at Lausanne, far from the German frontier for security reasons. Around 1946, Roche's management therefore had to find new applications for its ascorbic acid production facilities. They considered transforming them into a penicillin plant. This was quite understandable, as penicillin was the first microbial antibiotic, and a milestone in the emergence of biotechnology. At the US government's request during the Second World War, Roche Nutley had become involved in penicillin production.[68] Moreover, the first skilled bacteriologist joined Roche in 1946. But as prices for penicillin began to fall rapidly after the end of the Second World War, the existing equipment in Basel was considered to be too marginal to compete on the international market. The specific project of transforming the ascorbic acid production plants in Basel into a penicillin plant was therefore eventually cancelled, and the production of ascorbic acid restarted.[69]

65 R.I. Mateles, *Penicillin: A paradigm for biotechnology* (Chicago: Candida Corporation, 1998).

66 Bud, *The Uses of Life*, p. 100.

67 Quirke, *Collaboration in the Pharmaceutical Industry*, Part Two.

68 Peyer, *Roche*, 159.

69 HAR, PD.2.2.VIC - 103409 b: Dr A. Courtin. Meine Erinnerungen an die industrielle Entwicklung der Ascorbinsäuresynthese 1941-1952, 33. On Roche's reluctance to enter the business of antibiotic, see M. Bürgi and B. Strasser's contribution to this volume.

This association between ascorbic acid and penicillin underlines the important part played by ascorbic acid production in transforming synthesis from a purely chemical into a hybrid process. The biosynthetic technologies involved in ascorbic acid and penicillin manufacture prompted the foundation of a new fermentation laboratory at Roche Nutley in 1957, which aimed to improve the company's fermentation methods. During the 1960s, Roche took the decision to create the Roche Institute for Molecular Biology, which was dedicated to fundamental research in molecular biology, and was inaugurated in Nutley in 1968. The Institute – located at the interface between university and industry – had the objective of integrating molecular biologists, a community that had previously not felt very sympathetic towards the pharmaceutical industry, into the company's innovation process.[70]

Trends in academic research went in the same direction too. Although the discipline of biochemistry had been growing rapidly since the beginning of the century, a chair for biochemistry was introduced as late as 1956 at ETH. Within bacteriology, only one chair for agricultural bacteriology and one chair for hygiene and bacteriology had existed during the 1930s. In addition, whereas at ETH the bacteriological tradition was rooted in agriculture, biochemistry belonged to the department of chemistry. They were merged at ETH as late as 1962, by creating the degree of 'biochemistry and microbiology'.[71]

Conclusion

Research and Development are a contingent process. This becomes obvious if one looks at Roche's path to biotechnology. The path of the Swiss firm was quite unique, as it had no stake in vaccine production, and only

70 Idem.

71 http://www.ethistory.ethz.ch/rueckblicke/departemente/dbiol/weitere_seiten/dbiol_wfs_2_1 [cited 15 January 2006].

a very marginal one in antibiotics manufacture. Therefore, in the case of Roche, it was the fermentation step involved in ascorbic acid production that challenged the company's preference for pure chemical synthesis for the first time in the 1930s. The micro-organisms involved in the new process paved the way for a new industrial culture, combining chemical and biological production technologies. To adopt biosynthetic production methods and transform these 'local' skills from 'batch operations with a great deal of manual handling',[72] into explicit, standardized routine technologies that were transferable,[73] new forms of practices and alliances had to emerge – against a great deal of socio-technical resistance. This resistance consisted in a general dislike of fermentation technologies in the pharmaceutical industry, risk-minimizing strategies during the Great Depression, difficulties experienced between industrial and academic scientists, and between the different scientific disciplines required to establish new fermentation technologies. However, thanks to the production of l-ascorbic acid, Roche had become an important passage point[74] for academic and industrial chemists, chemical engineers, bacteriologists, microbiologists, biochemists and, later on, molecular biologists interested in pharmaceutical research and development. Microbes acted in this process as boundary objects,[75] which made these formerly separated social worlds meet and led to their integration into Roche's innovative processes.

72 HAR, PD.2.2.VIC – 101887: Vitamin C in USA. Abschrift ab Tonband. Dr J.J. Fuchs, December 1982, 1.

73 On 'standardized packages', see J.H. Fujimura, 'The Molecular Biological Bandwagon in Cancer Research: where social worlds meet', *Social Problems*, 35 (1988): 261–283; idem, 'Science: standardized packages, boundary objects, and "translation"', in A. Pickering (ed.), *Science as Practice and Culture* (Chicago: University of Chicago Press, 1992), pp. 168–211.

74 The concept of 'obligatory passage point', that is a point where anyone with a stake in the network has had to pass through in order to attain his goals, has been established by Michel Callon, in 'Eléments pour une sociologie de la traduction : la domestication des coquilles Saint-Jacques et des marins-pêcheurs dans la baie de Saint-Brieuc', *L'Année sociologique*, 36 (1986) : 169–208.

75 See S.L. Star and J.R. Griesemer, 'Institutional Ecology, "Translation" and Boundary Objects: amateurs and professionals in Berkeley's Museum of Vertebrate Zoology, 1907–1939', *Social Studies of Science*, 19 (1989): 387–420. On biotechnology as a 'boundary

The material-discursive trajectory of ascorbic acid production aptly illustrates how ascorbic acid produced new structures. Ascorbic acid production imposed its own rationale on the future modes of research and production at Roche. So, when the Swiss firm looked back on its biotech roots in the 1980s, the 'technical vitamin C synthesis' appeared as the 'classical example of a biotechnical procedure'[76] – thanks to the 'single-celled chemist', *Acetobacter suboxydans.*[77] Furthermore, in the 1980s Roche tried to transform ascorbic acid production into a purely biotechnological process. In 1988, Roche produced one tonne of ascorbic acid with an entirely microbiological procedure. This pure fermentation method, however, could not compete with the hybrid Reichstein procedure from an economic point of view. Therefore, this remained Roche's single tonne of ascorbic acid ever completely produced 'microbiologically, i.e. by fermentation'.[78]

In all these processes, our 'lead user', Tadeus Reichstein, saw himself as a 'subjectum'.[79] Maybe he was conscious of the historical contingency of his choices. For example, he would probably never have come to the idea of using the bacteria present in drosophila melanogaster's intestine to transform sorbitol into sorbose, if there had not been ready-at-hand recipes on how to catch and handle wild drosophila melanogaster – the geneticist's 'domestic animal' at that time. And, maybe he was aware that he was being subjected to a general trend towards the 'biochemicalization' of life – a trend that did not only create a demand for ascorbic acid, but was mirrored in the material objects, production devices, processes and technologies necessary for its production.

---

object' see R. Bud, 'Biotechnology in the twentieth century', *Social Studies of Science*, 21 (1991): 415–457.

76 Roche Magazin 21, Mai 1984, 19. See also the detailed interview with Tadeus Reichstein in the same volume, 10–15.

77 Roche Magazin 11, Dezember 1980, 7–10.

78 Roche-Nachrichten, 5.9.1988.

79 StABS, Abl. 1997/42, A 5.1. Antrittsvorlesung von Tadeus Reichstein an der E.T.H. Zürich. 21. Februar 1931. ‚Der Sinn der Alchemie', 13.

MICHAEL BÜRGI[1] AND BRUNO J. STRASSER[2]

14 Pharma in Transition: new approaches to drug development at F. Hoffmann-la Roche & Co, 1960–1980

Introduction

> [...] chemistry is at Roche's disposal just like a good wife, reliable, cooperative, maybe sometimes a little bit smiling with the knowledge that everything which had been realized came to a very great extent from her.[3]

The pharmaceutical industry transforms knowledge into drugs, and often into money. Today, drug innovation is based on knowledge produced by complex networks including pharmaceutical companies, biotechnology start-ups, and academic laboratories. It results from a combination of innovation strategies, for example research into the basic mechanisms of disease and drug action, using a series of molecular biology technologies such as

1 ETH Zürich, Institut für Geschichte, Technikgeschichte, Auf der Mauer 2 (ADM), CH-8092 Zürich, michael.buergi@bluewin.ch. My work has been supported by the Institut d'histoire de la médecine et de la santé, Université de Genève, and financed by the Roche Research Foundation and the Swiss National Science Foundation project No. 105311–109973/1. For stimulating discussions and critiques we would like to thank Viviane Quirke, Jean-Paul Gaudillière, and Sonja Matter, as well as Alexander Bieri, Bruno Halm, and Thomas Casutt of the Historisches Archiv Roche Basel for their kind support.

2 Yale University, Section for the History of Medicine, PO box 208015, New Haven, CT 06520–8015, USA, bruno.strasser@yale.edu

3 Historisches Archiv Roche (HAR) FE.0.4 – 103593 h: 'Continuity in Chemistry Research', memorandum submitted by Otto Isler, head of chemical research at Roche, 5.6.1969, p. 1.

recombinant DNA or monoclonal antibodies. The laboratory scientists involved often assume multiple roles in academia and industry.

This oversimplified picture of what has been called the 'biotechnological revolution' reflects the perception that drug development has entered a new era with distinctive social, economic, cultural and scientific characteristics.[4] In this paper, we investigate the transition into the biotechnology era of a major pharmaceutical company, F. Hoffmann-la Roche & Co (Roche),[5] from the perspective of its top research management, and focus specifically on drug development strategies. We describe *how* new approaches to drug innovation, emphasizing biological[6] research instead of chemical synthesis, were introduced in the 1960s. Our paper also tries to explain *why* Roche managers and scientists introduced these changes. We argue that at least four different factors motivated this transition: diminishing returns from previous modes of drug innovation, tighter FDA drug approval policies, isolation from academic biological research networks, and expansion of the management's time horizon towards long-term planning. This paper thus constitutes a contribution to the history of both the pharmaceutical industry and the management of scientific research in the twentieth century.

The standard picture of the 'biotechnological revolution' stands in sharp contrast with the description of two earlier drug innovation regimes: the extractive and the synthetic.[7] These regimes are best characterized as a

4 On biotechnology, see M.D. McKelvey, *Evolutionary Innovations: the business of biotechnology* (Oxford: Oxford University Press, 2000); M. Kenney, *Biotechnology: the university-industry complex* (New Haven: Yale University Press, 198); S. Krimsky, *Biotechnics & Society: the rise of industrial genetics* (New York: Praeger, 1991); L. Orsenigo, *The Emergence of Biotechnology: institutions and markets in industrial innovation* (London: Pinter Publishers, 1989); R. Loeppky, *Encoding Capital: the political economy of the Human Genome Project* (New York and London: Routlege, 2005).

5 The company changed its name from 'F. Hoffmann-la Roche & Co.' to 'Roche Holding AG' in 1989.

6 The Roche management and scientists used the terms 'biological' and 'biochemical' to designate a variety of laboratory approaches, ranging from metabolic studies to fundamental research in virus genetics.

7 V. Walsh, 'Paradigms in the evolution of life sciences research, and the changing structure of the innovation organization', in K. Grandin et al. (eds), *The Science-Industry Nexus: history, policy, implications* (Sagamore Beach: Science History Publications, 2004), pp. 189–221.

given combination of research strategies and institutional structures. The first came into being in the late-nineteenth century and relied on a strategy based on the chemical extraction of natural substances from plants and animals. It was mainly associated with the creation of small pharmaceutical firms. The second regime arose in the early-twentieth century, supplementing, but not replacing, the previous one. Its main strategy consisted in synthesizing structural variations of chemicals with known pharmacological effects and selecting the most promising variants through a screening procedure, i.e. testing the chemical compounds for pharmaceutical activity in animals, tissues or cells. It was generally associated with the establishment of in-house research laboratories, institutionalized relationships with academia, and the rise of multinational firms.

There exists no strong consensus regarding the origins, extent and precise characteristics of the historical transition towards the present mode of drug innovation. Defining a new innovation regime would therefore be premature. The most common explanation for this transition, however, is that it has been triggered by the invention of recombinant DNA technologies in the early 1970s, by changes in the market economy, in particular the availability of venture capital, and by amendments to intellectual property legislation in the United States. All of these developments were indeed important for the creation of start-up biotech companies, one of the key new players in this transition.[8] In the 1980s, the large drug companies began to adopt the new technologies for drug development, either by cooperating

---

We do not follow the author in her use of the term 'paradigm' because we do not feel that it conveys the institutional dimension we would like to stress, in combination with the cognitive dimension congruent with the Kuhnian origins of the term. In addition, the co-existence of paradigms stressed by the author differs from Kuhn's notion of paradigm replacement through scientific revolutions.

8 R. Teitelman, *Gene Dreams: Wall Street, academia, and the rise of biotechnology* (New York: Basic Books, 1989); S. Wright, 'Recombinant DNA technology and its social transformation, 1975–1982', *Osiris*, 2nd Series, 2 (1986): 303–360; G. Dutfield, *Intellectual Property Rights and the Life Science Industries: a twentieth century history* (Aldershot: Ashgate, 2003); D.J. Kevles, 'Diamond v. Chakrabarty and Beyond: the political economy of patenting life', in A. Thackray (ed.), *Private Science: biotechnology and the rise of the molecular sciences* (Philadelphia: University of Pennsylvania Press, 1998), pp. 65–79.

and eventually acquiring start-up companies, by incorporating new in-house capabilities, or by establishing collaborations with academic researchers.[9] Adopting a much broader perspective, the historian of science Robert Bud has stressed the century-long continuity of biotechnology as a vision and as a way of exploiting biological processes for production purposes.[10] Taking a middle ground, economic historians have underlined that only in the 1960s and 1970s did the amount of available fundamental biological knowledge make investments in biological research profitable for pharmaceutical firms, leading to changes in drug development strategies.[11]

Today one of the world's largest drug companies, Roche, has lived through the different transitions outlined above. Like many of its major competitors in the pharmaceutical sector, Roche was founded in the late-nineteenth century as a chemical firm. According to business historian Alfred D. Chandler Jr., Roche is currently a leader in genetic engineering and has played 'a significant role in building the infrastructure of the emerging biotechnology revolution'.[12] Thus it represents a good case study through which to explore the transformation of the pharmaceutical industry. Even in the early 1960s, i.e. before recombinant DNA, a profound change became apparent at Roche. The Roche management began considering new drug development strategies and introduced new approaches based on fundamental biological research. The main objective was no longer solely to understand the chemical and pharmacological properties of drugs, as in the synthetic regime, but also to explain the molecular mechanisms of drug action and the precise biological targets of drugs. In other words,

9 L. Galambos and J.L. Sturchio, 'Pharmaceutical Firms and the Transition to Biotechnology: a study in strategic innovation', *Business History Review*, 72 (1998): 250–278.

10 R. Bud, 'Biotechnology in the twentieth century.' *Social Studies of Science*, 21 (1991), pp. 415–457.

11 A. Gambardella, *Science and Innovation: the US pharmaceutical industry during the 1980s* (Cambridge: Cambridge University Press, 1995), pp. 22–26. For a similar quantitative argument see Walsh, 'Paradigms in the evolution of life sciences research', p. 193; also A.D. Chandler, *Shaping the Industrial Century: the remarkable story of the evolution of the modern chemical and pharmaceutical industries* (Cambridge and London: Harvard University Press, 2005).

12 Chandler, *Shaping the Industrial Century*, pp. 9–10, 15, quotations: pp. 241, 243.

Roche was undertaking a transition from a strategy of drug development centred on chemistry to another centred on biology. 'Rational drug design' became the motto for this new approach.[13]

Noting that this transition was already underway in the 1960s at Roche does not explain why it took place. Sciences can provide new opportunities for a company only if they are interpreted as opportunities by the company management and its scientific staff. Every company management has to 'read' its own scientific, economic, social and organizational situation and environment in order to decide what changes are to be made.[14] By focusing on the discussions among the Roche management in charge of research planning and on the actual research projects that were carried out in the laboratories, we can thus illuminate how and why fundamental biological research came to be viewed as promising or even necessary for drug innovation.

The transition towards biological research represented a particularly sharp turn for Roche. Indeed, the company had not previously engaged in work on vaccines and had only maintained a very limited research program on antibiotics. Furthermore, it did not rely heavily on biological modes of drug production. Thus, Roche hardly possessed any in-house experience and expertise with biological research, and its laboratory infrastructure was essentially that of a chemistry laboratory. In addition to this lack of internal resources, it possessed no contacts in the biological sciences to match its century-long networks of personal and institutional collaborations with academic chemists.

The situation at Roche regarding biological research and collaboration was quite typical for a chemistry-based pharmaceutical company. The

13 Nevertheless, the previous chemical strategies of drug innovation were far from irrational. Even in the nineteenth century, when the first synthetic drugs were developed, analogy with extracted drugs or synthetic dyestuffs gave some guidance. Synthesizing chemical variants of a molecule of known properties and testing their effects was indeed a rational approach. See M. Weatherall, *In Search of a Cure: a history of pharmaceutical discovery* (Oxford: Oxford University Press, 1990), pp. 36, 150.

14 K. Lipartito, 'Culture and the practice of business history', *Business and Economic History*, 24 (1995): 1–41 (5–6).

few biologists or medical researchers who worked in these companies were mainly involved in screening compounds synthesized by organic chemists for pharmacological activity or in testing drugs for toxicity. Other pharmaceutical companies, such as Eli Lilly or Merck, had established more extensive collaborations with academic biomedical laboratories, or even hired biomedical researchers as early as the inter-war period.[15] Most of them, however, were involved in developing improved biological production methods, not in investigating fundamental biological questions.

In some cases, such as the endocrine drugs, vaccines or antibiotics, the contribution of biological research was of a more fundamental nature. However, after the Second World War, with the establishment of the State as the main patron for scientific research and with increased funding for the academic biological sciences, biological researchers became less willing to work for industry and doing so lost much of its prestige.[16] Thus, even though the case of Roche might be particularly revealing with regard to chemistry-based companies, the transition we analyse represented a serious challenge to the pharmaceutical sector as a whole in the post-war period.

Our historical reconstruction of the transition undergone by Roche in the 1960s relies, among other sources, on the minutes of the Roche Research Management Group (RRMG) and the scientific reports submitted by the company's researchers to this managerial board. The RRMG was founded in 1956 and composed of a small nucleus of managers of the company's three main research sites in Switzerland, the United States, and Great Britain.

15 J.P. Swann, *Academic Scientists and the Pharmaceutical Industry: cooperative research in twentieth-century America* (Baltimore: John Hopkins University Press, 1988); N. Rasmussen, 'The Forgotten Promise of Thiamin: Merck, Caltech biologists, and plant hormones in a 1930s biotechnology project', *Journal of the History of Biology*, 32 (1999): 245–261; idem, 'The Moral Economy of the Drug Company-Medical Scientist Collaboration in Interwar America', *Social Studies of Science*, 34 (2004): 161–185; J.-P. Gaudillière, 'Better Prepared than Synthesized: Adolf Butenandt, Schering Ag and the transformation of sex steroids into drugs (1930–1946)', *Studies in History and Philosophy of Biological and Biomedical Sciences*, 36 (2005): 612–644.

16 N. Rasmussen, 'Of "Small Men", Big Science and Bigger Business: the Second World War and biomedical research in the United States', *Minerva*, 40 (2002): 115–146; Swann, *Academic Scientists and the Pharmaceutical Industry*, epilogue.

The group's original duty was to discuss commercial, chemical, and clinical aspects of research in order to coordinate research activities at Roche.[17] Its members belonged either to the general direction responsible for the whole firm including both the parent house in Basel and the subsidiaries, or they were part of one of the local executives in the USA or the UK. The RRMG's first chairman was the chemist Placidus A. Plattner, who had been managing director in charge of research since 1952.[18] Occasionally the RRMG meetings were also attended by Roche employees responsible for a particular area of research or by outside experts.[19]

The discussions taking place within the RRMG were exempt from the necessities of daily business as well as the constraints of public relations. The group's meetings were 'undisturbed by telephone calls and intrusions of outsiders', and its members would have 'the leisure and time to really let [their] hair down and work out the problems facing [them] continually'.[20] In addition, a number of scientific reports were submitted to the RRMG, in which researchers at Roche could express their opinions vis-à-vis management more extensively.[21] Understandably, the minutes of the RRMG and the reports submitted to this managerial board shed a different light on Roche's transition into the biotechnology era than do its published annual and business reports. They give unique insight into the vision and rationale

17 HAR FE.0.4 – 101129 a: Program of the first RRMG meeting, 25.–30.6.1956.

18 H.C. Peyer, *Roche. Geschichte eines Unternehmens 1896–1996* (Basel: Editiones Roche, 1996), p. 173. Other early members of the RRMG were the chemist Max Furter, member of the general direction since 1952 and in charge of manufacturing of specialities; the medical doctor and chemist Alfred Pletscher, head of medical research at Roche Basel since 1957 and Plattner's successor since 1967; the chemist Otto Isler, head of chemical research at Roche Basel; Lawrence D. Barney, president of Roche USA; the medical doctor Elmer L. Severinghaus and the chemist John Aeschlimann, both executives at Roche USA; and the chemist Alexander Morrison, head of research at Roche UK.

19 HAR FE.0.4 – 101129 a: Placidus A. Plattner to Roche President Albert Caflisch, 29.11.1957.

20 HAR FE.0.4 – 101129 a: Minutes of the RRMG meeting, 24.–30.4.1957, add. to p. 1.

21 HAR FE.0.4 – 101129 a: Placidus A. Plattner to Roche President Albert Caflisch, 29.11.1957.

of the management in charge of scientific research in a major pharmaceutical company during the second half of the twentieth century.[22]

Extraction and Synthesis at Roche

Before we turn to the transition undergone by Roche in the 1960s and 1970s, it is important to understand the research strategies and institutional arrangements on which the company relied in the first half of the twentieth century. In contrast to its local Swiss and German competitors, such as Ciba, Sandoz and Bayer, Roche did not evolve out of the chemical dyestuff industry.[23] Drug development at Roche was rooted in the chemical extraction of pharmacologically active natural products, a procedure that had been established in the early-nineteenth century.[24] From the inter-war period onwards, however, the company shifted towards chemical synthesis,

22 However, neither the minutes nor the reports are delivered completely in the Roche Archives. We therefore intend to outline the broader lines of discussion rather than to reconstruct the exact processes of decision making at Roche.

23 The German dyestuff manufacturer Bayer created its pharmaceutical division in 1888, Ciba also manufactured its first drugs before 1900 and created a 'Biologische Abteilung' in 1908. Sandoz produced analgesics beginning in 1895 and created its 'Pharmazeutische Abteilung' in 1917. G. Meyer-Thurow, 'The Industrialization of Invention: a case study from the German chemical industry', *Isis*, 73 (1982): 363–381; T. Straumann, *Die Schöpfung im Reagenzglas. Eine Geschichte der Basler Chemie* (Basel und Frankfurt a. M.: Helbing & Lichtenhahn, 1995); P.C. Waldmeier, 'Pharma-Forschung bei Ciba-Geigy und ihren Vorgängerfirmen CIBA und Geigy', in Athineos Philippu (ed.), *Geschichte und Wirken der pharmakologischen, klinisch-pharmakologischen und toxokologischen Institute im deutschsprachigen Raum* (Innsbruck: Berenkamp, 2004), pp. 858–863; H. Fritz, *Industrielle Arzneimittelherstellung. Die pharmazeutische Industrie in Basel am Beispiel der Sandoz AG* (Heidelberger Schriften zur Pharmazie- und Naturwissenschaftsgeschichte, 10) (Stuttgart: Wissenschaftliche Verlagsgesellschaft, 1992). See also A.S. Travis, *The Rainbow Makers: the origins of the synthetic dyestuff industry in Western Europe* (Bethlehem: Lheigh University Press, 1993).

24 Weatherall, *In Search of a Cure*, p. 20.

and by the 1960s it had become one of the world's largest manufacturers of synthetic drugs and vitamins. Drug development at Roche was now dominated by chemical synthesis; chemists occupied the leading positions in its management, and the company maintained close relationships with the chemical laboratories of Swiss universities. Alfred Pletscher, former head of research, summarized the company's development in the first half of the twentieth century by concluding that the 'traditional strength of Roche lies in synthetic chemistry'.[25]

Roche was founded in Basel in 1894 by the merchant Fritz Hoffmann and the apothecary Max Carl Traub as a spin-off of a small business that produced pharmaceuticals and commodity chemicals.[26] The growth of the company until the First World War relied mainly on sales of plant and animal extracts. The chemists Emil C. Barell and the apothecary Carl F. Schaegers, who both joined the company in 1896, initiated a research program on plant and organ extracts based on techniques they had learned when working at the university. Barell, who eventually became general director of Roche, had studied chemistry at the Swiss Federal Institute of Technology (ETH) in Zurich and the University of Bern. Schaegers had studied pharmacology in Munich and later at the University of Bern had worked together with the physician Theodor Kocher, who received the Nobel Prize in 1909, to investigate the extraction of the active substance of the thyroid gland. Since the late 1890s, Roche had sold a number of plant extracts, and from 1911 onward, the company launched a series of gland extracts, some of which were sold until after the Second World War.

The development of new drugs at Roche had often been guided by the experiences of pharmacologists and physicians working in academia, whom Roche's management and researchers had known personally. The company researchers themselves were in charge of improving the efficiency of the chemical extraction processes in order to lower production costs and

25 Pletscher, '25 Jahre Roche-Forschung', p. 25.

26 If not indicated separately, the following remarks on the history of Roche are based on Peyer, *Roche*.

achieve high levels of purity.[27] Before the First World War, Roche set up an analytical laboratory to test the extracted products. The pharmacological effect of most of the drugs launched by the company had already been known beforehand, and pharmacological research concentrated only on the toxicity and the effectiveness of the extracts. The cough syrup Sirolin for instance – one of Roche's first successful products, launched in 1898, and marketed until the 1950s – was composed of a chemically modified plant extract from a American tropical tree, a substance that had been assumed to be of therapeutic value as early as the mid-nineteenth century (Fig. 14.1).

During the inter-war period, Roche started manufacturing synthetic sedatives, vitamins and fine chemicals, and it strove to replace its extracted natural products with synthetic analogues. In particular fields such as pain killers, new synthetic compounds were considered less harmful than the conventional opium extracts, and the 1912 International Opium Convention restricting the use of opium forced the company to continue the search for alternatives.[28] In the early 1920s, the company launched its first semi-synthetic drug, the sedative Allonal.[29] Vitamins also became one of Roche's most important synthetic products. In the early-twentieth century, the concept of diseases due to deficiency of food factors present

27 Until 1907, Swiss patent law did not protect chemical processes, which allowed companies to imitate their foreign competitors, especially German companies. However, the 1907 legislation was also prompted by the fact that the Swiss chemical industry no longer relied on imitation, as it had done throughout the nineteenth century. An appraisal of the extent to which Roche imitated production processes during its early years of existence would require further research. N. Stettler, 'Erfindungen', in *Historisches Lexikon der Schweiz*, [electronic publication, www.dhs.ch], 6.12.2004; Straumann, *Die Schöpfung im Reagenzglas*, pp. 132–134; J. Tanner, 'The Swiss pharmaceutical industry: the impact of industrial property rights and trust in the laboratory, 1907–1939', in A.S. Travis et al (eds), *Determinants in the Evolution of the European Chemical Industry, 1900–1939* (Dodrecht: Kluwer, 1998), pp. 257–272.

28 Peyer, *Roche*, pp. 94, 99.

29 Allonal was composed of an analgesic compound called amidopyrine, which had already been synthesized in the nineteenth century, and allylisopropylbarbituric acid synthesized by the Roche chemist Ernst Preiswerk. Weatherall, *In Search of a Cure*, pp. 36–37; Peyer, *Roche*, p. 94.

Figure 14.1: The Roche speciality Sirolin, 1910. © Historisches Archiv Roche.

in trace amounts in normal diets had become widely accepted. Organic chemists succeeded in isolating the first vitamins, and nutritional scientists were enthusiastic about the newly discovered group of organic compounds. By the late 1920s, vitamins had become very fashionable and the subject of health propaganda in newspaper articles, radio broadcasts, and popular guides.[30] In 1933, Roche bought the patent rights for a synthetic pathway to vitamin C, and in the following decades, Roche purchased the rights to other synthetic vitamin pathways from university researchers, while contributing to the development of its own new synthetic processes.[31] Otto Isler for example, who joined Roche in 1936 after receiving his doctorate in chemistry at the ETH in Zurich and later became head of chemical research at Roche, contributed to the company's success in the vitamin business by discovering various vitamin synthesis pathways.[32] Eventually, Roche became one of the world's largest manufacturers of synthetic vitamins.

The development of chemical synthesis went hand-in-hand with an expansion of the company's in-house research capabilities. In contrast to their local and German competitors who started their business activity as dyestuff manufacturers in the nineteenth century, Roche did not dispose of extended facilities for research in synthetic chemistry. Therefore, new laboratory buildings were built as well as facilities for biological screening. Between 1924 and 1944, the number of researchers employed by Roche rose from 26 to 75.[33] As early as 1905, Roche had opened a branch office in

30 Weatherall, *In Search of a Cure*, chapter 7; S.M. Horrocks, 'The Business of Vitamins: nutrition science and the food industry in inter-war Britain', in H. Kamminga and A.Cunningham (eds), *The Science and Culture of Nutrition, 1840–1940* (Amsterdam and Atlanta: Rodopi, 1995), pp. 235–258 (238–239); R.D. Apple, '"They Need it Now". Science, advertising and vitamins 1925–1940', *Journal of Popular Culture*, 22 (1988): 65–83.

31 For more on this, see Beat Bächi's contribution to this volume.

32 M. Bürgi, *Die Anfänge der industriellen Vitaminproduktion. Fotografien aus dem Historischen Archiv Roche* (Basel: Historisches Archiv Roche, 2004); K. Bernhard, 'Otto Isler', in *Historisches Lexikon der Schweiz*, [electronic publication, www.dhs.ch], 29.7.2005.

33 By the end of the war, researchers counted for a little more than 2 per cent of the total personnel. Peyer, *Roche*, p. 139 and pp. 159–160.

New York, where a research department had been created five years later.[34] In 1929, the US branch office moved to Nutley, New Jersey. Its research department continuously expanded and, by the end of the Second World War, it employed more chemists than the site in Basel. A third but significantly smaller research infrastructure was established in Welwyn Garden City, England, in 1938. In the same period, the company had also settled in many other European countries and eventually in China, Japan and South America. In most of these places, Roche maintained production facilities with limited research capabilities in order to support production engineering.

After the creation of larger in-house research facilities, the chemists working at Roche began publishing in international scientific journals.[35] The focus on chemical synthesis and the increased scientific status of Roche researchers made it possible to develop new collaborations with some of the most renowned academic chemists working in Swiss universities (especially the ETH and the University of Zurich) such as Paul Karrer (Nobel Prize 1937), Leopold Ruzicka (Nobel Prize 1939), and Tadeusz Reichstein (Nobel Prize 1950). As far as clinical research was concerned, like other pharmaceutical companies, Roche relied on the work of private physicians who used and evaluated the company's products in their medical practices.[36] In some cases, toxicological problems did not appear until a drug had been widely used for medical treatment, and several compounds had to be withdrawn from the market.[37]

34 For the development of Roche USA see 'Roche in Amerika, 1905–1965', *Roche Zeitung*, 1 (1966): 2–19.

35 The Science Citation Index lists 16 articles published by the chemists Franz Elger, Markus Guggenheim and Ernst Preisewerk until 1944, and another 16 articles published by members of the laboratories at Roche Nutley; cf. also E. Gwinner and B. Dalle Carbonare, 'Roche Pharma Research from the Past to the Present', *Chimia*, 50 (1996): 514–518 (516).

36 Peyer, *Roche*, provides only very little insight into the history of clinical research at Roche; cf. also Swann, *Academic Scientists and the Pharmaceutical Industry*.

37 This was the case for the Allonal ingredient amidopyrine. Weatherall, *In Search of a Cure*, p. 37.

The impact of the Second World War on research at Roche was manifold. As early as December 1938, Roche's administrative board had decided to concentrate increasingly on the US and Great Britain. The board limited the scheduled expansion of the Basel research site in favour of the research sites in Nutley and Welwyn. The decision was partly due to the upcoming war in Europe. Furthermore, in May 1940, when Germany invaded France, Roche's general director Emil C. Barell left Switzerland and moved to Nutley. Several senior scientists from the headquarters in Basel accompanied him, and contributed to the development of Nutley's research department before 1945. During the war, Roche Nutley became the most important research site of the company. This corresponded to a business strategy of 'decentralized growth'. Roche's management tried to minimize risk by maintaining its business activities in European countries, including Germany, and in the United States. But by 1941, revenues at Roche Nutley had grown rapidly. By the end of the war, the US subsidiary accounted for over 50 per cent of the company's total revenues, which rose from approximately 65 million Swiss francs in 1939 to 177 million in 1945.[38] After the war, the company needed to improve the coordination of its increasingly decentralized research facilities, located in Switzerland, the United States and United Kingdom. The Roche Research Management Group was founded in 1956 to address this challenge. Wartime also strengthened the company's emphasis on synthetic vitamins. In its 1945 annual report, the administrative board of Roche stated that the 'effects of the war gave rise to an increasing demand for our products, especially for those suited to alleviate the pain that resulted from destitution and malnutrition'.[39] In other words, wartime food shortages and military needs had increased the demand for vitamins during these years. Among the ten top-selling products of Roche, vitamin-specialities registered the strongest growth in sales between 1941

38 L. Straumann and D. Wildmann. *Schweizer Chemieunternehmen im 'Dritten Reich'* (Veröffentlichungen der Unabhängingen Expertenkommission Schweiz – Zweiter Weltkrieg, Vol. 7) (Zürich: Chronos, 2001), pp. 177–186.

39 'Bericht des Verwaltungsrates der Firma Hoffmann-la Roche', 1945, without paging; cf. also Peyer, *Roche*, p. 175.

and 1945, with the sole exception of the synthetic compound Prostagmin, a drug used to treat muscular diseases.[40]

Between 1947 and 1963, the total revenues of Roche's pharmaceutical products grew from 154 to 930 million Swiss francs without any major acquisition.[41] Vitamins and sulphonamides, synthetic antibacterial agents, were Roche's best-selling products throughout the 1950s, and in the early 1960s the company launched another series of best-selling drugs, the synthetic benzodiazepine tranquillizers Librium and Valium. Sulphonamides were a class of synthetic dyestuff derivatives that had already been developed for medical use in the 1930s by researchers at the Bayer laboratories and the Pasteur Institute in Paris.[42] At the end of the 1940s, Roche launched its own sulphonamide, the lucrative Gantrisin.

Both the sulphonamide Gantrisin and the benzodiazepines Valium and Librium had been developed at Roche Nutley. The development of the latter illustrates the fact that, under the synthetic regime of drug development, new drugs were being developed without extensive knowledge about their biological mode of action. In 1955, the chemist Leo Sternbach, who had been an assistant of Leopold Ruzicka at the ETH in Zurich and who had joined Roche in 1940, succeeded in synthesizing the first synthetic benzodiazepine (chlordiazepoxide). The new compound showed a tranquillizing effect in animals. The company initiated clinical trials and finally brought Librium to market in 1960.[43] At the time when Sternbach and his colleagues developed Librium, Roche did 'not do a great deal of internal fundamental work on the biochemistry of mental disease'. Only a small group was kept active in this field 'in order to maintain contact with general progress'.[44] The reasons why the newly developed compound had a tranquillizing and anxiolytic effect were completely unknown, and of only marginal importance to the development of the drug. Indeed, the mode

40 Straumann and Wildmann, *Schweizer Chemieunternehmen im 'Dritten Reich'*, p. 270.

41 Information of Alexander Bieri, head of the Historisches Archiv Roche, 16.5.2006.

42 Weatherall, *In Search of a Cure*, pp. 150–153.

43 A. Bänninger et al., *Good Chemistry: the life and legacy of valium inventor Leo Sternbach* (New York etc: McGraw-Hill, 2004), pp. 49–52.

44 HAR FE.0.4 – 101129 a: Minutes of the RRMG meeting, 15.–23.10.1956, p. 4.

of action of benzodiazepine molecules on neurotransmitter receptors was only elucidated in the 1980s.[45]

Because benzodiazepines were a new class of pharmaceutical substances, Roche was able to assure large patent protection and develop a series of benzodiazepine compounds. By 1963, the new drugs had already realized the highest revenues of all Roche products; vitamins and sulphonamides came second.[46] From a medical point of view, the benzodiazepine tranquillizers were superior to their predecessors, e.g. the barbiturates, because they did not show any adverse side-effects and overdose was not lethal. Enthusiastically, the popular press in the United States called the tranquillizers 'Happiness Pills' or 'Emotional Aspirin'.[47] Valium became the world's best-selling 'middle-class drug', and the single most prescribed drug in the United States.[48] By the end of the 1960s, Roche obtained over 50 per cent of its total revenues from psychotropic drugs.[49]

During the same period, the company significantly enlarged its research capacities worldwide. In 1962, the total number of employees at the Roche research department in Basel reached 700.[50] Meanwhile, research at Roche Nutley had grown to a comparable extent, and eventually surpassed Basel in terms of expenditure.[51] Total research expenditure quadrupled between 1952 and 1962. Between 1958 and 1967, the contribution of Roche Basel's organic chemists consisted of approximately 2000 new compounds per year that were chemically synthesized or isolated and submitted for biological screening.[52]

45 Pletscher, '25 Jahre Roche-Forschung', p. 20–24.
46 HAR FE.0.4 – 101129 a: Minutes of the RRMG meeting, 13.–18.6.1963, p. 60.
47 S.L. Speaker, 'From "Happiness Pills" to "National Nightmare": changing cultural assessment of minor tranquillizers in America, 1955–1980', *Journal of the History of Medicine and Allied Sciences*, 52 (1997): 338–376 (347); M.C. Smith, *Small Comfort: a history of the minor tranquillizers* (New York etc.: Praeger, 1985).
48 D. Herzberg, '"The Pill You Love Can Turn on You": feminism, tranquillizers, and the valium panic of the 1970s', *American Quarterly*, 58 (2006): 79–103 (79, 80).
49 HAR FE.0.4 – 101129 c: 'RRMG 1969', undated presentation, p. 5.
50 Peyer, *Roche*, p. 187.
51 HAR FE.0.4 – 102852: Untitled memorandum of Kurt Feinstein, 17.6.1969.
52 HAR FE.0.4 – 103593 e: 'New Research Functions and Activities', report submitted by Prof. A. Studer and Dr Ch. Polzer, 14.5.1968, table 2.

The Shift towards Molecular Targets

By the mid-1960s, researchers and managers at Roche became increasingly dissatisfied with the traditional approach to drug development, i.e. chemical synthesis and biological screening. A number of scientific reports submitted to the Roche Research Management Group (RRMG) prompted discussions among its members on whether the company should reorient its research activities and spend more time and money on fundamental biological research instead of the traditional chemical approach. Fundamental biological research was expected to provide detailed knowledge of normal and pathological biological mechanisms and offer new opportunities for drug development. The stronghold of organic chemistry was beginning to be challenged within the company.

Biological research on the mode of action of particular drugs was not completely lacking at Roche in the previous decades, but it was maintained on a very limited scale. Until the end of the 1950s, the RRMG rarely discussed the need to reinforce biological approaches, such as antiviral research, and generally considered them as either too expensive or unpromising for the development of new drugs. Only in selected therapeutic fields, such as the biochemistry of mental diseases, did the RRMG decide to make limited efforts in fundamental biological research, but merely in order to monitor the progress of academic research, a common strategy among pharmaceutical companies.[53]

The Roche management also remained unwilling to engage in biological drug manufacturing. The RRMG refused to carry out research on biosynthetic antibiotics because Roche had 'more experience in chemical manufacturing than in fermentation'.[54] Some researchers therefore considered it too hazardous to enter the 'biosynthetic approach with all the

53 HAR FE.0.4 – 101129 a: Minutes of the RRMG meetings 15.–23.10.1956, p. 4, 16, 17, 19, and 24.–30.4.1957, p. 2, 20.–25.10.1957, p. 13.

54 HAR FE.0.4 – 101129 a: Minutes of the RRMG meeting, 15.–23.10.1956, p. 9.

risks and pitfalls involved in fermentation work'.[55] Similarly, in the field of viral diseases, the RRMG preferred the chemotherapeutic to the vaccine approach, even though the latter was considered to be 'most likely to produce immediate results'; a decision probably motivated by the fact that vaccines could not be manufactured synthetically, but also because research on vaccines would have overstrained the 'present personnel and facilities'.[56]

In the 1960s, the RRMG began to evaluate the prospects of 'rational' approaches to drug development, a term that first appeared in the RRMG minutes in 1967.[57] The topic had been raised in 1966 when scientists from the Roche laboratories in Welwyn submitted three reports on the 'future pattern of research' to the RRMG demanding considerable investments in biochemical and biological research.[58] All three reports stressed that the classical approach to drug development, combining chemical synthesis and biological screening, was increasingly dissatisfying. If for certain diseases the screening of potential drugs on animals was relevant to predict their action in the human body, the reports emphasized that no reliable screening tests existed for other diseases, such as psychiatric disorders. Furthermore, the sheer number of already existing drugs lowered the chances of finding new drugs by trial-and-error approaches alone. At this point, the chemist M. W. Parkes, researcher at Roche Welwyn and author of one of the reports, concluded that

> the drug industry must consider its position and revise its pattern of research. Since drugs are going to cost so much more to find, we must learn how to find them more reliably. [...]

55 HAR FE.0.4 – 103593 a: J. Herrero, M. Montavon, W. Mosimann, W. Rehm, G.E. Werner to the members of the RRMG, 17.5.1966.

56 HAR FE.0.4 – 101129 a: Minutes of the RRMG meetings, 25.–30.4.1960, p. 10, and 13.–18.6.1963, p. 2.

57 HAR FE.0.4 – 101129 b: Minutes of the RRMG meeting, 18.–23.5.1967, typescript entitled 'RRMG 1967', p. 69.

58 All three reports are entitled 'Future pattern of research', undated and submitted to the RRMG on the occasion of its 1966 meeting held between June 14 and 18. The authors are Alexander L. Morrison, head of research at Roche Welwyn, M.W. Parkes and R. F. Long. The reports are delivered in HAR FR.0.4 – 103593 a.

> Instead of starting with synthesis, stemming from a chemist's ideas, work should start from biochemical and biological investigation, based on the fundamental studies of others, and leading to a clarification of what the proposed drug is required to do and how it could do it, preferably in biochemical terms.[59]

Parkes underlined that the future development of research in the pharmaceutical industry would bring along the 'decline of the chemist from his present leading position'.[60] As a consequence, the report from Roche Welwyn claimed that the company should devote an increasing proportion of its research expenditures to biology and biochemistry.[61]

Two years later, the RRMG received a report evaluating drug development at Roche Basel during the past decade.[62] The report aimed at 'obtaining suggestions for the further procedure in the development of new drugs'. From 20.000 compounds tested during the period under investigation, 181 had been submitted to preliminary clinical trials and only seven had finally been introduced as specialities. None of the 181 molecules had been detected by 'submission of basically new compounds to a fixed, stereotyped battery of pharmacologic and chemotherapeutic tests', i.e. random screening. Instead, the compounds were the result of what has been called 'programmed screening', namely the molecular manipulation of chemical structures already known to have pharmacological action. But while random screening had obviously not been effective at all, molecular manipulation was not likely to generate entirely new drugs either. Furthermore, the report stated that successes like the discovery of the benzodiazepines could hardly be repeated and should be considered as resulting from an 'especially fortunate' combination 'of top scientists [...] and a large amount of luck'. The report therefore recommended to intensify 'basic research including

59 HAR FE.0.4 – 103593 a: 'Future pattern of research', undated report submitted to the 1966 RRMG meeting by M.W. Parkes, p. 4.

60 HAR FE.0.4 – 103593 a: 'Future pattern of research', undated report submitted to the 1966 RRMG meeting by M.W. Parkes, p. 1.

61 HAR FE.0.4 – 103593 a: 'The Future pattern of research', undated report submitted to the 1966 RRMG meeting by R. F. Long, p. 3.

62 HAR FE.0.4 – 103593 e: 'New Research Functions and Activities', report submitted by Prof. A. Studer and Dr Ch. Polzer, 14.5.1968.

biochemistry, immunology and experimental pathology' rather than 'a further increase in random screening', at least in fields where screening results had been disappointing. Other reports submitted to the RRMG argued in a similar way, highlighting the limits of the traditional approach to drug development and advocating new strategies to develop drugs more efficiently.[63]

Roche was far from being the only company facing difficulties in bringing new drugs to market. In the 1960s, the wave of new pharmaceutical products levelled off in general.[64] Roche badly needed new marketable compounds because the company had a very unbalanced product range and its sales were largely due to a single category of drugs: tranquillizers. For a member of the RRMG in charge of the commercial aspects of research at Roche, this situation entailed 'the well-known risk that too many eggs lie in the same basket, and that all the eggs could drop out once the basket gets a hole'.[65] The Roche management also had to worry about rising public concerns regarding psychotropic drugs. In the United States in particular, tranquillizer addiction and abuse had become a publicly debated issue throughout the 1960s.[66] The RRMG decided that the problem had to be addressed in the near future.[67]

In spite of the various arguments in favour of fundamental biological research, not all members of the RRMG were convinced. The chemist Otto Isler for instance insisted that:

> [...] chemistry is at Roche's disposal just like a good wife, reliable, cooperative, maybe sometimes a little bit smiling with the knowledge that everything which had been realized came to a very great extent from her. As you all know, Roche-earnings are

63 See for instance HAR FE.0.4 – 103593 e: 'Trends in Drug Research', undated report submitted by Dr Hines to the 1968 RRMG meeting, p. 1.

64 Chandler, *Shaping the Industrial Century*, p. 34.

65 HAR FE.0.4 – 101129 c: 'RRMG 1969', undated presentation, p. 5; cf. also HAR FE.0.4 – 101129 b: Minutes of the RRMG meeting, 18.–23.5.1967, typescript entitled 'RRMG 1967', pp. 28–29.

66 Speaker, 'From "Happiness Pills" to "National Nightmare"', pp. 348–353.

67 HAR FE.0.4 – 101129 b: Minutes of the RRMG meeting, 5.–8.6.1968, p. 5.

> due to sales of substances in the form of specialities or bulk and seldom or never in the form of theories or speculation.[68]

Isler considered the new biological approaches to drug development to be perhaps of theoretical, but hardly of practical interest. He therefore concluded that the future successes of Roche should rely on what he called 'conventional research', namely organic synthesis combined with biological screening. He admitted, however, that chemists should cooperate more closely with those in charge of biological and medical research. The chemist Alexander Morrison, head of research at Roche Welwyn and member of the RRMG, expressed similar provisos regarding new approaches to drug development, but for different reasons. He was convinced that in the future the economic prosperity of Roche would depend more on the drugs the company was already selling, than on the development of entirely new compounds. He therefore addressed the RRMG asking for a concentrated effort on the improvement of synthetic manufacturing processes in order to lower production costs and maintain the company's competitiveness.[69] However, Isler's metaphor of Roche and chemistry as husband and wife indicates that the position of chemistry at Roche had become clearly challenged – like the traditional marital model that had just come under fire in the counter-culture movements of the late 1960s.[70]

The field of antiviral research illustrates Roche's changing research priorities in the 1950s and 1960s. The development of an antiviral drug was pursued by Roche because its potential markets seemed even larger than those for antibacterial sulphonamides in which Roche was already heavily

68 HAR FE.0.4 – 103593 h: 'Continuity in Chemistry Research', memorandum submitted by Otto Isler, 5.6.1969, p. 1.

69 HAR FE.0.4 – 103593 b: 'Pattern of Future Research Activities in Welwyn', report submitted by Alexander Morrison, 9.5.1967, p. 3.

70 A. Marwick, *The Sixties: cultural revolution in Britain, France, Italy, and the United States, c.1958–c.1974* (Oxford: Oxford University Press, 1998), pp. 381–403; for Switzerland see E. Joris and H. Witzig (eds), *Frauengeschichte(n). Dokumente aus zwei Jahrhunderten zur Situation der Frauen in der Schweiz* (Zürich: Limmat Verlag, 1991), pp. 473–483.

involved.[71] In the 1950s, Roche had a modest antiviral research program that consisted in screening compounds randomly against a small number of viruses in tissue cultures and *in vivo*. However, between 1958 and 1967, the Basel research department only succeeded in preparing one antiviral compound for preliminary clinical trials, which finally did not show any satisfactory results.[72]

Because of this failure, the medical doctor and chemist Alfred Pletscher,[73] who had succeeded Placidus A. Plattner as managing director in charge of research and chairman of the RRMG, called for a 'new approach' to antiviral research.[74] Researchers at Roche repeatedly discussed the potential of the vaccine approach against viral diseases. But it would have entailed major investments or cooperation with a firm already well established in the vaccine business, since Roche did not have any experience in that field.[75] Virginius D. Mattia, a physician by training and president of Roche Nutley, outlined another antiviral approach, proposing that research should focus on viral nucleic acids.[76] As a result, Nutley formed a biochemical virology section within the department of chemotherapy, devoted to investigating the mechanisms of viral RNA replication in the host cell.[77] It was hoped that the results of this research would make it possible to develop a drug interfering with this mechanism. The rationale behind this research program on the molecular mechanism of viral infection was thus to identify appropriate molecular targets for an antiviral drug.

71 HAR FE.0.4 – 101129 a: Minutes of the RRMG meeting, 15.–23.10.1956, p. 16 and 25.–30.4.1960, p. 10.

72 HAR FE.0.4 – 103593 e: 'New Research Functions and Activities', report submitted by Prof. A. Studer and Dr Ch. Polzer, 14.5.1968, table 1.

73 R. M. Kunz, 'Alfred Pletscher', Experientia 44, 2 (1988), 91–93. (Title not mentioned before)

74 HAR FE.0.4 – 101129 b: 'RRMG Meeting 1967/(Notizen A. Pletscher)', 30.5.1967, p. 6.

75 HAR FE.0.4 – 103593 h: 'Antivirals', report submitted by Mr. Mosimann, 5.5.1969, p. 2; HAR FE.0.4 – 101129 a: Minutes of the RRMG meeting, 25.–30.4.1960, p. 10.

76 Peyer, *Roche*, p. 225.

77 HAR FE.0.4 – 103593 f: 'New Approaches to Antiviral Chemotherapy', report submitted by Roche Nutley, 21.5.1970.

In 1971, Roche Nutley organized an interdisciplinary meeting on antiviral and anticancer research. In his opening remarks, John J. Burns, biochemist and head of research at Roche Nutley, reaffirmed the need for a new approach to antiviral therapy, which Pletscher had mentioned a few years earlier but which had not yet been implemented systematically:

> It is evident, that random screening for antiviral and anticancer agents is a worthwhile approach and should not be abandoned. However, simultaneous basic research at the molecular level is indispensable if we are to comprehend the complexities of the biochemical and biological phenomena and use the acquired fundamental knowledge to design a rational program for antiviral and anticancer research.[78]

The participants at the meeting discussed viral RNA replication in the host cell, and other possible targets for antiviral drugs, such as virus penetration, virus assembly, or its release from the host cell. These targets began to be investigated by several research teams in Basel and Nutley in order to 'set up carefully controlled *in vitro* systems to measure the effect of chemical agents on any one or all of the steps involved in viral replication'.[79] A group of researchers in Basel, for example, started searching for an inhibitor of neuraminidase, i.e. an enzyme that plays a role in the virus' release from the host cell. It was assumed that finding a compound that inhibits the enzyme could stop the virus' proliferation in the infected organism.[80] It took, however, nearly three more decades before the first antiviral neuraminidase inhibitor was marketed. In 1999, GlaxoSmithKline and Roche both launched antiviral drugs that block the viral neuraminidase. Roche had not developed the compound in-house; instead it had purchased it from the biotechnology company Gilead Sciences Inc.[81] Even though Roche never developed its own neuraminidase inhibitor, this example illustrates how,

78 HAR FE.0.4 – 103593 i: 'Interdisciplinary Research Meeting. Antiviral Research, Anticancer Research', report submitted by Roche Nutley, 21.5.1971, introduction.

79 HAR FE.0.4 – 103593 i: 'Interdisciplinary Research Meeting. Antiviral Research, Anticancer Research', report submitted by Roche Nutley, 21.5.1971, p. 2.

80 HAR FE.0.4 – 103593 k: 'New Approaches to Antiviral Chemotherapy', report submitted by Mr. Mosimann, 5.5.1972, p. 2.

81 Cf. http://www.gilead.com

by the end of the 1960s, fundamental biological research had begun to play an increasingly important role for the company's new drug development strategies. This example also shows that while fundamental biological research was always viewed as an alternative to previous strategies, such as random screening, it was also seen as a way to reinforce them, for example by allowing the development of new test systems to assess the effects of drugs on specific molecular targets.

Raising Drug Safety Issues and the Need for Biology

In addition to the dissatisfaction with the effectiveness of conventional approaches to drug development, there was yet another reason why biological research became important at Roche during the 1960s. By 1963, the recently tightened drug approval policies in the United States provoked discussions among Roche managers and researchers on whether the company required more profound knowledge on the biological activity of drugs; knowledge that could neither be provided by synthetic chemistry, nor by animal testing alone.

In late 1962, the US Congress had passed the Kefauver-Harris Drug Amendments to ensure drug efficacy and greater drug safety. In the previous year, the German-based company Grünenthal had had to withdraw its thalidomide sleeping aid Contergan because of the strong suspicion that it was responsible for an alarming increase in malformations among newborn children in Europe. The FDA had not yet approved thalidomide, but the American company licensing the compound had already distributed the drug to over one thousand medical doctors for clinical investigations. Under the existing FDA laws and regulations, prescription drugs distributed for clinical investigations did not require FDA approval. The new Drug Amendments of 1962 considerably strengthened the control of clinical trials. From then on, FDA approval was necessary even for testing new compounds in clinical investigations. In early 1963, the FDA announced new regulations that obliged a company to provide the agency with comprehensive information about compounds before testing them in clinical trials.[82] The Roche Research Management Group stated in June 1963 that these new requirements made it 'very difficult to bring new drugs to the first clinical trials' in the United States. Furthermore, the allegations against Contergan had been confirmed in 1962 and drug safety was becoming a matter of growing public concern. In view of this 'increased sensitivity to toxicological problems', the RRMG wanted the company to respond with a policy of 'frankness including informative action', 'defensive measures' and 'fundamental research'.[83]

What kind of fundamental research did the Roche managers have in mind? The 1962 FDA legislation had its most significant impact on clinical investigations. Under the new regulations, clinical trials with new drugs had to be scientifically planned by qualified investigators and reported in detail to the FDA.[84] But the new Drug Amendments also affected the pharmaceutical industries' pre-clinical testing practices.[85] Besides toxicological data, the FDA now also required information on drug metabolites (i.e. the small molecules resulting from the drug's breakdown in the organism) and

82 R.E. McFadyen, 'Thalidomide in America: a brush with tragedy', *Clio Medica*, 11 (1976): 79–93; G.P. Larrick, 'How the Food and Drug Administration evaluates New Drug Applications', *The Journal of new Drugs*, 4 (1964): 63–74.

83 HAR FE.0.4 – 101129 a: Minutes of the RRMG meeting, 13.–18.6.1963, p. 41. In October 1962, the Swiss regulatory board IKS also planned to tighten its regulations, but the minutes of the 1963 RRMG meeting do not mention this initiative explicitly. See also CIBA Firmenarchiv Vg 1.01 Verwaltung: Minutes of the 'Pharma Geschäftsausschuss', No. 39, 25.10.1962, paragraph 3.

84 A.A. Daemmrich, *Pharmacopolitics: drug regulation in the United States and Germany* (Chapel Hill and London: University of North Carolina Press, 2004); cf. also G.P. Larrick, 'New Federal Regulations for the Control of New Drug Testing in Humans', *The Journal of new Drugs*, 2 (1962): 373–384.

85 Daemmrich, *Pharmacopolitics*, p. 27. Laboratory research on a drug's toxicological effects had, of course, already been part of drug development before 1963, and had been required by the FDA for market approval since 1938. See H.M. Marks, *The Progress of Experiment: science and therapeutic reforms in the United States, 1900–1990* (Cambridge: Cambridge University Press, 1997), pp. 71–97.

on the drug's interaction with other chemicals.[86] The RRMG discussed these requirements taking into account a number of internal scientific reports on that issue. In 1966, a report submitted by Roche Welwyn to the RRMG underlined that it was becoming increasingly difficult to bring a new drug on the market, since many countries were requiring 'detailed information on the biological activity, both desirable and undesirable, of all new compounds before they can be tested clinically'.[87] Such information could not be provided on the basis of chemical synthesis and biological screening alone. The authors of the reports therefore requested that the company should 'intensify fundamental research' as well as increase the 'proportion of pharmaceutical effort devoted to biology'.[88] Two years later, another report concluded that to produce merely 'a new compound that had a desirable pharmacological action in man' was no longer sufficient anymore. Instead, more knowledge would be needed 'about the mechanism of action of the drug' and 'a reasonable understanding of its metabolism'[89] – knowledge that had been completely irrelevant for the development of earlier Roche products, such as the above-mentioned benzodiazepines.

The fact that research on drug metabolism had become more important after the thalidomide tragedy, is further illustrated by the way Roche's local competitor Ciba reacted to the withdrawal of Contergan. Four days after Grünenthal had to stop selling its sleeping aid, the management of the pharmaceutical division at Ciba discussed the consequences this had for the company. In particular, the managerial board decided to prepare the defence of the Ciba speciality Doriden. Like Contergan, this compound was used as a sleeping aid and was suspected of having similar effects on the human foetus. But the Ciba management was hoping that its own drug

86 Larrick, 'How the Food and Drug Administration Evaluates New Drug Applications', p. 69.

87 HAR FR.0.4 – 103593 a: 'Future pattern of research', undated report submitted by Alexander L. Morrison to the 1966 RRMG meeting, p. 1.

88 HAR FE.0.4 – 103593 a: 'Future pattern of research' undated report submitted by R. F. Long to the 1966 RRMG meeting, p. 3.

89 HAR FE.0.4 – 103593 e: 'Trends in Drug Research', undated report submitted by Dr Hines to the 1968 RRMG meeting, p. 2, cf. also HAR FE.0.4 – 101129 b: Minutes of the RRMG meeting, 5.–8.6.1968, p. 2 and 4.

could eventually step into the market gap opened by the Contergan affair. This hope was based on a study exploring the metabolism of Doriden, which was carried out by the biochemist Karl Bernhard, professor at the university of Basel, and two of his collaborators working at Ciba.[90] In their study, they concluded that because of the rapid and complete elimination of its metabolites, Doriden displayed no side- or after-effects and had a low toxicity. In view of these results, the authors were convinced that metabolic studies, i.e. studies on the biological mechanism of action of drugs, would be increasingly important for drug development:

> The general trend towards greater safety and reliability in human medical therapy has created the need for more information on the mode of action and fate of drugs in the organism. There is now reason to hope that by means of detailed metabolic studies it will often be possible to discover in good time those harmful side-effects of a preparation which do not appear until the patient has been under treatment for some considerable period and which in many instances are due to a cumulative action or to the building up of toxic metabolites.[91]

The most significant consequences of the new FDA regulation on pre-clinical research at Roche were, however, the indirect result of their economic impact. In view of the tightened FDA approval policies for clinical trials, the Roche research managers concluded that the company needed more reliable pre-clinical research strategies in order to avoid costly clinical investigations for compounds with an uncertain potential. In the above-mentioned reports submitted to the RRMG in 1966, Alexander Morrison, head of research at Roche Welwyn, stated that in many cases *in vitro* and *in vivo* tests could give a reliable guide to the efficacy of the drug in humans. But in other cases, such as rheumatic or psychiatric diseases, researchers would not know whether 'the pharmacological test procedures bear any

90 CIBA Firmenarchiv Vg 1.01 Verwaltung: Minutes of the 'Pharma Geschäftsausschuss', No. 12, 28.11.1961, paragraph 1.

91 H. Keberle, K. Hoffmann, K. Bernhard, 'Metabolism of Glutethimide (Doriden)', *Experientia*, 18 (1962): 105–111 (105, 110).

relationship to the clinical condition'.[92] The Welwyn researchers therefore requested that the company intensify fundamental biological research, hoping this would increase the chances that a new chemical compound would actually work in humans. The stakes had simply grown 'too high to gamble on the selection from compounds virtually randomly submitted to tests of doubtful relevance'.[93] The RRMG agreed and approved the plans of Roche Welwyn to enlarge its biological research facilities.[94]

Public concerns about drug safety and the strong reaction of the US regulatory authorities resulting in new economic constraints thus constituted another incentive for Roche to reorient its scientific tradition focused on organic synthesis, and to devote an increasing part of its resources to biological research.

The Roche Institute of Molecular Biology and New Relationships with Academia

The turn toward biological targeting also manifested itself in significant institutional changes at Roche. In 1967, the general management in Basel approved plans of Roche Nutley to build the Roche Institute of Molecular Biology (RIMB), an institution entirely devoted to fundamental biological research. In the following year, Roche built a similar institution in Switzerland, the Basel Institute for Immunology (BII).[95] Even though both institutions counted for only a relatively moderate part of the company's

92 HAR FR.0.4 – 103593 a: 'Future pattern of research', undated report submitted by Alexander L. Morrison to the 1966 RRMG meeting, p. 1 and 2.

93 HAR FE.0.4 – 103593 a: 'Future pattern of research', undated report submitted to the 1966 RRMG meeting by M.W. Parkes, p. 4.

94 HAR FE.0.4 – 101129 b: Untitled memorandum of Kurt Feinstein, 21.6.1966, p. 1 and 2; Thomas et al., 'Fifty years of Welwyn research'.

95 Pletscher, '25 Jahre Roche-Forschung', see p. 46.

rising research expenditures,[96] they represented the introduction of a significantly different institutional culture. Indeed, these institutes offered academic working conditions where researchers were free to choose their research topics. This focus on fundamental, rather than applied, biological research, and the granting of academic working conditions, was perceived as a means to attract new scientific staff and develop close collaborations with biological researchers working in academia. (Fig. 14.2.)

Figure 14.2: John J. Burns, Virginius D. Mattia and Sidney Udenfriend, groundbreaking ceremony for the RIMB, 1968. © Historisches Archiv Roche.

96 According to a cost projection submitted to the RRMG in 1971, total research expenditures in Europe and Japan were supposed to rise from 60 million Swiss francs in 1967 to more than 270 million in 1976, whereof only 5.5 per cent were to be spent for the BII. Both institutions, the BII and the RIMB, employed approximately 200 persons by 1975. HAR FE.0.4 – 102852: 'Forschung: Fünfjahresvorschau', September 1971, and Peyer, *Roche*, p. 226.

The Roche Research Management Group had already discussed the establishment of 'an institution devoted to the study of one or more basic topics' in 1963.[97] Even though researchers at Roche had always been given a certain amount of opportunity to investigate problems of basic interest, no particular institutional setting had been created for this purpose so far.[98] Presumably, the discussions taking place at the RRMG meeting in 1963 had been influenced by the fact that, just a few weeks earlier, the local competitor Ciba had opened the 'Woodward Forschungsinstitut' in Basel. This institute consisted of a few laboratories housed in an industrial research building at Ciba. It was directed by Robert Woodward, a chemistry professor at Harvard University who was awarded the Nobel Price in 1965 for his work on the organic synthesis of natural products. Four to five times a year, Woodward came to Basel and held workshops with his collaborators at the Ciba institute. He also assumed the role of a scientific advisor for Ciba. Furthermore, the members of the institute were hired for a limited period only, much like postdoctoral fellows at American universities, and some of them eventually took over academic positions when they left Ciba.[99] But in contrast to the RIMB, the Woodward Institute clearly focused on chemical rather than biological research.

Four years after the first discussion had taken place within the RRMG, the project was finally realized after biochemists Sidney Udenfriend and Herbert Weissbach discussed plans to build a Roche institute for fundamental biological research in the United States with John J. Burns, head of research at Roche Nutley, in March and April 1967. They had known each other since the 1950s when working in the same laboratory at the NIH. In 1966 Burns had been appointed head of research at Roche Nutley in order to strengthen biological research within the company. The plans for the new institute were based on a rejected grant proposal for a Molecular

97 HAR FE.0.4 – 101129 a: Minutes of the RRMG meeting, 13.–18.6.1963, p. 41.

98 Cf. HAR FE.0.4 – 101129 b: Minutes of the RRMG meeting, 18.–23.5.1967, typescript entitled 'RRMG 1967', p. 31.

99 'Das Woodward Forschungsinstitut', special issue of *CIBA-Blätter*, 202 (March–April 1966); CIBA Firmenarchiv Vg 1.01: Minutes of the 'Direktionskomitee', No. 807, 4.6.1963, paragraph 7.

Pharmacology Institute at St. Louis University that Weissbach had submitted to the NIH in early 1967.[100]

Only a few weeks later, the company decided to build the RIMB, and the Roche management signed a charter stipulating that the institute would be 'wholly devoted to long range basic research designed to shed light on fundamental life processes'.[101] A number of early research projects at the RIMB were unrelated to any particular human diseases, such as those on the virus lambda, whose only victims are bacteria. Other projects aimed at understanding the basic mechanisms of protein synthesis and genetic regulation, or at determining the characteristics of biological macromolecules.[102] For example, members of the institute isolated and determined the amino acid sequences of neuropeptides, i.e. proteins, which play a role in the nervous system.[103] Other projects focused on interferon, i.e. a protein, which was expected to be useful in antiviral and anticancer therapy. This eventually led to the production of recombinant human interferon, Roche's first drug manufactured by genetically modified bacteria.[104] (Fig. 14.3)

Other industrial companies had created institutions for fundamental research as early as the inter-war period, among them the chemical company Du Pont or the pharmaceutical companies Merck and Squibb. By establishing the new facilities, these firms had hoped to increase the chances of recruiting first-rate academic scientists and to improve their relations with

100 H. Weissbach, 'Reflections on the Roche Institute of Molecular Biology after 20 years', in J. Drews and F. Melchers (eds), *Research at Roche* (Basel: Editions Roche, 1989), pp. 231–259; S. Udenfriend, 'A Short History of the Roche Institute of Molecular Biology', *BioEssays*, 7 (1987): 278–280.

101 Weissbach, 'Reflections on the Roche Institute of Molecular Biology', p. 238.

102 Roche Institute of Molecular Biology, *Annual Report*, 1969.

103 S. Udenfriend, 'Neuropeptides, A Personal History', in Rao S. Rapaka and N. Dhawan Bhola (eds), *Opioid Peptides: an update* (NIDA Research Monograph Series 87) (Rockville: National Institute on Drug Abuse, 1987), pp. 1–9.

104 Weissbach, 'Reflections on the Roche Institute of Molecular Biology', pp. 253–254; see also Roche Institute of Molecular Biology, *Annual Report*. 1970, p. 5, HAR FE.0.4 – 103593 i: 'Interdisciplinary Research Meeting. Antiviral Research, Anticancer Research', report submitted by Roche Nutley, 21.5.1971, and HAR FE.0.4 – 101129 e: Minutes of the RRMG meeting, 16.–18.6.1977, p. 18.

academia.[105] Similar arguments were voiced within the Roche management in favour of building the RIMB. The Roche Research Management Group estimated that the 'higher the standard of research' at Roche, the easier the recruitment of first-class workers.[106] However, in Switzerland, it was difficult to find researchers or research institutions to collaborate with in the first place, since the modernization of biology in the universities was just beginning to take place,[107] while in the United States, molecular biologists seemed little inclined to work for industry. Indeed, Herbert Weissbach identified a 'stigma of working for industry' among his academic colleagues.[108] Furthermore, working conditions for biological researchers at American universities were fairly attractive due to generous federal funding.

Roche therefore tried to make its new research institution as attractive as possible to academic biologists by granting working conditions similar to those prevailing in academia, and the company management thereby accepted features which were entirely 'foreign to the Roche organization'.[109] According to the charter signed by the Roche management, the RIMB scientists enjoyed 'independence in their choice and pursuit of research prob-

105 W.B. Carlson, 'Innovation and the Modern Corporation: from heroic invention to industrial science', in J. Krige and D. Pestre (eds), *Companion to Science in the Twentieth Century* (London and New York: Routledge, 2003), pp. 203–226 (218); Swann, *Academic Scientists and the Pharmaceutical Industry*, pp. 41–49; D.A. Hounshell, 'Continuity and Change in the Management of Industrial Research: the Du Pont Company, 1902–1980', in G. Dosi et al (eds), *Technology and Enterprise in a Historical Perspective* (Oxford: Clarendon Press, 1992), pp. 231–260.

106 HAR FE.0.4 – 101129 b: Minutes of the RRMG meeting, 18.–23.5.1967, typescript entitled 'RRMG 1967', p. 29.

107 HAR FE.0.4 – 101129 b: Minutes of the RRMG meeting, 18.–23.5.1967, typescript entitled 'RRMG 1967', p. 30, and 'Roche Forschung', an interview with Alfred Pletscher. *Roche Zeitung*, 1 (1968), pp. 43–48 (44); cf. also B.J. Strasser, *La Fabrique d'une nouvelle science: la biologie moléculaire à l'âge atomique (1945–1964)* (Florence: Olschki, 2006); M. Bürgi, 'Pharma, Politik und Polypeptide – Die Institutionalisierung der Molekularbiologie in Zürich, 1962–1971', *Traverse*, 3 (2005): 126–139; N. Stettler, *Natur erforschen. Perspektiven einer Kulturgeschichte der Biowissenschaft an Schweizer Universitäten 1945–1975* (Zürich: Chronos, 2002), pp. 181–192.

108 R.J. Bazell, 'Molecular Biology, "Corporate Citizenship and Potential Profit"'. *Science*, 174 (1971): 275–276 (275).

109 Weissbach, 'Reflections on the Roche Institute of Molecular Biology', pp. 242–243.

lems'. Their independence was also guaranteed architecturally. The institute was located in its own building, off the main industrial and research site, with a separate gate and roadway, and 24-hour access. Flexible work time, common in academic molecular biology laboratories, was also new to the Roche management. In addition, many of the Institute's members held formal appointments on nearby university faculties, such as Colombia University or City College of New York , and some served as executive editors of peer reviewed scientific journals such as *Molecular and Cellular Biology*. Even when advertising in the *New York Times* for the position of secretary to the new director, the RIMB stated it would 'give preference to applicants with work experience in an academic environment'.[110]

Figure 14.3: Sidney Udenfriend, John J. Burns, Robert B. Clark and Renato Dulbecco behind a model illustrating the molecular mechanism of protein synthesis in the cell, inauguration of the new RIMB building in 1971. © Historisches Archiv Roche.

110 Classified advertisement, New York Times (16 June 1968), p. W17.

The RIMB also offered a postgraduate education program and held international symposia on various topics of molecular biology. The second symposium organized in 1973, for example, was devoted to 'Genetic Recombination'.[111]

The strategy of creating an academic culture within the company seemed to have the desired effect. In 1969, Nutley reported to the RRMG:

> Excellent relations to academic institutions have been established. The academic world has received the whole project in such a good manner that an unexpected large number of distinguished scientists have applied for a job.[112]

In 1971, 128 scientists worked at the RIMB, and this number climbed up to 200 by 1975, many of them as postdoctoral fellows. According to the RRMG, it was the only industrial institution taking part in the 'Exchange-Visa-Program' of the US State Department intended for academic researchers.[113] Eventually, the institute hosted the internationally renowned scientist and Nobel Laureate Severo Ochoa. He came in 1974 and directed a research program at the RIMB until 1985. By the mid 1980s, the RIMB had seven members of the US National Academy of Science on its staff.[114]

In one instance, the RIMB even blurred the borders between Roche and the academic biological community to such a degree that it resulted in a lawsuit. In 1980, the University of California accused Roche of exploiting a cell line sampled from the bones of a patient in a Los Angeles hospital. This cell line had been important for the isolation of the interferon gene that enabled the biotech company Genentech in cooperation with the RIMB and the Roche industrial laboratories to produce the human recombinant interferon mentioned previously. In the early 1980s it was estimated that the annual sales of this protein would reach 3 billion US

111 Roche Institute of Molecular Biology, *Annual Report*, 1969 et seqq.

112 HAR FE.0.4 – 101129 c: Minutes of the RRMG meeting, 1969, typescript entitled 'RRMG 1969', p. 5.

113 HAR FE.0.4 – 101129 c: Minutes of the RRMG meeting, 11.–16.6.1970, p. 2 and Minutes of the RRMG meeting, 10.–15.6.1971, p. 5; Peyer, *Roche*, p. 226.

114 Weissbach, 'Reflections on the Roche Institute of Molecular Biology', p. 248; Udenfriend, 'A Short History', p. 280.

dollars in forthcoming years. Even though the University of California had not patented the cell line, it claimed that 'Roche "subverted" for profit the academic relationship' through which the cellular material was freely passed over to the RIMB. Robert Gallo, working for the US National Cancer Institute had given the cell line to Sydney Pestka who was in charge of interferon research at the RIMB. When the conflict between the University of California and Roche broke out, Gallo testified that he had regarded Pestka 'as a friend and researcher working at a non-profit institution, not as a corporation scientist'.[115]

Even though the RIMB was granted academic liberties, its researchers did not work in complete isolation from the rest of the company. At first, interactions between the institute's staff and the company researchers were only limited. Arnold Brossi, head of Chemistry at Roche Nutley, was rather sceptical about the usefulness of the RIMB. He seems to have been convinced that 'Roche was a successful company that had built its reputation primarily on the strength of its chemical division' and therefore 'questioned why the company should make a major financial commitment to an entirely new field that had no apparent relevance or products in mind'.[116] The senior researchers at Nutley were especially unconvinced about the RIMB. Nevertheless, closer contacts between the RIMB and the other research laboratories of Roche were eventually established, and a number of common research programs were launched.[117] To what extent Roche finally benefited from the research activities pursued at the RIMB cannot be evaluated here. But clearly, the new institute represented an important step in Roche's transition towards biological research strategies, allowing for frequent contacts between faculty in academia and industrial scientists at Roche, and towards the creation of collaborative networks in the biological sciences, similar to those which had existed for decades in the field of chemistry. The creation of the RIMB in 1967 thus contributed

115 N. Wade, 'University and Drug Firm Battle over Billion-Dollar Gene', *Science*, 209 (1980): 1492–1494.

116 Weissbach, 'Reflections on the Roche Institute of Molecular Biology', pp. 245–246.

117 HAR FE.0.4 – 101129 c: Minutes of the RRMB meeting, 11.–16.6.1970, p. 2; cf. also the annual reports of the Roche Institute of Molecular Biology.

to blurring the boundaries between academic and industrial research which have become a dominant feature of contemporary biotechnology.[118]

'Long-Range Planning'

The shift towards more fundamental biological research at Roche was not only the result of a perceived need by the Roche management for new knowledge to develop drugs, to answer regulatory needs or to connect with academic researchers. This shift also took place within a broader change in management culture at the Roche Research Management Group. Indeed, the RRMG was widening its time horizon and was consequently increasingly ready to consider investments in research whose outcome was hardly predictable. New theories of management that were emerging at the same time further reinforced this changing emphasis in research planning at Roche.

In the late 1950s, the opening remarks of the RRMG meetings still reflected a clear emphasis on short-term issues and the need to achieve rapidly commercially valuable research results. This urgency was partly due to the fact that the profits from vitamins had decreased after the Second World War and that Roche's growth rate threatened to fall behind that of its competitors.[119] On the other hand, Roche's emphasis on short-term research goals was a quite common strategy within the pharmaceutical industry at the time. According to an economic survey published in 1965 and relying on interviews with managers of pharmaceutical companies in the United States, research in this sector was not undertaken 'merely with the hope of

118 R. Loeppky, *Encoding Capital: the political economy of the Human Genome Project* (New York and London: Routlege, 2005), pp. 149–160.

119 See HAR FE.0.4 – 101129 a: Minutes of the RRMG meetings, 15.–23.10.1956, p. 1, and 24.–30.4.1957, add. to p. 1. The prize for 1 kilogram Vitamin C declined from 1140.– Swiss francs in 1936 to 102.– Swiss francs in 1950. Peyer, *Roche*, p. 175, 180 and 205.

some distant but unknown returns, but rather with the expectation that profitable gains will accrue within a reasonable period of time'.[120]

But this perspective was about to change. In 1960s, Placidus A. Plattner opened the RRMG annual meeting by saying:

> Since we are now in the fortunate position of having several promising substances as candidates for launching, the Group is able to devote more of its time to consideration of long range aspects.[121]

Plattner was referring to Roche's new benzodiazepine tranquilizer Librium; now that new drugs were ready for market, he considered it important to spend more time discussing long-term agendas.[122] The financial situation of the company further improved throughout the following decade, and by 1967 Roche had become the world's largest pharmaceutical company in terms of sales.[123] This favourable economic position made it possible to integrate long-term planning with the management of scientific research within the company, and by the mid 1960s, Roche Nutley had established an independent committee in charge of 'long-range planning' within its research department.

According to the editorial of the first issue of the eponymous journal, 'long-range planning' was neither a single technique nor a specific managerial strategy, but was aimed at 'developing attitudes, processes, and perspectives which make planning possible'.[124]

120 W.S. Comanor, 'Research and technical change in the pharmaceutical industry', *The Review of Economics and Statistics*, 47 (1965): 182–190 (190).
121 HAR FE.0.4 – 101129 a: Minutes of the RRMG meeting, 25.–30.4.1960, p. 1.
122 Cf. also HAR FE.0.4 – 101129 a: Placidus A. Plattner to the Roche Head Office, 9.5.1960.
123 Peyer, *Roche*, p. 213.
124 R. Perrin, 'Long Range Planning: the concept and the need', *Long Range Planning*, 1 (1968): 3–6 (5).

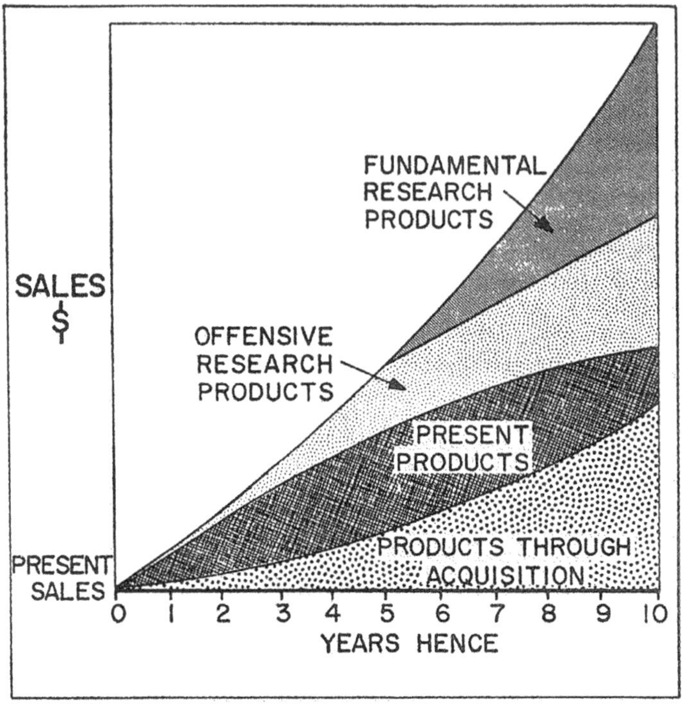

FIG. 1

Figure 14.4: Chart from a management guide for long-range research planning illustrating the need for fundamental research programs in industry, 1964.[125]

As far as managerial attitudes were concerned, the concept of 'long-range planning' was first of all a plea to take entrepreneurial risks, i.e. to plan the future even though it could never be properly predicted. Up to that point, unpredictability had been one of the key reasons why management had refrained from tackling issues beyond short-term planning.[126]

125 J.B. Quinn, 'Top-Management guides for Research Planning', in D.W. Ewing (ed.), *Long Range Planning for Management* (York etc.: Harper & Row, 1964), pp. 262–296 (385).

126 P.F. Drucker, 'Long Range Planning Means Risk-Taking', in Ewing (ed.), *Long Range Planning for Management*, pp. 7–20. On Drucker and management science, see S.P.

A first comprehensive reader for managers on 'long-range planning' had appeared in 1958. The second edition, published in 1964, contained an article on 'Top-Management Guides for Research Planning' written by James B. Quinn, professor of Business Administration at Dartmouth College. According to Quinn, the main purpose of research planning was to define the kind of research a company needed in order to achieve its long-term goals. He distinguished among three different kinds of research projects. In the first two instances, they could aim at defending products a company was already selling, or they could serve to develop foreseeable new products. Both of these approaches could be carried out with a high probability of success, but they might not enable the company to meet its economic goals. In order to do so, a company might need a third 'technology, which develops entirely new applications beyond the scope of those presently foreseen'. In these cases, Quinn argued, a fundamental research program was necessary even though its outcome could hardly be predicted.[127] The plea to take risks, which was a constitutive element of the concept of 'long-range planning', was therefore also an argument in favour of fundamental research (Fig. 14.4). The RRMG members learned more about this new management perspective in 1967, when an expert from Roche Nutley presented the concept of 'long-range planning' in more detail and distributed scientific papers from the Harvard Business School on that subject.[128] By that time, a number of corporations had already established formal long-range planning departments, and the Roche management itself considered the company's rising investments in fundamental research well in tune with this general trend towards long-term planning in corporate management.[129]

Waring, *Taylorism Transformed: Scientifc Management Theory since 1945* (Chapel Hill and London: University of North Carolina Press, 1991).

127 Quinn, 'Top-Management Guides for Research Planning', p. 386.

128 HAR FE.0.4 – 101129 b: Minutes of the RRMG meeting, 18.–23.5.1967, typescript entitled 'RRMG 1967', pp. 67–68.

129 'Auszug aus der Präsdidialadresse von Dr A.W. Jann, 10. Juni 1966'. *Roche Nachrichten* No. 3 (1966), without pagination; cf. also L.R. Michel, 'Long range Planning for Management', *Management Science*, 11, Series A, Sciences (1964): 210–211.

Conclusion

In his recent book on the history of twentieth century pharmaceutical industry, business historian Alfred D. Chandler Jr. traces Roche's position today back to what he calls the company's 'aggressive entry into the new biotechnology'[130] in the 1980s. Our investigation into the Roche Research Management Group (RRMG) during the two previous decades allows us to qualify this picture. The RRMG appears as an institutionalized place within the company where, even in the 1960s, the potentials of different scientific approaches to drug development were evaluated, legal constraints were interpreted, and future technological as well as institutional needs were determined. In this context, Roche decided to reorient its research priorities towards increased fundamental biological research. Thus, when Roche entered the new biotechnology in the 1980s, whether 'aggressively' or not, it had already undergone a major scientific and institutional transition towards a biological mode of drug innovation.

In this paper, we have tried to demonstrate how and why the management and the researchers at Roche began reassessing the company's traditional approach to drug development based on organic synthesis and a reliance on close relationships with chemistry departments in academia. We have argued that four main factors explain why the company adopted a new approach to drug development, aimed at understanding the molecular mechanisms of drug action and identifying the precise biological targets of drugs. First, the previous chemical approaches were criticized by managers and scientists because of decreasing results, opening the door for claims that fundamental biological knowledge would allow for more productive modes of drug development. Second, the tightened FDA drug approval policies required more profound pre-clinical knowledge on the mechanism of action of drugs, which could be provided by biological research. In addition, the increased costs of clinical trials made it even more important to carefully evaluate the chances of success at an early pre-clinical stage. Third, given

130 Chandler, *Shaping the Industrial Century*, p. 258.

that Roche had decided to emphasize biological research, it focused particularly on fundamental research as a means to attract academic scientists and overcome its isolation from the academic biological research community. And finally, having adopted the new principles of long-range planning and accepted the risks inherent to planning an unpredictable future, the management was ready to consider investments in fundamental biological research even though its economic value was still unforeseeable.

The link between the transition undergone by Roche in the 1960s and the biotechnology era still remains to be explored. How important, for example, was Roche's new in-house biological expertise when recombinant DNA technologies became available and opportunities emerged to collaborate with biotech start-ups? The RIMB might have played an important role in that respect. Roche monitored advances in molecular genetics attentively through its researchers at the RIMB and their close contacts with the academic molecular biology community. In the late 1970s, the RIMB also served as an interface between Roche and the new biotech start-ups such as Genentech, with which Roche scientists collaborated for the isolation of the interferon gene, before Roche acquired Genentech in 1990.

Besides the links to contemporary biotechnology, another area which requires more investigation is clinical research. In retrospect it has become clear that the new FDA laws and regulations of the early 1960s had a major impact on clinical research. It is therefore surprising how little the Roche management and scientists discussed a reconfiguration of the company's relationships with clinics and physicians – at least in the sample of sources this article is based upon. In the discussions we outlined, improving clinical research was not considered as an alternative way to overcome problems in developing new marketable compounds. More archival research would be needed in order to evaluate what importance clinical research had for the Roche management when it was confronted with the various problems outlined in this article.

Other important aspects still need further investigation. As far as Roche is concerned, more attention could be paid to the impact of the new generation of managers on the company's research strategies, and in particular of the American managers who started to fill top positions traditionally occupied by Swiss executives. Such a trans-national perspective could be further explored by a comparison of Roche with other companies, for instance the Swiss-based Ciba, which was considerably less engaged in the United States than Roche until the 1970s, or the large American pharmaceutical companies which had acquired more experience with biological research than had Swiss and German companies during the inter-war period.[131] A trans-national perspective is also essential to address more carefully the impact of the social, economic and especially legal environments on pharmaceutical research, which we could only cover briefly. Drug patent laws, drug safety regulations, and drug advertisement limitations constitute powerful forces that shape the opportunities for drug development. In addition to examining the contexts of pharmaceutical activities, another topic that remains to be explored are the links between the industry's search for biological targets for new drugs and the earlier practices of biological research, in particular biological screening, toxicological research, and the improvement of biological production methods. This century-old biological expertise might have constituted important material, intellectual and institutional resources when pharmaceutical companies embarked upon fundamental biological research on a larger scale in the 1960s.

Finally, this case study raises new questions about the periodization of the transformation of the pharmaceutical industry in the twentieth century. Our paper suggests that the introduction of a new research strategy – starting from the (biological) target, not the (chemical) bullet – constituted an important transition for the management and for the scientists at Roche. This perspective, taking into account the management, scientific, as well as institutional practices, opens a middle ground between the *longue durée* of biotechnology as a way to exploit living processes industrially, and the short-term accounts of the 'biotech revolution' centred on the introduction of recombinant DNA technologies.[132]

131 Swann, *Academic Scientists and the Pharmaceutical Industry.*

132 For a similar account see J.-P. Gaudillière, 'Introduction: Drug Trajectories.' *Studies in History and Philosophy of Biological and Biomedical Sciences*, 36 (2005): 603–611.

ANNE-LAURE SAIVES, MEHRAN EBRAHIMI
AND ROBERT H. DESMARTEAU, WITH THE COLLABORATION
OF CATHERINE GARNIER[1]

15 Knowledge Creation Dynamics and the Growth of Biotech Firms in Quebec[2]

Introduction

The biopharmaceutical industry is not new to Canada. More specifically, the role Quebec has played in the field of health care goes back to the early-seventeenth century with the establishment of Hôtel-Dieu in 1639 as the first Canadian hospital. Throughout the nineteenth century, Montreal universities began to teach medicine. The establishment of a Wyeth and Brother (of Philadelphia) subsidiary in Montreal in 1883, followed by the founding of a new company in 1899 by Charles Frosst marked the beginning of the industry in Quebec. Shortages in medicines and medical supplies from overseas (Europe) due to the First World War subsequently incited the development of innovating pharmaceutical activities in Montreal (Quebec) as well as in the rest of Canada. While many foreign, mostly European, companies (Rhône-Poulenc, Ciba, Schering, Sandoz, Roche, etc.) established themselves in Quebec, it was the Canadian firm Ayerst that founded the

1 The authors acknowledge financial support from the Social Sciences and Humanities Research Council of Canada (SSHRC) (GEIRSO – *Major Collaborative Research Initiatives* (MCRI)) and FQRSC – *Professor-Scholar start-up program* (Fond Québécois de recherche sur la société et la culture).

2 An earlier version of this paper has been presented at the IAMOT conference, May 2005. The authors acknowledge David Holford, PhD candidate (HEC Montreal), and Viviane Quirke, from Oxford Brookes University, for their help for the translation and revision of this paper.

first pharmaceutical biological laboratory in Montreal in 1925. While the Canadian biopharmaceutical industry achieved world-wide renown in 1922 due to the large scale production of insulin at the Connaught laboratories (a drug for the treatment of diabetes discovered earlier that year by F.G. Banting and C.H. Best at the University of Toronto), Quebec has also been the centre of numerous important biomedical research activities. Some of these activities included: the work of Hans Selye on psychological stress begun in 1932 at McGill University; the studies in tuberculosis conducted by Armand Frappier, founder of the Montreal Institute of Microbiology and Hygiene in 1938 (later renamed the Armand Frappier Institute in 1975); and the investigations on the brain and neurological mechanisms by Wilder Penfield (founder of the Montreal Neurological Institute in 1934). The institutions founded by these pioneers constitute the cornerstones of Quebec bioclusters today.[3]

At the beginning of the last century, the rivalry or emulation resulting from two distinct university networks (francophone, anglophone) and the adapted health care services (multiethnic) significantly contributed towards Montreal becoming a world class research centre in life and health sciences – as well as attracting numerous pharmaceutical manufacturing companies. As Germain and Rose have explained:

> Montreal is the primary health and life-sciences research center in Canada [...]. The city has a large and closely integrated complex of university hospitals and specialized research centers, some of which are among the world leaders in several fields of clinical research and practice [...]. This research network's capacity to produce leading-edge specialists is enhanced by the unique access to both French- and English-language international networks of medical expertise.[4]

As a result of successive government strategies aimed at promoting the local development of health sciences and biotechnologies,[5] the Montreal

3 J. Niosi et al., *Biotechnologie et Industrie au Québec* (Montréal: Transcontinental, 2002).

4 A. Germain and D. Rose, *Montréal, the Quest for a Metropolis* (Chichester and London: John Wiley & Sons, 2000), p. 139.

5 L. Picard, *Rapport du comité consultatif au comité ministériel sur le développement de la région de Montréal* (Montréal: Gouvernement du Canada, Nov. 1986).

region became internationally renowned as a 'biotech cluster'. A key part of the Quebec biotechnological research apparatus was created in 1987 by the National Research Council via the establishment of the Biological Research Institute (BRI). At the turn of the twenty-first century, Genome Quebec complemented and completed this apparatus in terms of genomic and proteomic research. Furthermore, the two patent laws of 1987 and 1993, which extended Canadian pharmaceutical patent lives from 7 to 20 years, further encouraged the development of pharmaceutical and biotechnological R&D activities.

Nevertheless, the announcement of the closure of BioChem Pharma's[6] research laboratories (seen as Quebec's champion in biomedical research), which had been acquired just two years earlier by the UK pharmaceutical company Shire, was a harbinger of changing development modalities within this industry. In today's global markets and corporate 'financialization',[7] continual restructuring characterizes the current crisis of the world's pharmaceutical industry.

In the last few years, the pharmaceutical industry, especially within North America, has become more and more alarmist about its future.[8] The threats invoked include: the drop in R&D productivity,[9] the slow evolution of marketable targeted therapies, the reduction in exclusivity periods linked to patent protection and competition between the original drugs and their generic versions, and the astronomical increase in R&D costs, resulting in part from stricter rules and regulations governing clinical trials (according to various sources, 880 m US$ to 1.7 billion US $ and 12–15 years of efforts), at a time when fewer pharmaceutical products meet or surpass the 'blockbuster' level (approximately 500 m$ sales per year). The industry is facing the complexity of its own mission, that is the (highly costly and time-consuming) discovery and (short-term)

6 Biotechnological firm founded in 1986 by Drs Belleau, Bellini et Dionne, who were at the origin of the famous 3TC/Epivir treatment against HIV/AIDS.

7 Stockhammer Engelbert, 'Financialisation and the slowdown of accumulation', *Cambridge Journal of Economics*, 28 (September 2004): 719–741 (p. 720).

8 PhRMA, *Pharmaceutical Industry Profile 2003* (Washington D.C., 2003).

9 With the discovery of 0.45 new compounds per year, it is insufficient to insure the current growth rate of 7–8 per cent.

marketing of innovative drugs. Whereas the profit margins observed on stock markets are still at levels superior to the average of other sectors,[10] those margins are decreasing.

In the same period, the pharmaceutical sector has experienced an expansion of new technologies, i.e. a 'third wave' of biotechnologies,[11] whereby the latter are succeeded by new ones at an ever increasing rate. Combinatorial chemistry techniques and the High Throughput Screening (HTS) of molecules have helped to rationalize the traditional drug discovery process. More recently, biotechnologies resulting from advances in computer technologies and genetics have reshuffled the cards of innovation between new players (such as new biotechnology firms, i.e. biopharmaceutical firms, bioinformatics firms, genomics services providers, etc.). According to the knowledge-based theories of the firm,[12] those biotech companies are currently focusing on managing invention by making judicious choices of intellectual property to be protected and traded, and of various financial alternatives to grow (and create value from their technological platforms).

What are the consequences of these evolutions on the emerging bio-industrial sector in Quebec? In this paper, we will answer this question by studying the technological and organizational development cycle of these

10 M.-H. Depret and A. Hamdouch, 'Innovation networks and competitive coalitions in the pharmaceutical industry: The emergence and structures of a new industrial organization', *European Journal of Economic and Social Systems*, 14 (2000): 229–270.

11 New biotechnologies refer to the 3rd generation of biotechnologies. 1st generation: pharmaceuticals (ex: penicillin (1942)) based on microbial theory (development of pharmaceutical compounds to fight infectious diseases); 2nd generation: biopharmaceuticals (ex: insulin (1982), Human Growth Hormone (1985), vaccine for Hepatitis B, Interferon (1986), factor VIII (1989)), based on replacement/substitution theory and Recombinant DNA; 3rd generation: pharmacogenomics (functional genomics, proteomics), for the development of products for the treatment of complex genetic diseases (ex: Herceptin) and personalized and predictive medicine.

12 On the knowledge-based view of the firm (KBV), see R.M. Grant, 'Toward a Knowledge-Based Theory of the Firm', *Strategic Management Journal*, 17 (Winter 1996): 109–122. For the knowledge-creation model, see I. Nonaka and H. Takeushi, *The Knowledge-Creating Company: how Japanese companies create the dynamics of innovation* (New York/Oxford: Oxford University Press, 1995).

firms. Using the hypothesis that techno-scientific knowledge creation is the driving force behind the creation of value within the bio-pharmaceutical industry, we examine the possibility of combining the creation of techno-scientific knowledge and the constraints of the financial governance that drives the contemporary bio-industrial system.

By analysing the concept of the technological life cycle or development cycle starting with a brief literature review on innovation management, knowledge management and strategic analysis as applied to the biotechnology sector, we will formulate our research questions around the issue of understanding 'high-technology' organizational transformations, that is, techno-scientific knowledge creating firms, during their different stages of development as in the case of biotechnology companies. We will then present the exploratory method used, and the empirical results from a multi-factorial analysis of the field of study, that is, the biotechnology system of Quebec. We will highlight a number of paradoxes that derive from these observations, and will propose a new framework for understanding the 'development modes' of these biotechnology companies.

Theoretical Framework and Research Questions

Biopharmaceutical knowledge creation and innovation

Hamdouch et al. have described the challenges facing the pharmaceutical industry.[13] In Canada and in North America, the cost increases related to R&D, the expiry of patent protection on numerous blockbuster drugs, or the reduction of these periods of exclusivity, along with the lengthening of

13 A. Hamdouch and M.-H. Depret, 'La gouvernance des jeunes entreprises innovantes: un éclairage analytique à partir du cas des sociétés de biotechnologies', *Revue Finance Contrôle Stratégie*, 7 (2004): 67–94; A. Hamdouch and D. Perrochon, 'Formes d'engagement en R&D, processus d'innovation et modalités d'interaction entre firmes dans l'industrie pharmaceutique', *Revue d'Économie Industrielle*, 93 (2000) : 29–50.

the discovery process, as well as competition from generic drugs have only further complicated the strategic choices of pharmaceutical firms. Pignarre has argued that the increasing complexity of the innovation process is a consequence of the growth of *evidence-based* medicine, that is to say, the generalization and imposition of clinical trials so as to demonstrate safety and effectiveness of new medicinal preparations prior to bringing them to market.[14] Moreover, the new technologies resulting from advances in genetic combinations and bio-informatics (structural and functional genomics, proteomics, etc.) have multiplied the competencies required to develop potentially profitable knowledge-based products. Similarly, Hamdouch et al. have observed that the pharmaceutical industry no longer gravitates around a few large integrated companies, but rather is characterized by the diversity, the evolving character and the entangled modes of interactions (horizontal, vertical and transversal) linking the different types of agents that today make-up the industrial bio-pharmaceutical system.[15] In particular, the strategy of 'pre-emption' adopted by 'large pharmaceutical companies' consisting in gaining access as soon as possible to the knowledge embodied in patents and promising candidate drugs so as to renew their product pipelines,[16] leads them to linking up specifically with biotechnology firms.

This strategy emerges also within other value creating systems such as the bio-food or environmental value chains. In fact, biotechnology firms dispose of distinctive technological competencies that raise them to the rank of partners of pharmaceutical companies, in co-operative agreements that resemble 'coopetition'[17] rather than sub-contracting (Figure 15.1).

14 P. Pignarre, *Le Grand secret de l'industrie pharmaceutique* (Paris: La Découverte/Poche, 2004).

15 Hamdouch and Perrochon, 'Formes d'engagement en R&D'; Hamdouch and Depret, 'La gouvernance des jeunes entreprises innovantes'; see also C. Parolini, *The Value Net* (Chichester: John Wiley & Sons, 1999).

16 Hamdouch and Perrochon, 'Formes d'engagement en R&D', p. 42.

17 If both the elements of cooperation and competition are visible, the relationship between the competitors is named coopetition. M. Bengtsson and S. Kock, '"Coopetition" in business networks – to cooperate and compete simultaneously', *Industrial Marketing Management*, 29 (September 2000): 411–426 (p. 415).

Within such agreements, their mission consists of new techno-scientific knowledge creation.

Within these high-technology sectors, based on technological innovation ('technology-based' and 'science-driven'),[18] the strategic resource of biotechnology firms is knowledge (in the sense used by KBV)[19] or more precisely the capacity to create, absorb and monitor knowledge within networks of collaboration and innovation.[20] These firms have integrated the reticular and partnering dimensions of value creation, as recently stipulated by Chesbrough, with the idea that the dynamics of 'open' innovation is bi-directional: 'Open Innovations means that valuable ideas can come from inside or outside the company and can go to market from inside or outside the company as well'.[21]

Equally numerous are the authors within the field 'knowledge management' who have adopted this reticular vision of knowledge creation in the new economy founded on knowledge; an economy of relationships necessary to accede to and grow the expertise which is at the source of value creation. Hence, the creation and circulation of knowledge involve processes that are first and foremost social processes permitting the articulation

18 A.-L. Saives, M. Ebrahimi, R.H. Desmarteau and C. Garnier, 'Les logiques d'évolution des entreprises de biotechnologie', *proceedings of the XIVth AIMS conference* (Association Internationale de Management Stratégique), Angers, 6–9 June 2005; A.-L. Saives, R.H. Desmarteau and D.A. Seni, 'Vers une conception des bio-industries', *Économies et Sociétés, série 'Dynamiques agroalimentaires'*, 27 (2005): 957–968.

19 Grant, 'Toward a Knowledge-Based Theory of the Firm'; Nonaka and H. Takeushi, *The Knowledge-Creating Company*.

20 W.M. Cohen and D.A. Levinthal, 'Absorptive capacity: a new perspective on learning and innovation', *Administrative Science Quarterly*, 35 (1990): 128–152; W.W. Powell, 'Learning From Collaboration: knowledge and networks in the biotechnology and pharmaceutical Industries', *California Management Review*, 40 (1998): 228–240; G. Pisano, 'The Governance of Innovation: vertical integration and collaborative arrangements in the biotechnology industry', *Research Policy*, 30 (June 1991): 237–250; idem, 'Knowledge, integration, and the locus of learning; an empirical analysis of process development', *Strategic Management Journal*, 15 (Winter 1994): 85–100.

21 H. Chesbrough, *Open Innovation: the new imperative for creating and profiting from innovation* (Boston, MA: Harvard Business School Publishing, 2003), p. 43.

of explicit, formal, and classified knowledge, along with tacit knowledge.[22] Within organizational strategies, it becomes essential to create and foster contexts for socialization so as to reinforce the cycle of tacit, conceptual and operational knowledge around shared metaphors. If the organization, as a collectivity of individuals, is considered as a place of socialization combining tacit and explicit knowledge, it is also necessary to re-situate these individuals within the centre of organizational preoccupations.

Thus, there is an important qualitative jump between the individual and organizational finality with respect to the invention (techno-scientific knowledge creation) and innovation (oriented towards commercialization of knowledge) phase that is driven by this new pharmaceutical industrialization system. If this techno-scientific knowledge is the source of competitiveness amongst the biopharmaceutical firms of tomorrow, how is this knowledge generated on the one hand, and activated on the other? What transformations occur and operate between intangible intellectual resources such as patents or techno-scientific and technological knowledge creation, and specific financial assets of bio-technology firms?

Development stages of biotechnology firms

With the advent of the new scientific paradigm founded on genomics,[23] the literature on the management of innovation, considered to be most abundant within the biotechnology sectors, often describes the bio-pharmaceutical system as revolving around two distinct homogeneous groups, which occasionally merge into each other, namely, large pharmaceutical corporations and companies dedicated to biotechnology. However, this literature does not open up the black box of their respective organizations, and even less of their organizational practices related to the successive

22 Nonaka and Takeuchi, *The Knowledge-Creating Company*.

23 N. Lacetera, 'Corporate Governance and the Governance of Innovation: the case of the pharmaceutical industry', *Journal of Management and Governance*, 5 (2001): 29–59; G. Pisano, *The Life Sciences Revolution: a technical primer* (Boston, Mass.: Harvard Business School Publishing, 6 August 2002).

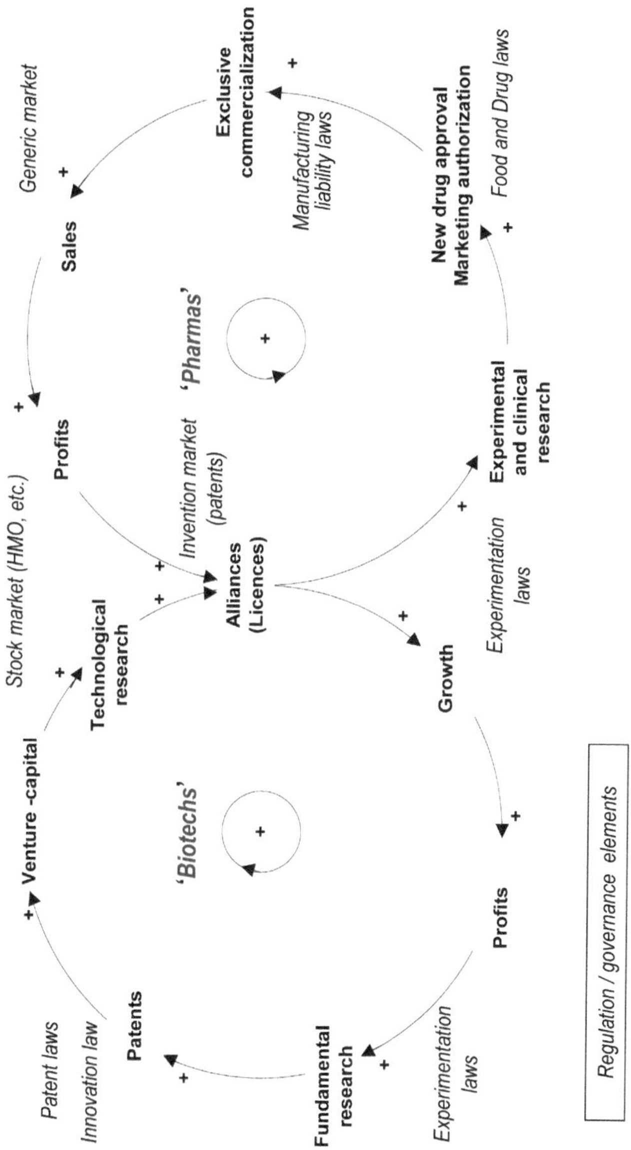

Figure 15.1: The virtuous cycle of growth of bio-pharmaceutical activities.

stages of their growth, as defined by Greiner or Churchill and Lewis for example.[24]

The 'metamorphosis' models of development[25] are built on various criteria, and consist in up to five stages of development.[26] The first stage is usually characterized by a very *informal organization* with a *low level of specialization* as the enterprise is managed by a single entrepreneur. In the following stage, the *formal structure* of the organization emerges. The *task specialization* as well as the *centralization of decisions* with a *functional structure* start to increase. The third stage is one of *delegation*. The formal structure of the firm is more clearly hierarchical, but the decision modalities evolve towards a more participative model. The next two stages (*coordination* and *collaboration*) concern large enterprises, where the high degree of formalization of the divisionary or matrix structure requires reinforced means of coordination.

Even if the bio-industries are considered to have entered into a 'consolidation' phase,[27] the most widespread representation of bio-technology firms, especially in Quebec, is that they remain small in size, are in an emergent phase, are under-capitalized, experience the full risk associated with the development and/or clinical validation cycle of their products, unless they are able to enter into a virtuous cycle of growth brought on by determining factors of success which are: access to intellectual property

24 L.E. Greiner, 'Evolution and Revolution as Organizations Grow', *Harvard Business Review*, 50 (July–August 1971): 37–45 (reprinted in 1998, *HBR*, 76, May–June, pp. 55–66); N.C. Churchill and V.L. Lewis, 'The Five Stages of Small Business Growth', *Harvard Business Review*, 61 (May–June 1983): 30–50.

25 A. Godener, 'PME en croissance : peut-on prévoir les seuils organisationnels ?', *Revue Internationale PME*, 15 (2002): 39–63.

26 Those criteria might consist of: management style, organizational structure, managerial focus, control systems, characterization of the direction (capital propriety, strategic objectives, etc.). Greiner, 'Evolution and Revolution as Organizations Grow'; Churchill and Lewis, 'The Five Stages of Small Business Growth'.

27 The bio-industries, of industrial sectors that emerged from the commercialization of biotech innovations are today considered to be entering a phase where leaders have consolidated their position, whereas weaker organizations have been forced to reorganize themselves or disappear. We are here referring to a stage of the *industry* life cycle (and not to the firm life cycle).

(patents and licenses), to venture-capital, and to strategic alliances and collaborations for the commercialization of their knowledge.[28]

Because biotech firms remain numerous and very small, the current and known business models of biotechnology firms are therefore centred towards the management of inventions within the first phases of a *financial* development cycle. This financial cycle consists in a succession of stages (pre-start-up, start-up, growth). These are marked by the judicious direction of R&D projects towards bringing a product or process to a potential market,[29] by the choice of intellectual property to be protected (typically via patents) and by the choice of financial resources from institutional start-up funds to a series of venture capital investments and the eventual raising of funds on the stock market via an IPO. Typically, an established researcher within the academic field comes up with a university research project (pre-start-up). This constitutes an 'incubated' start-up,[30] which evolves towards a semi-private firm for the purpose of satisfying a potential market need. This project may become more or less condemned in the quest for these financial resources. It should be noted that evolution takes place within the first of the five growth phases as described by Greiner,[31] whereby the initial phase of 'creativity' involves the creation of both the 'product' (or service) and its market.

Hamdouch and Depret are among the first to have attempted to characterize this progressive organization of the firm under the different 'regimes of governance' at each stage of its evolution (start-up, growth and routine

28 J. Niosi, 'Alliances are not enough. Explaining rapid growth in biotechnology firms', *Research Policy*, 32 (2003): 737–750.

29 J. Fisken and J. Rutherford, 'Business models and investments trends in the biotechnology industry in Europe', *Journal of Commercial Biotechnology*, 8 (Winter 2002): 191–199.

30 'Business incubator' is a notion in entrepreneurship and economic development. 'The businesses share administrative services, such as clerical and reception staff, computer and copying equipment, bookkeeping help, and conference rooms, in addition to renting space. Management assistance is usually available, and financial assistance sometimes is provided'. F.L. Fry, 'The Role of Incubators in Small Business Planning', *American Journal of Small Business*, 12 (Summer 1987): 51–61 (p. 51).

31 Greiner, 'Evolution and Revolution as Organizations Grow'.

types of governance).[32] However, their description of the three phases (pre-start-up, start-up, growth) in terms of age indicators, critical size, modes of investment, modes of intellectual property protection, etc. reflects a very *financial* vision of company growth, which has little in common with the company founder (who is very often a researcher).

In fact, 'biopharmaceutical' biotechnology firms are often managed by scientific entrepreneurs, who are said to suffer from a lack of competencies with respect to making the most enlightened strategic and organizational choices.[33] It is precisely at the intersection between these different points of views – of the researcher, the manager, and the investor – that our questions are situated (Figure 15.2).

For the researcher, who is in search of 'credit',[34] at the 'pre-start-up' stage or rather the 'pre-entrepreneurial' stage, the invention activity is inscribed within a program of academic research more or less determined by the state of scientific and technological knowledge at the time, and is constrained by legal barriers (such as the patents of other researchers) and by financial resources (varying from public subsidies, government grants to private contracts). At this stage, the researcher involved in 'start-up activities' makes repeated attempts to reproduce his research program in the form of many different projects, which may accidentally (but less and less so) meet the pre-occupations of industrial partners. How does this 'revolution' take place,[35] from its first phase, governed by the researcher who has embarked on an innovation-oriented activity, to the second phase, governed by venture-capital investors, and then to the third phase of the research routinization within the so-called 'surviving' or innovative firms,[36] which are essentially governed by financial interests? More precisely, from a managerial point of view, our research questions are here the following:

32 Hamdouch and Depret, 'La gouvernance des jeunes entreprises innovantes'.

33 Niosi et al., *Biotechnologie et Industrie au Québec.*

34 B. Latour and S. Woolgar (French Transl.), *La Vie de laboratoire* (Paris: La Découverte, 1988).

35 Greiner, 'Evolution and Revolution as Organizations Grow'.

36 Churchill and Lewis, 'The Five Stages of Small Business Growth'.

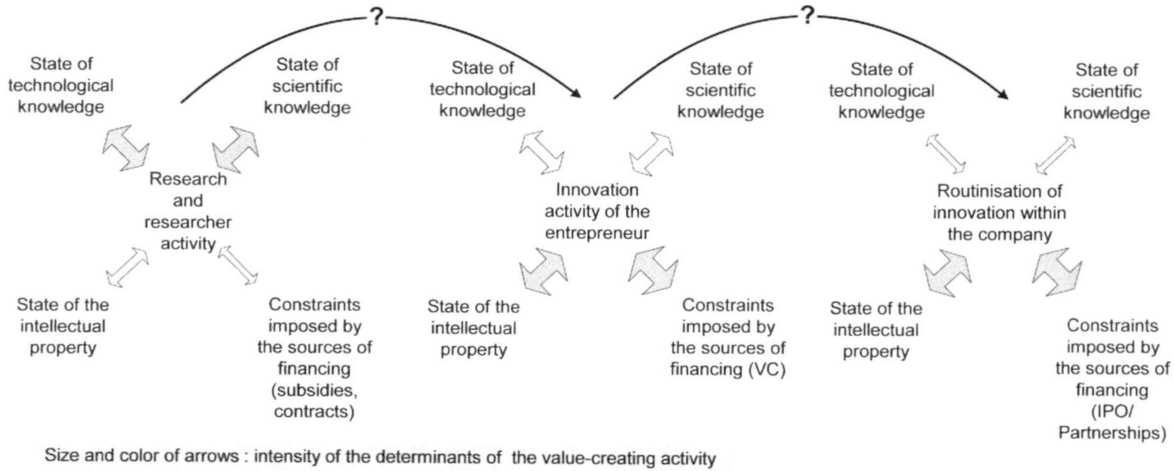

Figure 15.2: The stages and constraints of the development of biotechnological innovation.

1) What are the different development modes of high technology and knowledge-creating organizations, such as biotechnology firms?
2) What are the transformation mechanisms of these high-biotech organizations as they pass from one mode of development to another?
3) What are the managerial implications of these transformations?

Methodology and Statistical Results

Description of sampling of the firms and of the variables analysed

After more than 10 years of existence, the internationally renowned industrial cluster of Quebec has attained relative maturity[37]. Both the creation and the restructuration of firms are now a permanent reality. The Quebecan biocluster therefore contains a significant density and diversity of actors well suited to the study of these theoretical questions.

Indeed, the bio-industrial firms of Quebec make up the principal biotechnological cluster within Canada, along with Toronto and Vancouver.[38] 73 per cent of the total estimated population of companies from the bio-industries is from Quebec, with 60 per cent of these firms practising their activities in the human health sector. To analyse the stages of development of biotechnology companies within this system, we will utilize the data obtained from semi-directive interviews, which were carried out with a number of firms from the bio-industries of Quebec (that is, more than 52 per cent of its estimated population of biotech companies).

During this investigation, a series of quantitative and qualitative variables were used to code for the contents of over 110 semi-directed interviews,

37 A.-L. Saives and R.H. Desmarteau, 'Comprendre l'ancrage territorial d'une bio-industrie : le cas de la grappe bio-pharmaceutique de la région de Montréal (Canada)', *Revue Internationale PME*, 18 (2005): 75–108.

38 Niosi et al., *Biotechnologie et Industrie au Québec*.

conducted with the leaders of these firms of all sizes.[39] These variables were based on criteria that had been used by Niosi in an earlier investigation of biotechnology firms, conducted in Canada in collaboration with StatCan, and widely utilized in France as well.[40]

More specifically, a series of 28 nominal variables were employed in a multi-factorial analysis with multiple correspondences and an ascending hierarchical classification of data. These 28 variables make up a series that defines the stages of evolution and modes of governance within firms according to the three following dimensions:

1) *the firm's organization* (origin of spin-off), embededness (location and anchorage factors), size (age, number of employees), governance type and financing modalities (public research funds, 'seed-money', venture-capital, IPO), and core competences (mechanisms of revenue generation such as patents, licences, exportation);
2) *the firm's strategy* (strategic vision in a 5 years perspective);
3) *the project manager's perception* of the difficulties encountered by these companies vis-à-vis the process of innovation (obstacles to growth) and financing of innovation.

For greater clarity, two other series of variables (20 continuous variables and approximately fifty nominal variables) were preserved and treated statistically.[41]

39 The pharmaceutical companies (and large bio-food companies) were excluded from the present analysis since they are partners to the innovation process of the biotechnology firms.

40 Niosi, 'Alliances are not enough'; J.-P. Boissin, D. Catherine, F. Corolleur, R. Coronini, S. Lemarié, V. Mangematin, M. Trommeter, 'Structure de gouvernance dans les PME high-tech ; caractéristiques des dirigeants, structure de contrôle et croissance de l'entreprise', *Rapport au Commissariat Général au Plan, convention 1999/9* (December 2002).

41 For details, see A.-L. Saives et al., 'Les logiques d'évolution des entreprises de bio-technologie.

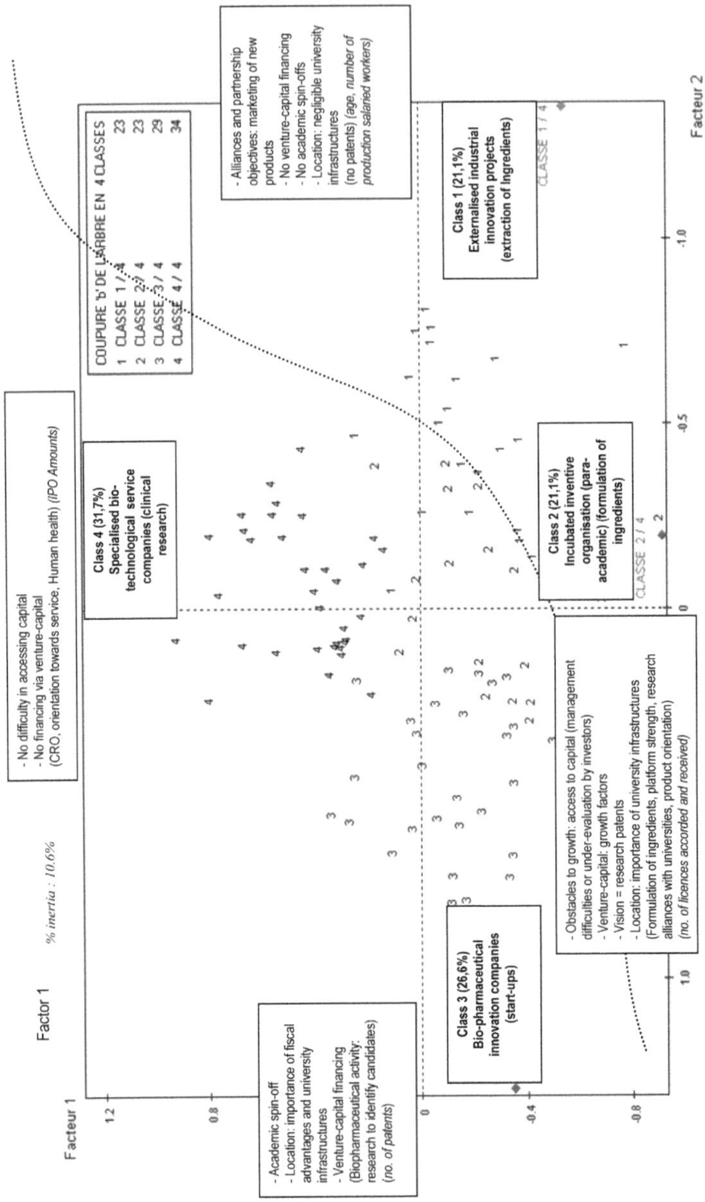

Figure 15.3: A multi-factorial classification chart of biotechnology firms (Quebec, Canada)

Exploratory statistical observations

Figure 15.3 represents the factorial chart obtained according to the two dominant axes of our multi-factorial analysis.[42] On the horizontal axis, we find the organizational dimensions related to *context of the creation of the company*, as well as the *temporal dimension* of the development cycles of the firm.

On the left, this axis shows the enterprises that originate from academic spin-offs, that is to say from the transfer of technology from universities benefiting from a fertile environment of knowledge creation (the availability of local university research infrastructures) supported by funding mechanisms that are typical of 'pre-start-up' (research income tax credits, first round of risk-capital). Conversely, towards the right, we find older firms, which do not finance themselves via venture-capital and do not necessarily originate from academic spin-offs. These firms are oriented towards the development and marketing of new products rather than towards pure innovation resulting from the creation of new knowledge.

The vertical axis of this chart distinguishes between firms according to their governance model, and according to their mechanisms of revenue generation. At the bottom of the chart are situated the firms that are in search of venture-capital, and for which it is difficult to obtain adequate capital as long as they have not achieved the conditions of success required by the investors (the patent, as a gauge of credibility; strategic management, as a gauge of the ability to position themselves on the market in the long term). At the top, in contrast, are situated the firms that do not have any particular funding concerns due to their attractiveness to financial and stock markets (e.g. because of the guaranteed return on investment achieved by providing profitable biotechnological services specialized in clinical research).

The classification that was conducted in an increasing hierarchical fashion allows us to distinguish between four different groups of

42 Percentage of inertia on axis 1: 10.6 per cent, on axis 2: 9.8 per cent, total for five axes: 39.3 per cent.

companies[43] as described in table 1. As in Torres's work,[44] our empirical research enables us to identify 'contrasting cases' so as to test the robustness of the small and medium enterprise (SME) concept. We therefore observe four contrasting cases of biotechnology SMEs, which make it possible to reflect on the modes of organizational development of these high-tech and 'science-based' companies. These four types of firms can be characterized as follows:

The industrial company – externalized innovation project (company spin-off)

This first group of companies (with an entrepreneurial industrial culture) comprises firms that bear little or no relation to the innovation dynamics of academic spin-offs seen within the biopharmaceutical sector. They consist of SME's (close to ten years old on average), which have built their activities from financial sources other than venture-capital (often issued from other company headquarters) and from technological expertise in a specific bio-process. Their aim is to market new products to auto-finance their industrial development as well as to create value. Their technological platforms include the physicochemical and biological extraction of functional ingredients from marine biomass, high-throughput screening applied to natural plant extracts, accelerated culture in bioreactors and purification of plant cells, isolation of natural phyto-chemical fractions, bacterial fermentation, ultra-filtration of peptides, etc. Their products are conceived in partnership with university and/or public research centres, since they have limited financial means for R&D. These firms display a managerial behaviour inspired by the classical productive model which aims towards the marketing of tangible industrial products. We find them essentially in the physicochemical and biological extraction and functional ingredient

43 Quotient (intra inertia/total): 29 per cent.

44 O. Torres, 'Pour une approche contingente de la spécificité de la PME', *Revue Internationale PME*, 10 (1997): 9–43.

formulation sectors, which generate short to medium term sales on the human nutrition markets (nutraceuticals, functional foods, etc.).

The inventive incubated organization

This second class groups together young inventive companies (approximately five years old on average) which run against difficulties associated with the managing of intellectual property. These difficulties delay the innovation process, which would otherwise attract more substantial investments (and investors) to further develop their technological competencies. These competencies include the culture of lactic ferments, the enzymatic hydrolysis for isolation of peptides, the formulation of protein mix, etc. These companies are usually embedded within an innovative environment made up of widely solicited public research institutions (universities and public laboratories) and service infrastructures (including incubation within an academic or technopole environment). These companies also possess a traditional strategic vision, focusing on the commercialization of tangible industrial products (for example, the bio-material and functional ingredient sectors, etc.). At this point in time, these companies generate little or no revenue related to their principal innovation activity, distinctive competencies linked to cognitive and organizational capabilities of their human resources or their technological know-how.

These companies are closely linked to academic circles (typically, they publish academic work in collaboration with their universities of origin) so as to tap into strategic techno-scientific knowledge that is complementary to their own. On average, these firms generate the highest number of patents, and have no difficulty in attracting venture-capital and stock market Keeping in mind the early stage of development they had reached during our investigation, they felt that in the medium term they would participate in fusion-acquisition activities, and also redistribute their intangible assets (portfolio and pipeline of patents, licenses etc.) according to their key technological competencies.

	Class 1	*Class 2*	*Class 3*	*Class 4*
	Externalized industrial innovation project (company spin-off)	Incubated inventive organization (techno para-academic research)	Bio-pharmaceutical innovation start-up	Specialized bio-technological service companies
	(21.1% of firms)	*(21.1% of firms)*	*(26.6% of firms)*	*(31.2% of firms)*
Project	Activity related to extraction of biological ingredients**	Extraction/Formulation of ingredients***	In vitro and in silico research for bio-pharmaceutical candidates*** Scientific publications**	Clinical Research**
Organisation			University spin-offs***	No academic spin-offs ***
	Research alliances* Strategic alliances*			No research alliances***, neither with universities**
	Other sources of financing*** Difficulty in accessing capital due to management problems***	Obstacles to growth: IP protection*** Cost/delays for approvals***	Venture Capital = growth factor*** No difficulty in accessing venture-capital***	No difficulty to access capital***
	University and public laboratory infrastructures: negligible location factors*** Fiscal advantages: negligible location factor***	University and service infrastructures: significant location factors*** Public laboratory infrastructures: significant location factor**	University infrastructure and fiscal advantages: significant location factors ***	University infrastructures: significant location factor **
	No patents *** Strength: quality of management*	Distinctive technological plate-form* Patents*	Patents* Distinctive technological plate-form*	IP = not an obstacle to growth** No distinctive technological plate-form**
Strategy	Market: human nutrition*** 5 year strategic vision: marketing of products** Objectives of alliances: marketing of new products*	Product orientated **	No exporting*** 5 year strategic vision: fusion*	Exportation**

Table 15.1 (a): Description of 4 classes of biotechnology firms (Quebec, Canada)
(According to the order significance of the descriptive variables)

	Class 1	*Class 2*	*Class 3*	*Class 4*
	Externalized industrial innovation project (company spin-off)	Incubated inventive organization (techno para-academic research)	Bio-pharmaceutical innovation start-up	Specialized bio-technological service companies
	(21.1% of firms)	*(21.1% of firms)*	*(26.6% of firms)*	*(31.2% of firms)*
Continuous illustrative variables	*– Age superior to the mean – No of salaried production employees superior to the mean – No of salaried sales employees superior to the mean – No of patents inferior to the mean – No of salaried R&D employees inferior to the mean – Sales and revenues very much superior to mean*	*– Age inferior to the mean – No of salaried Quebec employees inferior to the mean – R&D expenses inferior to the mean*	*– No of patents superior to the mean – No of salaried R&D employees superior to the mean – R&D expenses superior to the mean – Age inferior to the mean – No of approved and/or received licenses superior to the mean – Amounts raised by IPO superior to the mean*	*– No of salaried clinical research employees superior to the mean – No of years listed on stock market superior to the mean (10 years) – Sales and revenues superior to the mean – No of managers superior to the mean – Age superior to the mean – No of salaried Quebec employees superior to mean*

**** Significant variable to a baseline inferior to 1%; ** Significant to a baseline inferior to 5%; * Significant to a baseline inferior to 10%*

Table 15.1 (b): Description of 4 classes of biotechnology firms (Quebec, Canada)
(According to the order significance of the descriptive variables)

The specialized biotechnology service firm

The last class consists of a group of older companies (approximately ten years old) originating from more purely entrepreneurial creations that supply specialized services (clinical research, diagnostic products, laboratory products and chemicals) to bio-industrial companies as well as biopharmaceutical research organizations. These companies are either not particularly innovative, or else have turned their innovative process into a repetitive routine. Their innovations have sometime become generic in nature, and these companies rarely have a distinctive technological platform. They develop by means of internal growth and the internationalization of their markets. They benefit from investor confidence going back several years (40 per cent of them are listed on stock markets), and generate levels of business that are superior to the average by means of a management team (administration and finance, sales management, management of regulations) whose size is superior to the mean of the sampled companies.

At this stage, we would like to point out that a dashed line curve was arbitrarily drawn for illustration purposes in Figure 15.3 so as to show the relationship between the multi-factorial results obtained from the various types of biotechnology firms and the traditional S-curve of the cycle of development of companies and technologies. We pursue the interpretation of these empirical results in the light of our initial theoretical questions in the next and final section of this paper.

Discussion

The different 'modes of development' of high-tech companies

As mentioned earlier, Hamdouch and Depret are among the first to have characterized the types of management of young biopharmaceutical companies by distinguishing between three governance 'regimes', which are associated with the financial maturities of these firms.[45] In the same way, the qualitative data related to each one of the classes of companies allows us to propose three development modes within the growth 'cycle' of knowledge intensive biotechnology firms. These three 'modes' are qualified as 'pre-entrepreneurial', 'entrepreneurial', and 'managerial', as shown in Table 15.2. However, we must specify here that, as we have not yet completed a longitudinal study of these firms, we are not in a position to assert that the three stages follow each other sequentially.

These stages seem to differ according to the nature of the organization's project itself. In the first *pre-entrepreneurial* case, a researcher or research service provider from industry or a public research institution is responsible for a project to discover universal scientific laws, thus perpetuating the exploration (in the sense used by March)[46] and the creation of scientific knowledge with both an individual (that of 'credit' in the sense used by Latour and Woolgar)[47] and a collective goal (the production of common goods/knowledge). He acts with a relative degree of freedom within a community at the cutting-edge of scientific knowledge, and within an academic bureaucracy where key scientific competencies are organized and embodied in research teams. In this case, knowledge creation is evaluated by peers through publications and patents.

In the second *entrepreneurial* stage (represented by the companies of types 1 and 2), the researcher-entrepreneur is responsible for a technological project (to solve a specific problem) with the aim of attaining relative independence from the academic bureaucracy to which he had belonged, for example. However, his success in achieving this individual goal depends on the degree of autonomy he receives from the interested parties (stakeholders). Governance is therefore shared between the founder, the investors (often venture-capital), and advisory groups (board of directors, scientific board, and innovation and expertise networks) of the company committed to the commercial integration and exploitation of scientific knowledge

45 Hamdouch and Depret, 'La gouvernance des jeunes entreprises innovantes'.

46 J. March, 'Exploration and exploitation in organizational learning', *Organization Science*, 2 (February 1991): 71–87.

47 Latour and Woolgar, *La Vie de laboratoire*.

within the knowledge economy, characterized by complex, multi-disciplinary and multi-directional innovation. The intangibles involved (human capital, social capital, etc.) make it difficult to measure the creation of value by these firms. Hence, 'the investors under-value our company!' can often be heard from the leaders of Quebec biotechnology firms in search of venture-capital (especially type 2 in our classification).

At the '*managerial*' stage, the project carried out by the organization, directed by managers but essentially governed by the shareholders, becomes a financial project. Without a doubt, the third type of firm identified in our classification offers the clearest illustration of this. The scientist, often closely involved in the founding of the company, now has a more specific role, that is when he is not excluded from strategic decisions. The goal of collective profit serves the interests of individuals (shareholders or partners). The objective of the organization consists in valorizing techno-scientific knowledge at every stage, through a speculative exchange of segmented strategic knowledge (patents, licences on therapeutic molecules having attained the stages of phase I, II or III clinical trials, etc.).[48]

In this contemporary speculative economic 'corporate governance', the key competency of the organization resides in the management of its portfolio of intangible assets (pipeline of patented products, expertise) within a commercial network of value creation. Innovation is made routine and/or is fragmented for the benefit of a short-term race oriented towards the commercialization of scientific knowledge and technical know-how. Today, this race is the source of many debates, and its short-term outlook is highly criticized.

The passage of a firm from one stage to another is neither systematic nor strictly demonstrated (without any longitudinal data). Nevertheless, from the qualitative data we have collected, we notice that the transformation of firms from one mode of development to another generates two states of 'rupture' in the sense of organizational crisis as evoked by Greiner:[49] 1) a *teleological rupture* for the scientific researcher becoming an entrepre-

48 'Research outputs' according to Chesbrough, *Open Innovation*.

49 Greiner, 'Evolution and Revolution as Organizations Grow'.

neur, which is often referred to,[50] and linked to the commercial orientation of the invention process; and 2) a *creativity crisis* linked to the bureaucratization of the innovation process within the company as it becomes managerial.[51] We describe in greater detail what we understand by ruptures in the following sub-section.

How to reconcile the development modes of high-tech knowledge creating firms?

A) The 'teleological rupture':

A teleological rupture (between pre-entrepreneurial and entrepreneurial stages) would consist of the apparent distance between the two distinctive missions and objectives of the researcher, committed to the development and perpetuation of a common cognitive patrimony and in search of credit on the one hand, and on the other hand, the entrepreneur who, depending on the entrepreneurial authors, is in search of independence or perennial autonomy, as well as profit that is able to ensure economic and material power. To conciliate these two (caricaturized) representations of research activity (academic or targeted), it would seem useful to return to the work conducted by Latour and Woolgar. In their work conducted during the 1970s, the sociologists clearly identify the motivations of the researchers and the transformation of the different managed capitals within a laboratory (the dotted circle in Figure 15.4).

Hence, according to Latour and Woolgar, a *given* intellectual capital (proto-knowledge) leads to the proposal of scientific arguments, which, once published in the form of scientific articles, can be read and appreciated by peers who in turn enrich the *recognition* capital of the researcher, which is indispensable for the securing of research subsidie*s* to finance

50 P. Tambourin, P.-B. Joly and J.-P. Dupuy, 'Les traditions françaises à l'épreuve des biotechnologies', *Les Amis de l'École de Paris, compte-rendu de la séance du 6 octobre 2003 (Rédaction E.Bourguinat)* (École de Management de Paris, 2003).

51 Pignarre, *Le Grand secret de l'industrie pharmaceutique.*

	Mode	Pre-entrepreneurial	Entrepreneurial	Managerial
PROJECT	Project	Scientific project	Technological innovation project	Financial project
	Main actor	Service provider-researcher	Entrepreneur-researcher	Manager
	Founder	Heavy involvement	Shared implication	Targeted implication
	Objective	Perpetuity (of scientific knowledge creation)	Independence	Profit
	Aim	Credit (Intellectual legitimacy and recognition (glory, social power))	Economic recognition (economic power, material recognition)	Economic power and recognition
	Individual – collective link	Individual aim with a collective vocation	Individual aim with an individual vocation	Collective end-result with an individual vocation
ORGANISATION	Culture	'neo-academic'	business	capitalization
	Work Division	By knowledge	By inputs	By inputs and outputs
	Structure	Simple professional bureaucracy	Simple and/or organic	Mechanist Bureaucracy
	Governance	Scientific community (peers)	Promoters (venture-capital, angels)	Institutional and private investors
	Embeddedness	Community of scientific knowledge	Innovation and expertal network +/– glocal	Value creating system
	Core competence	Capacity of making unexpected links between ideas	Identification and exploitation of technological opportunities, integrated within a specific platform	Management of tangible and intangible assets (intellectual property, expertise relationships)
STRATEGY	Mission	Knowledge creation, knowledge exploration	Knowledge integration and knowledge exploitation Renewal of Economy	Commercialization and intensive diffusion of strategic knowledge
	Position in the innovation cycle	Invention	Innovation	Industrialization / Merchandising

TELEOLOGICAL RUPTURE
Commercial orientation of the discovery process into a process of invention

CREATIVITY CRISIS
Bureaucratization of the innovation process

Table 15.2: The development modes of biotechnology firms.

the *technical* capital (equipment) of the laboratory, permitting the collection of new data and information, thus initiating a new cycle generating recognition-credit.[52]

We would like to emphasize here that the transformation of the organization from the *pre-entrepreneurial* to the *entrepreneurial* mode depends on the appearance of a source of 'credit' other than publications, that is the patent, and of the gradual transformation of the notion of 'credit'. The search for 'credit' on the part of researchers was found in the studies conducted by Latour and Woolgar in their double quest for *recognition* (individual attribution of intellectual glory) and *credibility* (in the sense of reliability, reputation of competence). Based on our observations, the notion of 'credit' takes on a new meaning, with the gradual transformation of the patent from an instrument of knowledge diffusion, into a tool for reassuring investors and therefore an exclusive source of financial revenues once exploited under license(s) (Figure 15.1). Commercializing the potential fruits of the patent under a regulatory framework altered by the recent laws on innovation, in the context of a shortage of funding for public and semi public research[53], therefore increases the possibilities of intellectual and material recognition, which are the drivers of entrepreneurship.

B) The creativity crisis:

Another paradox appears when innovations are turned into routines ('routinization') as a result of the short-term vision adopted by firms at the 'managerial' stage. The long biopharmaceutical research and development time periods (it takes approximately ten years to complete pre-clinical and clinical phase I, II and III research, for example) contrast sharply with the short investment cycles (one to three years for venture-capital) and shareholder expectations. The race to generate revenues from intangible assets as the patents on *blockbuster drugs* expire (Figure 15.1) reinforces the financial

52 Latour and Woolgar, *La Vie de laboratoire*.

53 The need for venture-capital was expressed by our interviewees when they said: 'it's the only source of financing available!', 'it's a necessary evil!'

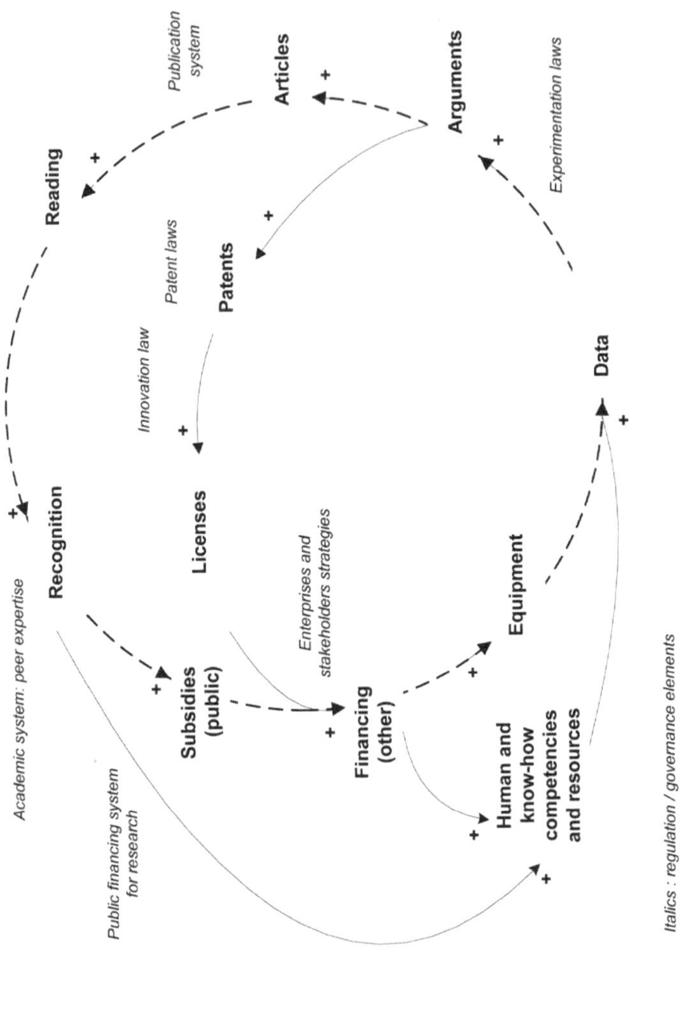

Figure 15.4: The transformation of capitals generated from academic research activities.

choices made by management. Paradoxically, these choices carry little risk compared with the prime mission of the financial world.

In the short term, the choices involved in creating new markets include improving on existing products (follow-up drugs, me-too products) or the commercialization of profitable products, such as diagnostic products or laboratory chemicals (type 4 companies within our sampling), rather than the creation of truly new medications ('novel' medicines).[54] Close to half of the pharmaceutical products commercialized in 1997 already existed ten years earlier, and fewer than 10 per cent prescription drugs had less than 2 years of existence.[55] Close to two thirds of new medicines approved by the FDA between 1989 and 2000 used active ingredients already on the market.[56] According to Pignarre, with the standardization of pharmaceutical research methods (HTS, combinatorial chemistry) resulting from a traditional industrial approach to increasing productivity, combined with the division of cognitive activities between pharmas and biotechs and the creation of rules and regulations for the marketing of products (clinical trials), the biopharmaceutical industry has been suffering from a 'bureaucratization' of its innovation processes, which has impeded its creativity.[57]

Conclusion

The exploratory analysis of knowledge intensive biotechnology firms within Quebec has enabled us to identify three development 'modes' involved in their growth cycle (*pre-entrepreneurial*, *entrepreneurial* and *managerial*

54 Pignarre, *Le Grand secret de l'industrie pharmaceutique.*

55 A. Hamdouch and M.-H. Depret, *Nouvelle économie industrielle de la pharmacie : structures industrielles, dynamiques d'innovations et stratégies commerciales* (Paris: Elsevier, 2001).

56 NIHCM, *Changing Patterns of Pharmaceutical Innovation* (May 2002).

57 The division of cognitive activities between pharmas and biotechs has had the effect of segmenting, parcelling and specializing knowledge, rather than encouraging its circulation.

modes). Most of the companies sampled are situated in the transition between the second and third modes of development. Moreover, two ruptures mark the three development phases of these companies: a teleological rupture, and a crisis of creativity. They are both inscribed within a dominant current of 'corporate governance'. Information, expertise and knowledge being the engines of 'value creation', how does one grasp these two ruptures, and relate them to the financial view of value creation?

The creativity of the innovation process, or the intensity of knowledge creation, is intimately linked to the individual or group that generates them. How, then, does one integrate a knowledge economy, whose logic is based on continuity and the individual human being, within a financial economy, which is nowadays based on immediateness and speculation?

Epingard pertinently reminds us that immaterial investment consists in an economically useful detour in relation to production, but that its outcome is intangible and can only be observed indirectly, just like the knowledge and skills that constitute human capital. As a result, there is an uncertainty that is partly irreducible, not only about productive assets, but also about their value.[58] It is precisely this uncertainty that poses a problem for financial management with respect to evaluating immaterial assets during the economic activities of a technological company. Currently, there exists no appropriate method to evaluate the effects of the knowledge-creation process, since such an evaluation must integrate non-measurable elements, such as motivation on the part of the experts, the convergence between individual objectives and ambitions and those of the company, the articulation between individuals and knowledge, and the potential of a cognitive network of companies.

58 'l'idée qu'il y a bien, comme pour l'investissement matériel, détour économiquement utile à travers un acte de production, mais dont le résultat, l'output est impalpable, non directement observable, à l'image des savoirs et aptitudes qui constituent le capital humain. Il en résulte une incertitude partiellement irréductible, à la fois sur la réalité de l'actif productif ainsi constitué et sur sa valeur'. P. Epingard, *L'Investissement immatériel* (Paris: CNRS Éditions, 1999), p. 153.

The commercialization of 'scientific' projects, which have thus become 'financial' projects, might lead to the reduction of knowledge and know-how into codifiable and quantifiable information, as well as information that is dis-embodied from the holder of knowledge. It might also lead to the transformation of an optimizing, holistic organizational logic into a maximizing logic of project fragmentation aimed towards creating financial value. But the dynamics of knowledge creation and diffusion within the knowledge economy fundamentally differs from commercialization. If we grant knowledge its full significance, it consists of a 'continuum that extends from non-representable (the presentiment) to interpreted information, and which integrates, as an essential element, the individual as well as collective dimension of tacit knowledge'.[59]

In other words, what characterizes the knowledge economy is that knowledge becomes incorporated in a durable manner within projects, individuals and the organization. Financial logic has evolved in recent years so as to considerably modify its perception of these two elements (the individual and the organization). But the limits of this logic rapidly become apparent:

> An important part of codified knowledge results from the conversion of tacit knowledge, that is to say non-conveyable outside the actions of those who possess it because it is inscribed within a human being. Through the codification of knowledge, man exploits the unique faculty he has, in the words of Leroi-Gourhan, of being able to 'place his memory outside himself'.[60]

To help the manager adapt to the unavoidable reality of an economy dominated by finance, it is necessary to pursue the debate concerning these

59 'un continuum qui s'étend du non représentable (pressentiment) à l'information interprétée, et qui intègre, en tant que composante essentielle, la dimension tacite des savoirs, individuels comme collectifs'. Epingard, *L'Investissement immatériel*, p. 155.

60 'Une part importante des connaissances codifiées provient de la conversion de savoirs tacites, c'est-à-dire non exprimables en dehors de l'action de celui qui les détient parce qu'enchâssés dans leur support humain. À travers la codification des savoirs, l'homme exploite cette faculté unique qu'il possède de pouvoir, selon l'expression de A. Leroi-Gourhan, 'placer sa mémoire en dehors de lui-même'. Epingard, *L'Investissement immatériel*, p. 159.

managerial stakes by adopting new approaches to knowledge management, which value the creation of both internal and external sites of exchange and of socialization, that is to say, at the boundary of the organization.[61]

These are precisely the theoretical and managerial issues raised by the 'network business model' that is being promoted today within the academic and the pharmaceutical business literatures.[62]

61 Nonaka and Takeushi, *The Knowledge-Creating Company*; P. Senge, *La Cinquième Discipline* (Paris: First Editions, 1991); G. Von Krogh, I. Nonaka, and M. Aben, 'Making the most of your Company's Knowledge: a strategic framework', *Long Range Planning*, 34 (2001): 421–439; I. Nonaka and R. Toyoma, 'The Theory of the Knowledge-creating firm: subjectivity, objectivity and synthesis', *Industrial and Corporate Change*, 14 (2005): 419–436.

62 Y. Pesqueux, *Organisations : modèles et représentations* (Paris: PUF, 2002); JSB Intelligence, *Emerging Business Models in the Pharmaceutical Industries – Strategic Analysis of the Market, Future Revenue Models and Key Players* (2005), 80p.

Notes on Contributors

BEAT BÄCHI studied History, Sociology, and Philosophy at the University of Zurich. His doctoral thesis, carried out at the Swiss Federal Institute of Technology, examined the history of synthetic vitamin C. He is currently working on regulatory science at the Institute for Science and Technology Studies at the University of Bielefeld, focusing on the history of the production of boundary values for dangerous chemical compounds in the Federal Republic of Germany, 1950–1985. His most recent publication is: *Vitamin C für alle! Pharmazeutische Produktion, Vermarktung und Gesundheitspolitik* (Zürich: Chronos, 2009).

MICHAEL BÜRGI is a PhD student at the University of Basel and is currently working at the Institute of History, ETH Zürich. His work focuses on the history of the connections between the Swiss pharmaceutical industry and academia. He is co-editor of a book on local practices of knowledge production in nineteenth- and twentieth-century Switzerland: *Lokale Naturen – 150 Jahre Thurgauische Naturforschende Gesellschaft* (Frauenfeld, 2004).

WILLIAM J. BUXTON is Professor of Communication Studies at Concordia University in Montreal. He is author of *Talcott Parsons and the Capitalist Nation-State: political sociology as a strategic vocation* (University of Toronto Press, 1985), co-author of *The Politics of Knowledge and Information: American philanthropy and Canadian libraries* (McGill University, The Centre for Research on Canadian Cultural Industries and Institutions and the Graduate School of Information and Library Studies, 1998), co-editor of *Harold Innis in the New Century: reflections and refractions* (McGill-Queen's University Press, 1999), and editor of *Patronizing the Public: American philanthropy's transformation of culture, communication, and the humanities* (Lexington Books, 2009).

TONY CORLEY is a member of the Centre for International Business History, Henley Business School at the University of Reading, He has published extensively on the history of the oil, consumer products and pharmaceutical industries, and on the origins of British foreign direct investment and multinational enterprises. His books include histories of Huntley & Palmers and of the Burmah Oil Company. A further work, *From Pills to Pharmaceuticals: the Beecham Enterprise, 1848–2000*, is to be published in 2010. In addition to writing many articles and conference papers, he has contributed nearly 100 entries to the *Oxford Dictionary of National Biography* (2004), and was recently awarded an honorary DLitt by Reading University.

ROBERT H. DESMARTEAU is Professor of Corporate Strategy at the Department of Strategy, Social and Environmental Responsibility, in the School of Management at the University of Quebec at Montreal (UQAM). He is also the Director of the Executive MBA program of this school. In 1997, he completed a doctorate in radical strategy from UQAM after thirteen years of hospital management. In Quebec, he has advised more than thirty organizations, both private and public, in their strategic planning. His research and publications are on the business models and the corporate governance of biotechnology firms, the modelling of bio-clusters, and the governance of healthcare organizations.

MEHRAN EBRAHIMI is Professor of Management of Technology at the Department of Management and Technology, School of Management, University of Quebec at Montreal (UQAM). His previous affiliations include HEC Montreal, and Sigmapi Corporation. He coordinates the Group for Enterprise Management Studies in the Aeronautical Sector (GEMS-Aero) at UQAM. He has written articles on knowledge management in high-tech organizations, knowledge creation processes, and on the interference between the 'financialization' of the economy and management practices, especially in the aeronautical, telecom and biopharmaceutical sectors. Recent books (in French) include: *La Mondialisation de l'ignorance* (Quentin Édition, 2001), *Nouvelle économie, nouveaux enjeux de formation* (Quentin Édition, 2002).

FRANK HUISMAN is Professor in the History of Medicine at the University Medical Center Utrecht and at the Descartes Center for the History and Philosophy of the Sciences of Utrecht University. He has published on medical historiography, quackery, and the cultural authority of medicine. Recently, he co-edited with John Harley Warner: *Locating Medical History: the stories and their meanings* (Johns Hopkins University Press). He was area editor of the five-volume *Dictionary of Medical Biography* (Greenwood Press). He is currently working on a book exploring the transformation of the Dutch health care system between 1880 and 1940.

MARION A. HULVERSCHEIDT is Research Assistant at the Institute for the History of Medicine, Charité, Berlin. Her research interests include the history of tropical medicine, especially the research area of malaria chemotherapy. Her recent publications include: 'German Malaria Experiments with Humans, Supported by the DFG until 1945', in W.U. Eckhart (ed.), *Man, Medicine and the State: the human body as an object of government sponsored medical research in the 20th century* (Stuttgart: Franz Steiner Verlag, 2006), pp. 221–235; 'Chinin als Malariatherapeutikum – prekär und unverzichtbar?', in B.Wahrig, H. Stoff, A.v. Schwerin, V. Blaz (eds), *Precarious Matters/Prekäre Stoffe* (Berlin: MPIWG Preprint, 356, 2008), pp. 83–98; '1942- Die klinische Prüfung des Sontochin. Arzneimittelforschung in Krieg', in N. Eschenbruch. N. Balz, U. Köppel, M.A.Hulverscheidt (eds), *Arzneimittel des 20. Jahrhunderts. Historische Skizzen von Lebertran bis Contergan* (Bielefeld: transcript Verlag, 2009), pp. 143–166.

AXEL C. HÜNTELMANN is Postdoctoral Research Fellow at the Bielefeld Graduate School in History and Sociology, working on the cultural history of growth, 1770–1970. Recently he has worked on the German Imperial Health Office, 1876–1933; on a biography on Paul Ehrlich; and on serum and vaccine regulation in the German Empire and in France. His publications include: 'Das Diphtherie-Serum und der Fall Langerhans', in *MedGG*, 24 (2006): 71–104; *Hygiene im Namen des Staates. Das Reichsgesundheitsamt 1876–1933* (Göttingen, 2008); 'Dynamics of Wertbestimmung', in *Science in Context*, 21 (2008): 229–252; together with Johannes Vossen and Herwig Czech he has edited a collection of articles on German Health

Offices: *Gesundheit und Staat. Studien zur Geschichte der Gesundheitsämter 1870–1950* (Husum, 2006).

MARIE-THÉRÈSE LUSSIER is Professor in the Department of Family Medicine and Emergency Medicine at the Medical School of the University of Montreal. She is also member of the Research Team on Primary Care at the Centre for Health and Social Care in Laval. Her research mainly consists in analysing communication during primary care encounters. She has recently become interested in the problem of chronic disease discussed during medical consultations. She teaches graduate students and interns in family medicine, as well as in continuous education. An expert on Medical Communication, she sits on several committees whose task it is to establish teaching programmes and other activities on the subject.

MANON NIQUETTE is Associate Professor in the Department of Information and Communication at Laval University (Québec, Canada), and researcher of the Interdisciplinarity Studies and Social Representations Group (Groupe d'études sur l'interdisciplinarité et les représentations sociales, GEIRSO). She has published on social interaction in science museums and world fairs, on the history of women in public relations, and on social marketing. Her current research is on pharmaceutical advertising and, more specifically, on the representations of drug uses in drug promotional material.

HEATHER MUNRO PRESCOTT is Professor of History at Central Connecticut State University in New Britain. Her research interests include history of women's health, history of childhood, and disability history. Her first book, *A Doctor of Their Own: the history of adolescent medicine* (Harvard, 1998), received the Will Solimene Award of Excellence in Medical Communication from the New England Chapter, American Medical Writers Association. She is also the author of *Student Bodies: the impact of student health on American society and medicine* (University of Michigan, 2007).

CAY-RÜDIGER PRÜLL is Lecturer at the Medical Faculty of the University of Freiburg, Germany. His work focuses on the history of pathology, pharmacology and military medicine in the nineteenth and twentieth centuries, with a particular interest in the social and cultural history of medicine and the relationship between medicine and politics. His publications include: 'Part of a Scientific Masterplan? – Paul Ehrlich (1854–1915) and the Origins of his Receptor Concept', *Medical History*, 47 (2003): 332–356; 'Caught between the Old and the New – Walther Straub (1874–1944), the Question of Drug Receptors and the Rise of Modern Pharmacology', *Bulletin of the History of Medicine*, 80 (2006): 465–489.

CLAUDE RICHARD is Research Associate on the Research Team on Primary Care at the Centre for Health and Social Care in Laval. His research has been mainly concerned with the development and testing of tools for analysing communication (MEDICODE and the French version of the *Roter Interaction Analysis System*). Recently, he has become particularly interested in the discussions about medication that occur during medical consultations. He has co-authored with Marie-Thérèse Lussier the volume (in French): *La Communication professionnelle en santé* (ERPI, January 2005). Together, they have written more than 50 articles on the topic of professional health communication, including for the journals *Canadian Family Physician* (Canada) and *L'Actualité médicale* (Québec).

ANNE-LAURE SAIVES is Professor of Strategic Management at the Department of Management and Technology, School of Management, University of Quebec at Montreal (UQAM). She coordinates the Group for Enterprise Management Studies in Bio-industries (GEMS-Bio) at UQAM and is an Associate Researcher on the CRSH-program on the 'pharmaceutical drugs chain' (GEIRSO, Research Group on Interdisciplinary Studies and Social Representations) and the National Institute for Functional Foods and Nutraceuticals (INAF, Laval University). She has published articles on biotech clusters, the territorialization strategies of biofood/biotech firms, and business models of biotechnology firms. Her main book (in French) is: *Territoire et compétitivité de l'entreprise* (L'Harmattan, 2002).

MARÍA JESÚS SANTESMASES is Senior Research Fellow at the Department for Science, Technology and Society (Departamento de Ciencia, Tecnología y Sociedad), Centre for Human and Social Sciences (Centro de Ciencias Humanas y Sociales), CSIC, Madrid. She works on the history of post-World-War-Two developments in experimental biology and medicine, more recently on antibiotic uses and production and on practices in human genetics. Her recent publications include her book: *Severo Ochoa, de músculos a proteínas* (Madrid, 2005); a special issue of the *Journal of the History of Biology*, 29 (2006), on *Radiobiology in the Atomic Age*, co-authored with Angela Creager; and an edited volume on twentieth-century research in Spain with Ana Romero, *Cien años de política científica en España* (Madrid, 2008).

JONATHAN SIMON is *Maître de conférences* (Lecturer) at the Université Lyon 1, where he teaches history and philosophy of science. Working principally on the history of pharmacy and its relationship to chemistry, his most recent book is: *Chemistry: the impure science*, co-authored with Bernadette Bensaude-Vincent. He has studied and taught in Paris, Sydney, Strasbourg and Berlin. In Berlin, he collaborated with Axel Hüntelmann on a project funded by the DFG and led by Volker Hess on a comparative history of serotherapy in France and Germany.

SERGIO SISMONDO teaches philosophy and sociology at Queen's University, Canada. He is the author of a number of general and theoretical works in Science and Technology Studies, including *An Introduction to Science and Technology Studies* (2nd Edition, 2009). His current research is on political economies of pharmaceutical knowledge, focusing especially on publication planning and other aspects of the industry's use of research for marketing.

BRUNO J. STRASSER is Assistant Professor in the Program for the History of Science and Medicine at Yale University. He is the author of a book on the history of molecular biology in Switzerland: *La Fabrique d'une nouvelle science: la biologie moléculaire à l'âge atomique, 1945–1964* (Florence, 2006). His work focuses on the history of the biomedical sciences in the twentieth century, especially molecular biology, bioinformatics, biotechnology, and the pharmaceutical industry. He is currently writing a book on the history of biomedical collections and databases.

JOHN P. SWANN has been an historian at the US Food and Drug Administration since 1989. His publications have focused on the history of drugs, biomedical research, the pharmaceutical industry, and regulatory history. The study in this volume is drawn from a project he is currently involved in, a history of obesity and its pharmaceutical therapy. To find out more about FDA history, including a guide to research resources, see: http://www.fda.gov/AboutFDA/WhatWeDo/History/default.htm.

DOMINIQUE A. TOBBELL is Assistant Professor in the History of Medicine at the University of Minnesota. Her research focuses on the history of pharmaceuticals, health policy, and academic medicine. Her forthcoming book, *Pills, Power, and Policy: how drug companies and physicians resisted federal reform in Cold War America*, describes how the American drug industry and key sectors of the medical profession became allies against federal reform. Her other publications include: 'Allied Against Reform: pharmaceutical industry-academic physician relations in the United States, 1945–1970,' *Bulletin of the History of the Medicine*, 82 (2008): 878–912; '"Who's Winning the Human Race?" Cold War as Pharmaceutical Political Strategy,' *Journal of the History of Medicine and Allied Sciences* (July 2009).

Index

FSC
www.fsc.org
MIX
Papier | Fördert
gute Waldnutzung
FSC® C083411

Kolibri 360 GmbH
Ferdinand-Jühlke-Straße 7
99095 Erfurt, Deutschland
produktsicherheit@kolibri360.de